ENCYCLOPEDIA OF ORGANIC CHEMISTRY

ENCYCLOPEDIA OF ORGANIC CHEMISTRY

Vol. 6

By
Sananda Chatterjee
DAE, DCA, CMCNet
Scientific Consultant
Institute for Natural Sciences
Kolkata
(West Bengal)

DISCOVERY PUBLISHING HOUSE PVT. LTD.
NEW DELHI-110 002

Published by:
Tilak Wasan
DISCOVERY PUBLISHING HOUSE PVT. LTD.
4383/4B, Ansari Road, Darya Ganj
New Delhi-110 002 (India)
Phone : +91-11-23279245, 43596064-65
Fax : +91-11-23253475
E-mail : parul.wasan@gmail.com
discoverypublishinghouse@gmail.com
web : www.discoverypublishinggroup.com

***First Edition:* 2012**

ISBN: 978-81-8356-875-3 (Set)

Encyclopedia of Organic Chemistry

Printed at:
Shree Balaji Art Press
Delhi

Preface

Organic Chemistry is that branch of chemistry, which deals with the reactions related to the life form, or one of the source of the life form, carbon (we can say the end of it).

This carbon mainly conjugates with different elements such as nitrogen, hydrogen and oxygen sometimes metals and other non-metals also). Organic chemistry studies mainly the extraction, structure, properties, of these compounds.

In this book chapters deals with the different methods of purification, extraction of compounds, some processes were used and devised by Alchemists.

The extraction preservation of natural dyes are examples of alchemical work, these been discussed in the chapter "dyes indicators and pigments".

Dyes is the mal-pronunciation of the damathi word dyum meaning colour, it was, the ancient Egyptians who use different natural pigments to extract different colours.

In this book the organic compounds has been mainly classified into two broad groups, the aromatic and the aliphatic compounds. The extraction of alcohol is definitely an alchemical process'.

The destructive distillation of wood is one of the finest processes devised by the alchemists for the preparation of methanol is one of the important topics of this book.

Looking ahead in to the modem world, the world of future, the subject comes in vision is the biochemistry, the base of which is organic chemistry, the chapter containing carbohydrates, and the chapter, chemical compounds of biological and biochemical interest is the best element of focus for the students stepping foot in the advanced staircase of biochemistry.

The last chapter, chapter 41, discusses the different organic reactions and their suitably mechanisms, this will help those students who are very much afraid of organic formulas and reactions. This chapter also deals with the chirality one of the interest of Dr. Palash Gangopadhyay (author's friend).

Organic Chemistry is not a mere subject, it is the rhythm of life tuned in the essence of carbon, which looks black but carries the dazzleness of diamond in its heart.

SANANDA CHATTERJEE

Contents

Preface

30. Diazo and Diazo Compounds 1327

Introduction and Brief History—Diazotization of Benzene—Nomenclature and Types —Benzene Diazomium Chloride—Preparation—Structure—Physical Properties—Chemical Properties—Diazo Compounds : Diazotates—Reactions of Diazomium Compounds—The Comparative study between nitrosamines and diazo – compounds—Decomposition of Diazomium Salts and Compound—The Photo Sensitivity of Diazo Compounds—Reaction of Diazo Compounds with Sulphamic Acid—Scope of the Reaction—Substitution reactions of diazonium ions—The reaction with naphthalen-2-ol—Some More Diazo Compounds—Hydrazine—Introduction—Synthesis—Physical Properties—Chemical Properties—Biological Effects—Uses of Hydrazine—Derivatives of Hydrazine—Phenylhydrazine—Physical Properties—Chemical Properties—Nitroso and Hydroxylamine Derivatives of Hydrocarbons—Diphenyl-hydroxylamine—Azoxy Compounds—Introduction—Azoxybenzene—Introduction—Preparation—Physical Properties—Chemical Properties.

31. Aromatic Alcohol Aldehydes and Ketones 1370

Introduction and Brief History—Diazotization of Benzene—Nomenclature and Types —Benzene Diazomium Chloride—Preparation—Structure—Physical Properties—Chemical Properties—Diazo Compounds : Diazotates—Reactions of Diazomium Compounds—The Comparative study between nitrosamines and diazo – compounds—Decomposition of Diazomium Salts and Compound—The Photo Sensitivity of Diazo Compounds—Reaction of Diazo Compounds with Sulphamic Acid—Scope of the Reaction—Substitution reactions of diazonium ions—The reaction with naphthalen-2-ol—Some More

Diazo Compounds—Hydrazine—Introduction—Synthesis—Physical Properties—Chemical Properties—Biological Effects—Uses of Hydrazine—Derivatives of Hydrazine—Phenylhydrazine—Physical Properties—Chemical Properties—Nitroso and Hydroxylamine Derivatives of Hydrocarbons—Diphenyl-hydroxylamine—Azoxy Compounds—Introduction—Azoxybenzene—Introduction—Preparation—Physical Properties—Chemical Properties.

32. Aromatic Acids **1398**

Introduction—Benzoic Acid—Introduction—Preparation—Physical Properties—Chemical Properties—Benzoyl Chloride—Introduction—Preparation—Physical Properties—Chemical Properties—Benzoyl Peroxide—Introduction—Preparation—Uses—Chemical Properties—Quinic Acid—History—Physical Properties—Phthalic Acid—History—Preparation—Other Process—Physical Properties —Chemical Properties—Isophthalic Acid—Reactions—Terephthalic Acid—Physical and Chemical Properties—Derivatives—Physical Property of Teraphthalic Acid—Salicylic Acid—Introduction—Preparation—Laboratory Preparation of Salicylic Acid—Manufacture—Physical Properties—Chemical Properties—Uses—Anisic Acid—Introduction—Physical Properties—Uses—Tyrosine—Introduction—Preparation—Biosynthesis—Physcial Properties—Uses—Tyrosol—Introduction—Preparation—Physical Properties—Chemical Properties and Uses—O-Hydroxycinnamic Acid—Introduction—Physical Properties—Chemcial Properties—Orsellinic Acid—Introduction—Preparation—Physical Properties —Uses—Lecanoric Acid—Evernic Acid—Kynurenic Acid—Physical Properties—Gamma-Hydroxybutyric Acid—Chemical Properties—Uses—Phenylacetic Acid—Introduction—Preparation—Physical Properties—Chemcal Properties—Uses—Mandellic Acid—Introduction—Preparation—Lab Preparation—Physcical Properties—Chemical Properties and Uses—Vanillyl Mandelic Acid—Aromatic Sulphonic Acids—Nomenclature—Benzene Sulphonic Acid—Introduction—Preparation—Mechanism—Manufacture—Physical Properties—Chemical Properties—Reactions—Uses—P-toluenesulfonic Acid—Preparation—Physical Properties—Uses.

33. Aromatic Compounds with Condensed Nuclei **1476**

Introduction—Naphthalene—History—Preparation—Industrial Preparation of Naphthalene —Structure of Naphthalene—Physcial Properties—Chemical Properties—Uses—Derivatives—Isomerism of Naphthalene Derivatives—Anthracene—Introduction and Occurrence —Physical Properties—Chemical Properties—Derivatives of Anthracene—Anthraquinone—Introduction—Natural Occurrences—Preparation—Physical Properties—Chemical Properties—Purpurin--Introduction—Physical Properties—Anthrapurpurin—Physical Properties—Flavopurpurin—Physcial Properties—Phenanthrene—Introduction—Preparation—Physcial

Properties—Chemcial Properties—Substitution Product of Phenanthrene—Hydroxy Phenanthrene—Morphol—Synthesis of Morphol—Dimethyl Morphol—Morphenol—Introduction—Phenanthraquinone and Its Derivatives—Nitroderivatives of Phenanthraquinone—Some More Compounds with Condensed Nuclei—Retene—Physcial Properties—Chrysene—Physcial Properties—Fluoranthene—Physical Properties—Tetracene—Physical Properties—Hexacene—Physical Properties—Benzpyrine—Physical Properties—Carcinogenic Compounds—Stranles Ring and Condensed Ring Structures.

34. The Carbohydrates **1528**

Introduction—Nomenclature —Classification of Carbohydrates—Monosaccharides—Introduction—General Properties of Monosaccharides—Glucose—Introduction—Synthesis—Properties of Glucose—Some Detail Physical Properties of Glucose—Structure of Glucose—Chemical Properties —Osazone and the Conversion of Aldoses and Ketoses—Further Structural Details of the Glucose Molecule—Analysis of Glucose—Fructose—Introduction—Synthesis—Physical Properties—Chemical Properties and Constitution —Galactose—Mannose—Arbinose—Ribose—Physical Properties—Ribulose—Xylose—Physcial Properties—Oligosaccharides—Sucrose—Synthesis of Sucrose—Physical Properties—Chemical Properties—Sugar As a Macronutrient—Use in Human Nutrition—Maltose—Physical Properties—Lactose—Raffinose—Melezitose—Glycogen—Synthesis—Breakdown—Starch—Introduction—Synthesis—Physical Properties—Chemical Properties—Uses—Dextrin—Inulin—Uses—Cellulose—History—Uses.

CHAPTER

30

Diazo and Diazo Compounds

Introduction and Brief History

Azo comes from the French word Azoo - means without animal - the element which does not support animal life - the element is nitrogen (azote - nitrogen).

Diazotization of Benzene

The process by which an aromatic primary amine is converted to a diazonium compound. The preparation and reactions of diazonium salts were discovered in 1858 The diazonium compounds were first discovered by Peter. Griess and were the basis of the synthetic dye industry and the development of other industrial chemistry in Europe. In diazotization, sodium nitrite is added to a solution of the amine in aqueous acid solution at 0-5°C. Reaction of the amine with nitrous acid gives a nitrosamine. Tautomerization and loss of water lead to the diazonium ion, which is stabilized by delocalization of the positive charge at the ortho and para carbon atoms of the ring.

NH_2 HONO N=O NH H^+ $-H_2O$ N≡N+ N– N+ (1)

Nitrosamine Diazonium ion

Nomenclature and Types

Specific Classes of Compounds

Azo, azoxy, diazo, and related compounds

Diazenes. The parent structure HN = NH is named "diazene", and the groups derived from it, i.e., HN = N– and –N=N–, are named systematically as diazenyl and diazenediyl, respectively

Azo compounds. Compounds with the general structure R—N=N—R', where R and may be alike or different, are known generically as "azo compounds". However, these compounds may be more systematically named substitutively as derivatives of the parent hydride diazene.

Examples:

$H_3C—N=N—CH_3$
Dimethyldiazene
(traditionally Azomethane)

$C_6H_5—N=N\text{-}C_6H_5$
Diphenyldiazene
(traditionally Azobenzene)

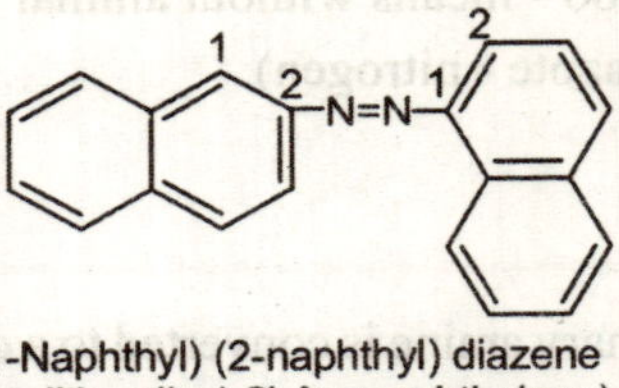

(1-Naphthyl) (2-naphthyl) diazene
(traditionally 1,2'-Azonaphthalene)

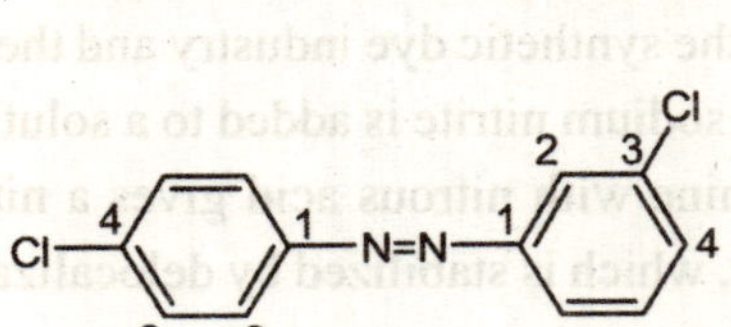

(3-Chlorophenyl) (4-chlorophenyl) diazene
(traditionally 3,4'-Dichloroazobenzene)

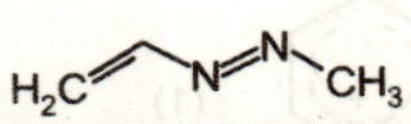

Methyl(vinyl) diazene
(traditionally Etheneazomethane)

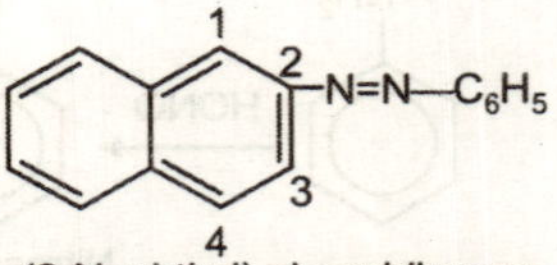

(2-Naphthyl) phenyldiazene
(traditionally Naphthalene-2-azobenzene)

A monoazo compound with the general structure R—N=N—R' in which R is substituted by a principal characteristic group is named on the basis of the parent hydride, RH, substituted by an organyldiazenyl group, R—N=N—. If both R and are substituted by the same number of the principal characteristic group, a multiplicative name may be used.

Examples:

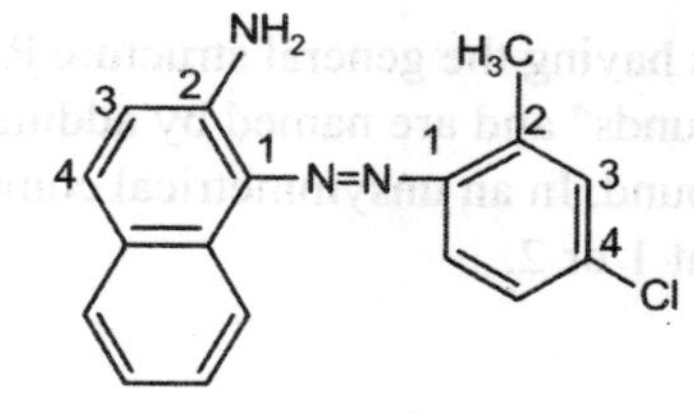

1-[(4-chloro-2-methylphenyl)diazenyl] napthalene-2-amine

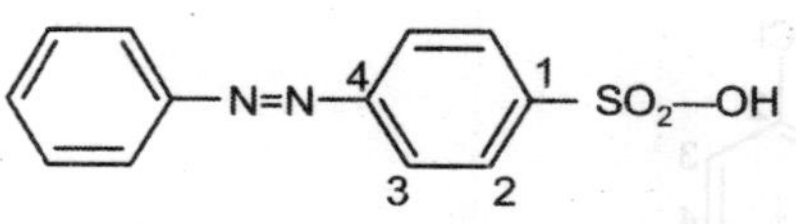

4-(Phenyldiazenyl) benzene-sulfonic acid [traditionally 4-(Phenylazo) benzenesulfonic acid]

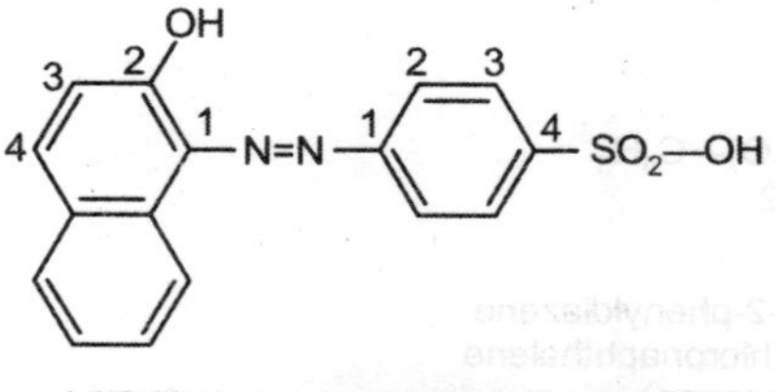

4-[(2-Hydroxy-1-naphthyl) diazenyl] benzenesulfonic acid [traditionally 4-[(2-Hydroxy-1-naphthyl) azo] benzenesulfonic acid

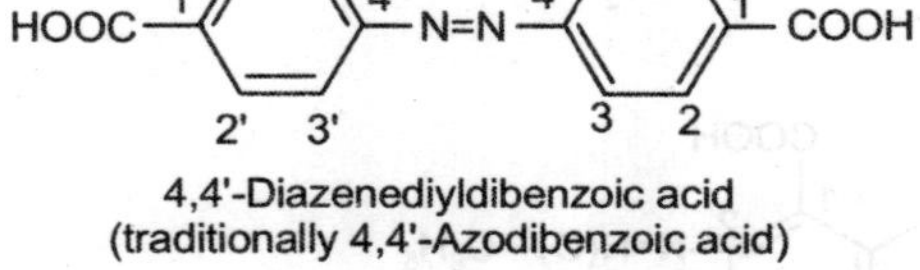

4,4'-Diazenediyldibenzoic acid (traditionally 4,4'-Azodibenzoic acid)

Bisazo compounds and more complex analogues are named on the basis of the parent structure "diazene", in the absence of a more preferred parent compound.

Examples:

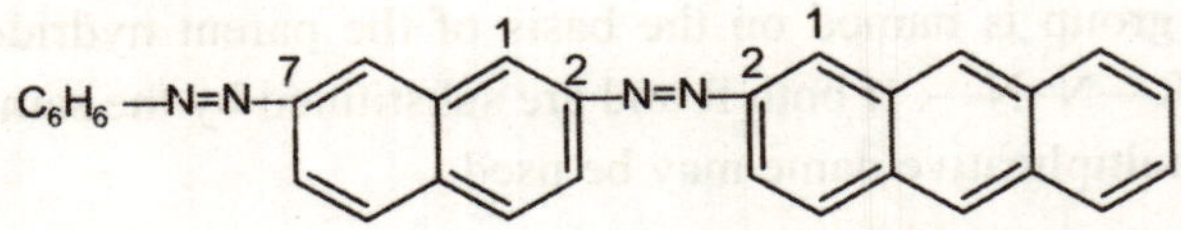

C_6H_6—N=N— (7-naphthyl-2)—N=N—(2-anthryl)

(2-Anthryl) [7-phenyldiazenyl)-2-naphthyl)] diazene
(traditionally Anthracene-2-azo-2'-naphthalene-7'-azobenzene)

Azoxy compounds. *N*-Oxides of azo compounds having the general structure R—N_2 (O)—R or R-N_2(O)—R' are known generically as "azoxy compounds" and are named by adding the separate word "oxide" to the name of the corresponding azo compound. In an unsymmetrical compound, the position of the azoxy oxygen atom is expressed by the locant 1 or 2.

Examples:

C_6H_5—N_2(O)—C_6H_5

Diphenyldiazene oxide
(traditionally Azoxybenzene)

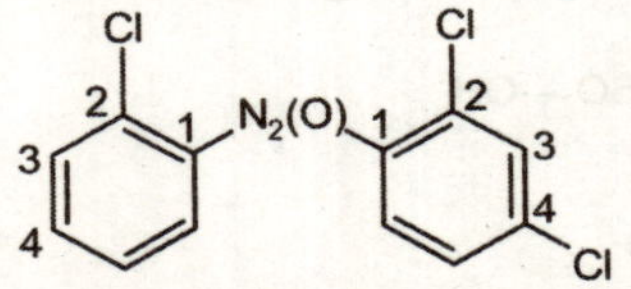

(2-Chlorophenyl) (2,4-dichlorophenyl)
diazene oxide (the position of the
oxide is unknown) (traditionally
2,2',4-Trichloroazoxybenzene)

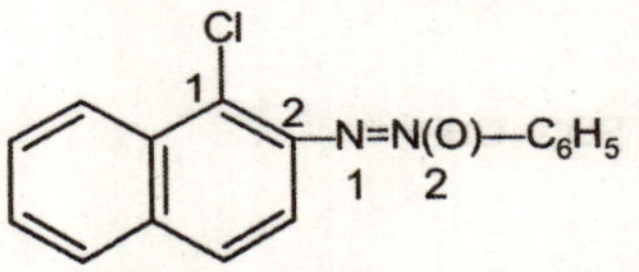

1-(1-Chloro-2-naphthyl)-2-phenyldiazene
2-oxide (traditionally 1-Chloronaphthalene
-*NNO*-azuxybenzene)

An azoxy compound in which the general structure is R—N=N(O)—R' or R—N(O)=N—R', in which R is substituted by a principal characteristic group, is named on the basis of the parent hydride, RH, substituted by the R'-azoxy group in which the position of the oxygen atom is denoted by the prefix *NNO-*, *ONN-*, or *NON-*, as appropriate.

Example:

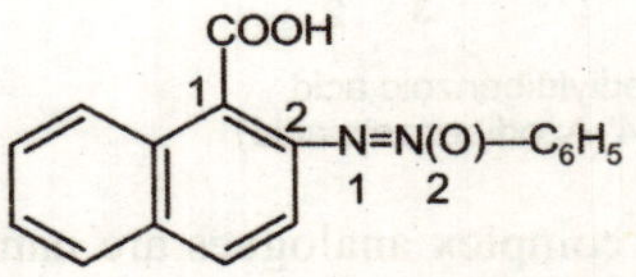

2-(Phenyl-*ONN*-azoxy)
-1-naphthoic acid

Diazonium compounds. Compounds with the general structure R—$N_2^+X^-$ are named by adding the suffix "-diazonium" to the name of the parent hydride RH followed by the name of the ion as a separate word.

Examples:

$H_3C\diagup\diagdown N_2^+Cl^-$

Ethane diazonium chloride

$C_6H_5—N_2^+Cl^-$

Benzenediazonium chloride

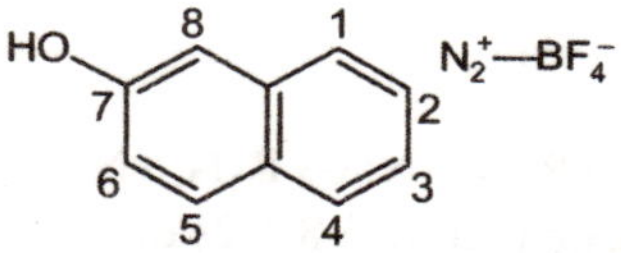

7-Hydroxynaphthalene-2-diazonium tetrafluoroborate

Azo compounds with the general structure R-N=N-X should be named as derivatives of the parent structure diazene, HN=NH.

Examples:

C_6H_5–N=OH
Phenyldizenol
(traditionally Benzene diazohydroxide)

C_6H_5–N=N–O^-Na*
Sodium phenyldiazenolate
(traditionally Sodium benenediazoate)

C_6H_5–N=N–$SO_3^-Na^+$
Sodium phenyldiazene sulfonate
(traditionally Sodium benzenediazoate)

Diazo compounds. Compounds containing a group N_2 attached by one nitrogen atom to one carbon atom are named by adding a prefix "diazo-" to the name of the parent hydride.

Examples:

CH_2N_2
Diazomethane

N_2CH—CO—O—C_2H_5
Ethyl diazoacetate

Azides

Compounds containing a group N3 attached through one nitrogen atom to a parent hydride structure are named (a) in substitutive nomenclature by adding the prefix "azido-" to the name of the parent hydride RH, or (b) in functional class nomenclature by citing the class name "azide" as a separate word following the name of the group R.

Examples:

$C_6H_5—N_3$

(*a*) Azibobenzene

(*b*) Phenyl azide

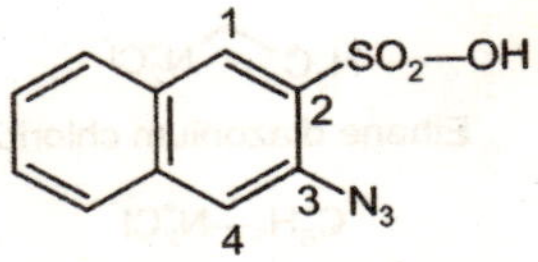

(a) 3-Azidonaphthalene-2-sulfonic acid

Isodiazenes

Compounds with the general structure $R_2N—N: \leftrightarrow R_2N^+{=}N^-$ may be named substitutively as derivatives of the parent "radical" hydrazinylidene (see R-5.8.1.2) or on the basis of the trivial name isodiazene.

Example:

$$(CH_3)_2N: \leftrightarrow (CH_3)_2N^+ = N^-$$

Radicals

Monovalent radicals. A radical formally derived by the removal of one hydrogen atom from any of the mononuclear parent hydrides of the carbon family of elements, or from a terminal atom of a saturated unbranched acyclic hydrocarbon, or from any position of a monocyclic hydrocarbon ring is named by replacing the "-ane" ending of the systematic name of the parent hydride by "-yl".

Examples:

$•CH_3$ Methyl

$CH_3—CH_2—CH_2•$ Propyl

$•GeH_3$ Germyl

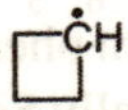

Cyclobutyl

A radical formally derived by the removal of one hydrogen atom from any position of any other parent hydride, is named by adding the suffix "-yl" to the name of the parent hydride, eliding the final "e" of the name of the parent hydride, if present.

Examples:

• SH

Sulfanyl

$•NH_2$

Azanol

Aminyl

(traditional name amino)

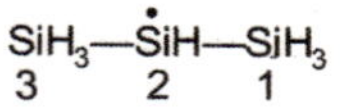

$SiH_3—\dot{Si}H—SiH_3$
3 2 1

Trisilan-2-yl

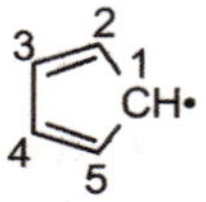

Cyclopenta-2,4-dien-1-yl

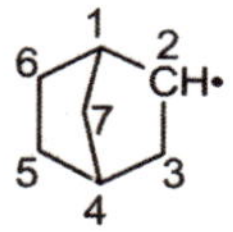

Bicyclo[2.2.1] heptan-2-yl

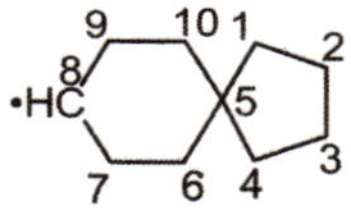

Spiro [4.5]decan-8yl

As exceptions, the nmes of the radicals HO•and HOO•are "hydroxyl" and "hydroperoxyl", respectively.

Divalent and trivalent radicals. Divalent radical centres formally derived by removal of two hydrogen atoms from the mononuclear parent hydrides CH_4, NH_3, and SiH_4 may be named respectively "methylene" or "carbene"; "azanylidene", "nitrene", or "aminylene"; an "silylene". Derivatives of these parent hydride radicals are named substitutively.

The use of these names does not imply a specific electronic configuration. If needed, such a distinction should be made by using a separate word or descriptive phrase, such as singlet and triplet.

Examples:

$$(C_6H_5)_2\dot{\underset{\cdot}{C}} \text{ and } (C_6H_5)_2C:$$

Diphenylmethylene (preferred)
Diphenylcarbene

$$CH_3\dot{\underset{\cdot}{N}} \text{ and /or } CH_3N:$$

Methylazanlidene
Methylnitrene
Methylaminylene

$$C_6H_5-CH_2-\dot{\underset{\cdot}{Si}}H \text{ and/or } C_6H_5-CH_2-CH_2-\ddot{Si}H$$

Benzylsilylene

$$CH_3-CO-\dot{\underset{\bullet}{N}} \text{ and/or } CH_3-CO-N:$$

Acetylazanylidene
Acetylnitrene
Acetylaminylene

Other bivalent radical centres and trivalent radicals centres formally derived by removal of two or three hydrigen atoms from the same skeletal atom of a parent hydride may be described by adding a suffix "-ylidene" or "-ylidyne" following the procedure prescribed above for the suffix "-yl"

Examples:

$CH_3-\dot{C}H\cdot$ and/or $CH_3-CH:$

Ethylidene

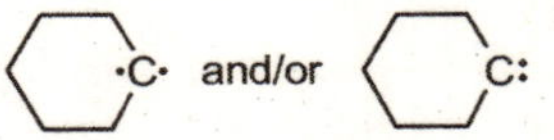

Cyclohexylidene

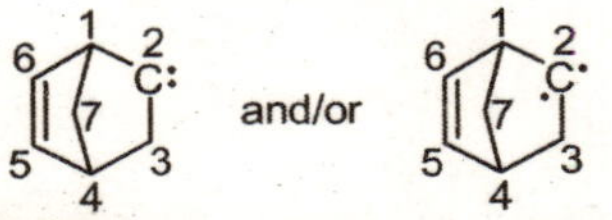

Bicyclo[2.2.1]hept-5-en-2-ylidene

Polyradicals containing two or more radical centres, formally derived by the removal of two or more hydrogen atoms from each of two or more different skeletal atoms of a parent hydride, are named by adding to the name of the parent hydride combinations of the suffixes "-yl" for a monovalent radical centre, "-ylidene" for a divalent radical centre, and "-ylidyne" for a trivalent radical centre, together with the appropriate numerical prefixes indicating the number of each kind of radical centre. The final "e" of the name of the parent hydride, if present, is elided when followed by the letter "y".

Examples:

$\cdot HN-NH\cdot$ (2 1)

Diazane-1,2-diyl
Hydrazine-1,2-diyl

$\cdot CH_2-\dot{C}H-CH_2\cdot$ (3 2 1)

Propane-1,2,3-triyl

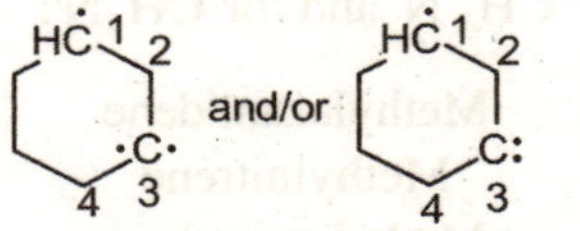

Cyclohexan-1-1yl-3-ylidene

$CH_3-\dot{\underset{\bullet}{C}}-CH_2-\dot{\underset{\bullet}{C}}-CH_3$ and/or $CH_3-\dot{\underset{\bullet}{C}}-CH_2-\ddot{C}-CH_3$

and/or $CH_3-\ddot{C}-CH_2-\ddot{C}-CH_3$

Pentane–2,4–diylidene

$\dot{C}-CH_2-\dot{C}H-CH_2-\dot{C}H$ and/or $:\dot{C}-CH_2-\dot{C}H-CH_2-\dot{C}H$ and/or
$\cdot\dot{C}\text{-}CH_2-\dot{C}H-CH_2-\ddot{C}H$ and/or $:\dot{C}-CH_2-\dot{C}H-\dot{C}H_2-\ddot{C}H$
Pentan-3-yl-1-yliden-5-xlidyne

The presence of a radical centre in a substituent to be cited as a prefix is expressed by the prefix "ylo-" denoting the removal of a hydrogen atom.

Example:

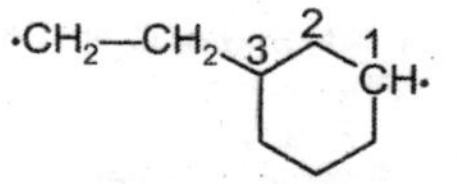

$\cdot CH_2-CH_2$ (3-position of cyclohexane ring, $\dot{C}H\cdot$ at 1)

3-(2-Yloethyl)cyclohexyl

Radical centres on characteristic groups. Acyl radicals, i.e., radicals with at least one chalcogen or nitrogen atom attached to the radical centre by a (formal) double bond, which may be considered to be formally derived by the loss of the hydroxy group from acid characteristic groups, are named by replacing the "-ic acid" or "-carboxylic acid" ending of the name with "-yl" (occasionally "-oyl") or "-carbonyl". Alternatively, acyl radicals may be named on the basis of an appropriate parent hydride using prefixes such as "oxo-", "thioxo-", "imino-", etc.

Examples:

$CH_3-[CH_2]_4-\dot{C}S$

Hexanethioyl
1-Thioxohexyl

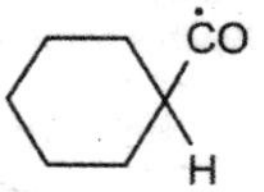

(cyclohexane ring bearing $\dot{C}O$ and H)

Cyclohexanecarbonyl
Cyclohexyloxomethyl

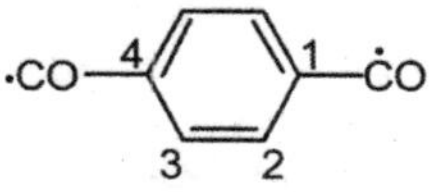

$\cdot CO-C_6H_4-\dot{C}O$ (positions 4, 1; ring positions 3, 2)

Terephthaloyl
1,4-Phenylene bis(oxomethyl)

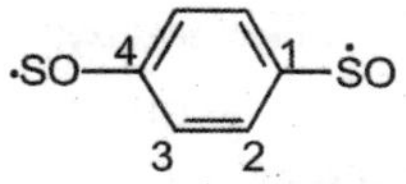

$\cdot SO-C_6H_4-\dot{S}O$ (positions 4, 1; ring positions 3, 2)

Benzene-1,4-disulfinyl
1,4-Phenylenebis(oxo-λ^4-sulfanyl

$(CH_3)_2\dot{P}O$

Dimethylphosphinoyl
Dimethyloxo-λ^5-phosphanyl

A radical derived formally by the removal of one or two hydrogen atoms from an amine, imine, or amide characteristic group may be named by adding a suffix "-aminyl", "-iminyl", or "-amidyl", to the name of the parent hydride for monovalent radicals and as a substituted nitrene for bivalent radicals. Alternatively, such radicals may be named substitutively on the basis of the parent hydride "azane".

Examples:

$CH_3-\dot{N}H$
Methanaminyl
Methylazanyl
(traditionally methylamino)

$HCO-\dot{N}H$
Formamidyl
Formylazanyl

$C_6H_5-\ddot{N}$ and/or $C_6H_5-N:$
Phenylazanylidene
Phenylnitrene
Phenylaminylene

Polyradicals with radical centres identically derived located on two or more amine, imine, or amide characteristic groups are named by applying the principles for nomenclature of assemblies of identical units using the multiplicative prefixes "bis-", "tris-", etc.

Examples:

$H\dot{N}-CH_2-CH_2-\dot{N}H$
Ethylenebis (aminyl)
Ethylenebis (azanyl)

$\cdot N=C=N\cdot$
Methanediylidenebis (aminyl)
Methanediylidenebis (azanyl)

$\ddot{N}-CO-[CH_2]_4-CO-\ddot{N}$ and/or $\ddot{N}-CO-[CH_2]_4-CO-N:$
and/or $:N-CO-[CH_2]_4-CO-N:$
Hexanedioylbis (azanylidene)
Hexanedioylbis (nitrate)
Hexne dioylbis (aminylene)

A radical derived formally by the removal of the hydrogen atom from the hydroxy group of an acid or hydroxy characteristic group is named on the basis of a composite parent radical name formed by combining the appropriate prefix derived from the parent hydride or acid residue attached to the chalcogen atom with the term "oxyl", "peroxyl". etc.

Examples:

$CH_3—O\cdot$
Methoxyl

$ClCH_2—CO—O\cdot$
Chloroacetoxyl

$\cdot O—O—HC$(cyclobutane)$CH—O—O\cdot$
Cyclobutane-1,3-diyldiperoxyl

Chalcogen analogues are named on the basis of parent radicals, such as "sulfanyl" and "selanyl".

Examples:

$C_6H_5-S.$
Phenylsulfanyl

$C_6H_5-CO-Se.$
Benzoylselanyl

BENZENE DIAZOMIUM CHLORIDE

Preparation

Preparation From phenylamine or aniline (*which can be made by reduction of nitrobenzene*)

reagents sodium nitrite hydrochloric acid and aniline

conditions keep below 10°C

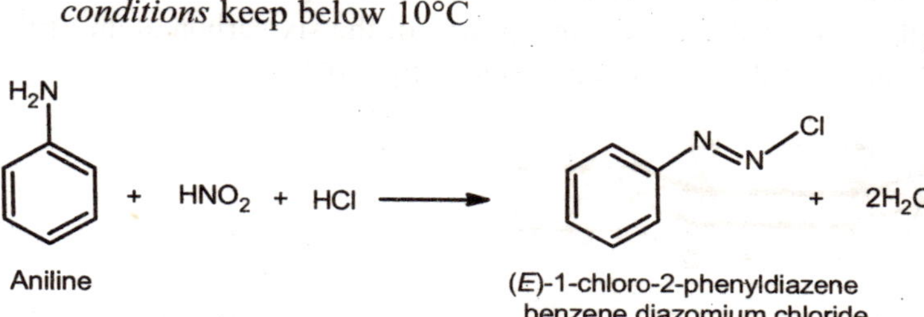

equation:

procedure

Aniline is allowed to react with hydrochloric acid and then a little sodium nitrite is added the reaction temperature is kept under 10ºC

nitrous acid is unstable and is made *in situ* from sodium nitrite

$$C_6H_5NH_2 + NaNO_2 + 2HCl \xrightarrow{<10°C} C_6H_5N_2^+Cl^- + NaCl + 2H_2O$$

The solution is kept cold to slow down decomposition of the diazonium salt

$$C_6H_5N_2^+Cl^- + H_2O \xrightarrow{>10°C} C_6H_5OH + HCl + N_2$$

Structure

To explain the structure of benzene diazomium chloride, when we look at its structure it has 3 p orbitals containing central N atom, one of then overlaps with sp^2 orbital of the cyclohexal ring of benzene. The other two p-orbital superpose with the mirror image of the second N-atom to form two N-atoms. The non-bonded electron pair stays on the free section of the second N-atom (see fig. 30.1)

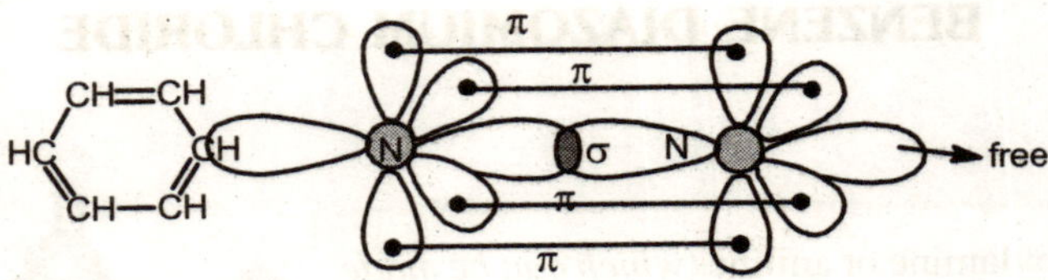

Fig. 30.1. Showing the Structure of Benzene diazomium chloride

Now these p-orbitals which are available, interact with those of benzene ring, therefore welcomes the delocalization of electrons. Now these delocalized electrons pearls all the six carbon atom of the benzene ring and as well as the two N-atom of the diazomium ion (see fig. 30.2).

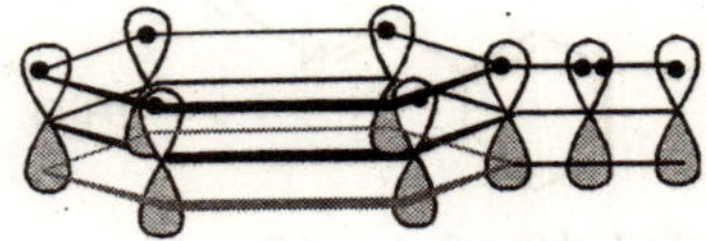

Fig. 30.2. Showing the delocalization of electrons in benzene diazomium chloride

The C – N bond as in the fig. 30.2 exhibits the double strength.

Isomerism in azo (-N=N-) and R_2C=N-X compounds

Organic (or inorganic) compounds of the structure **R-N=N-R'** (e.g. aromatic azo dyes) can exist as cis and trans isomers in just the same way as alkenes, where R or R' = H, alkyl, aryl group etc. **R can be different or the same as R'**. Stereochemically, the lone pairs on the nitrogen effectively behave as an atom bonding pair of electrons in determining the **trigonal planar orientation of the -N= bonds and the lone pair of electrons on the nitrogen**. Three groups of electrons around an atom X, always give a trigonal planar arrangement around the central atom >X-. **The double bond, N=N or C=N, ensures that too high an energy is required for ready rotation about the double bond**. The cis and trans forms will have different physical properties such as melting/boiling points.

Examples of -N=N- systems:

(1) **(cis)** R–N̈=N̈–R' and (2) **(trans)** R–N̈=N̈–R'

(all R-N=N-R' bond angles are about 120°)

Examples of >C=N- systems

Carbonyl compounds like aldehydes and ketones undergo condensation reactions of the type

$$\mathbf{RR'C{=}O + H_2N{-}X \Rightarrow RR'C{=}N{-}X + H_2O}$$

where **R is different to R'** and = **H, alkyl** or **aryl** etc. **geometrical isomers can occur.**

and when **X= H** (ammonia), **alkyl** (aliphatic primary amine), **aryl** (aromatic primary amine), **OH** (hydroxylamine), **NH_2** (hydrazine), **$NHC_6H_3(NO_2)_2$** (2,4-dinitrophenylhydrazine)

(3) (cis wrt R') $\begin{matrix} R' & & X \\ & C=N & \\ R & & \end{matrix}$ and (4) (trans wrt R') $\begin{matrix} R' & & \\ & C=N & \\ R & & X \end{matrix}$

(you can think of R' as a higher ranking group wrt R)

When **R = R'** i.e. (5) $\begin{matrix} R & & X \\ & C=N & \\ R & & \end{matrix}$ **geometrical isomerism is not possible**.

(all >C=N-X angles are about 120°)

Physical Properties

Molecular Formula	= $C_6H_5ClN_2$
Formula Weight	= 140.5703
Composition	= C(51.27%) H(3.59%) Cl(25.22%) N(19.93%)
Molar Refractivity	= 38.35 ± 0.5 cm^3
Molar Volume	= 119.0 ± 7.0 cm^3
Parachor	= 299.5 ± 8.0 cm^3
Index of Refraction	= 1.557 ± 0.05
Surface Tension	= 40.0 ± 7.0 dyne/cm
Density	= 1.18 ± 0.1 g/cm^3
Polarizability	= 15.20 ± 0.5 $10^{-24}cm^3$
Monoisotopic Mass	= 140.014126 Da
Nominal Mass	= 140 Da
Average Mass	= 140.5703 Da

Chemical Properties

Benzene diazonium chloride undergoes two main types of reaction

- **SUBSTITUTION OF THE DIAZONIUM GROUP** (nitrogen expelled)

OH *reagents* water **(hydrolysis)**

equation

$$C_6H_5N_2^+Cl^- + H_2O \xrightarrow{>10°C} C_6H_5OH + HCl + N_2$$

reagents potassium iodide solution

conditions warm

equation $C_6H_5N_2$+ Cl^- + KI ——> C_6H_5I + KCl + N_2

• **COUPLING REACTIONS** the nitrogen atoms are retained

Phenols *reagents* **phenol** and **sodium hydroxide**

conditions **alkaline** solution below 10°C

equation

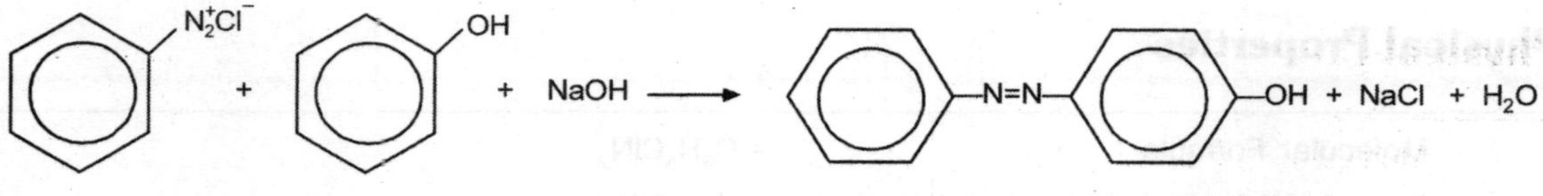

$N_2^+Cl^-$ + OH + NaOH ——→ –N=N– –OH + NaCl + H_2O

(4-hydroxyphenol)azobenzene
Yellow

Diazo Compounds : Diazotates

The group $C_6H_6N_2$ is the common in all the diazomium NH_4 by replacing on hydrogen by the phenyl group and three by a tetravalent nitrogen, like ammonium it forms unstable hydroxide.

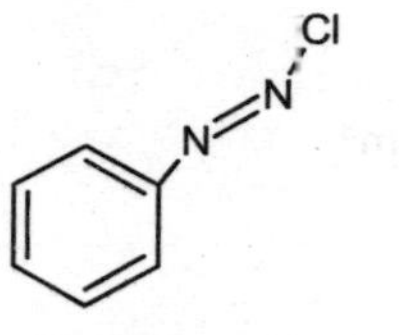

1-chloro-2-phenyldiazene
benzene diazomium chloride

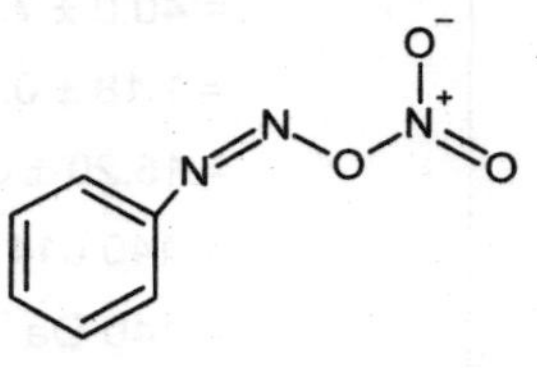

1-(nitrooxy)-2-phenyldiazene
benzene diazomium nitrate

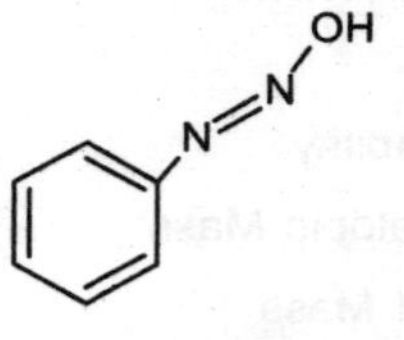

1-hydroxy-2-phenyldiazene
benzene diazomium hydroxide

The salts when dry are sensitive to shock and explode on warming, and the hydroxide can be liberated as an oily liquid, but the salts are always used in aqueous solution. The reaction is generally similar to primary aromatic amines.

We are going to read about it in this chapter

The formula as proposed by Kekule $C_6H_5N{:}N.Cl$ does not satisty the strong ionization of diazomium salts, and has been in favour of Blomstand's formula and after this structure has a similarity with $NH_4^+Cl^-$

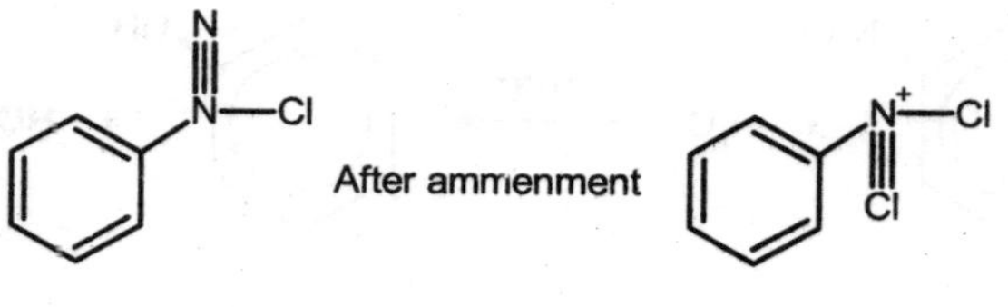

After ammenment

The groping – N : N – is however, used for the non-ionized diazo – compounds, such dizoaminobenzene, also for the azo compounds and again for totally different class of diazotates.

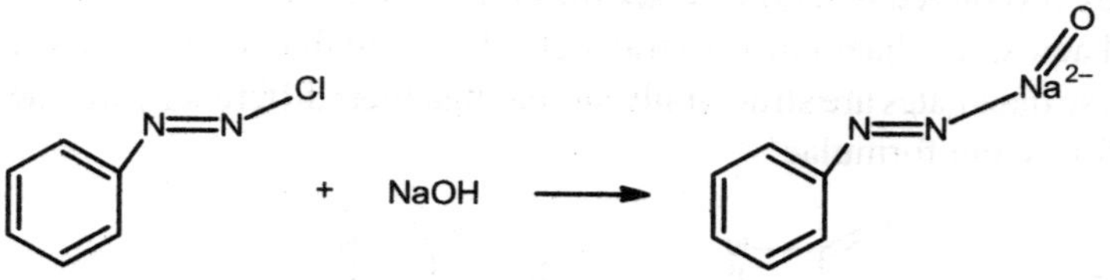

(E)-1-chloro-2-phenyldiazene + NaOH ⟶ oxo[(E)-phenyldiazenato-kN[2]]sodate(2-)

This is a reversible reaction, when excess hydrochloric acid is added. The double bond in the N = N bond brings a kind of geometric isomerism, the *syn* – diazotates which are first formed being less reactive than the *anti* – diazotates, but both can be reduced to hydrazines. The *syn*-diazotates can be coupled to alkaline solutions of phenols to form azo dyes.

Reactions of Diazomium Compounds

When a diazomium salt is treated with alkalies it is converted into a metallic salt or diazotates with a general formula of Ar—N=N—O—X where Ar is the aromatic group and X is the metal

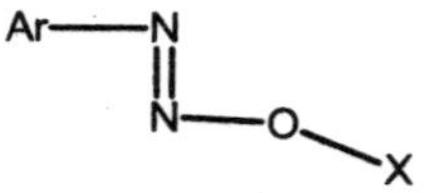

If we see the reaction below the dilemma will be over.

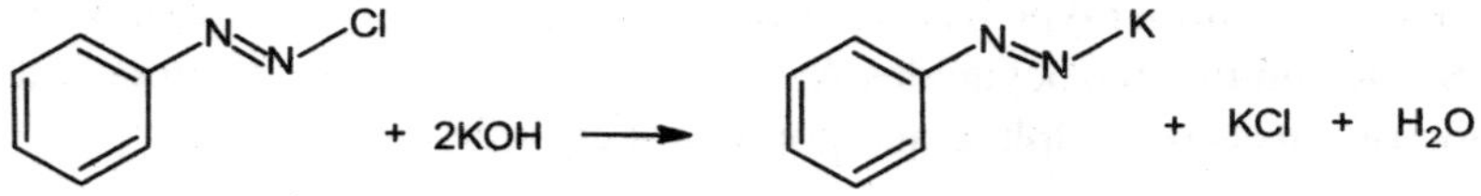

$C_6H_5N{=}NCl$ + 2KOH ⟶ $C_6H_5N{=}NK$ + KCl + H_2O

Potassium (E)-phenyldiazenide

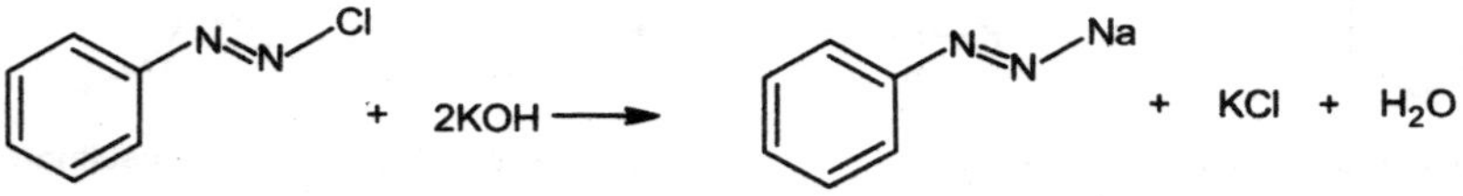

$C_6H_5N{=}NCl$ + 2KOH ⟶ $C_6H_5N{=}NNa$ + KCl + H_2O

Sodium (E)-phenyldiazenide

These diazotates can exist in two isomeric modifications, which are colourless and possess many similar properties. Both are readily reduced to hydrazines and with benzoyl chloride yield nitrobenzanilide. On oxidation both are converted into nitroamine salts e.g.

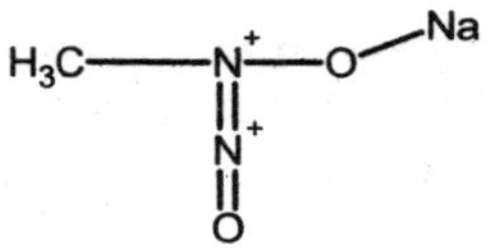

$H_3C—N^+(=N^+=O)—O—Na$

They differ mainly in the relative speeds in which they undergo reactions, for example, the reactable diazotates that are formed couple with phenols in alkaline solutions to give azo dyes, where as the stable diazotates obtained by prolonged reactions of alkalies fail to give this reaction or react very slowly, with the action of mineral acids, the diazotates are transferred back to diazomium salts. This was established by **Hantzsch**, that these diazotates are structurally similar that their differences are due to stereoisomerism, as illustrated in the following formulae.

Syn diazotate labile and couples readily

Anti - diazotate stable and couples slowly or not at all

According to this both syn and anti diazotates are to be regarded as stereo isomeric oximes of nitrobenzene.

Experiments show that both forms are actually produced by the interaction of hydroxylamine and nitrobenzene.

Nitrosobenzene + $H_2N—OH$ $\xrightarrow{KOH}$ Potassium phenyldiazenolate + H_2O

In this reaction labile syn –forms are first obtained, which then pass into the stable anti – modification.

Stereoisomerism of a similar type has also been found to exist in the case of the diazo-sulphonates, $Ar - N == N - SO_3K$ and the diazocyanides $Ar - N == N - CN$. These are obtained from diazomium salts by the action of potassium sulphite and potassium cyanide respectively.

benzene diazomium chloride + $K—SO_2—O—K$ ⟶ potassium (*Z*)-phenyldiazenesulfonate + KCl

benzene diazomium chloride + KCN ⟶ (*Z*)-phenyldiazenecarbonitrile + KCl

The stable anti- diazocyanides are converted into ready to and reactive syn forms by exposure to light. Confirmation of the conclusion of Hantzsch that the stable compounds are of the anti – configuration has now been obtained by dipole moment determinations. The values being small compaired with those of the syn forms.

The Comparative study between nitrosamines and diazo - compounds

Primary Nitrosoamine and anti diazohydroxides – The group – N_2OH in the compounds R – N_2OH is tautomeric, functioning either as an anti diazohydroxides (- N== N – OH) as a primary nitrosamine structure (- NH – NO). According to Hantzsch all the metallic salts R – N_2O or to be regarded as more frequently happens as primary nitrosamines R – NH –NO Hence, in the latter case, during the conversion of the salt (diazotates) into the hydrogen compound an intramolecular rearrangement takes place

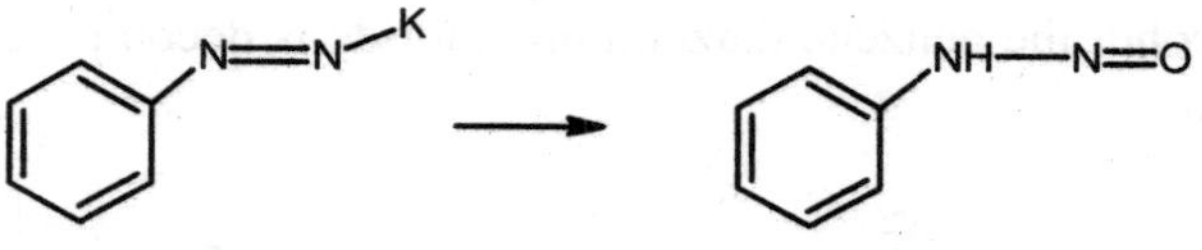

potassium (*E*)-phenyldiazenide → *N*-nitrosoaniline

Nitroamines behave as pseudo-acids reacting with alkalies to form salts of the anti-diazotate structure. The anti diazotates are thus closely related to the nitrosoamines.

The isolation of both isomeric forms of the hydrogen compound form one and the same tautomeric substance has been effected in a few instances e.g. int eh case of 2, 4, 6 trinitrobenzene

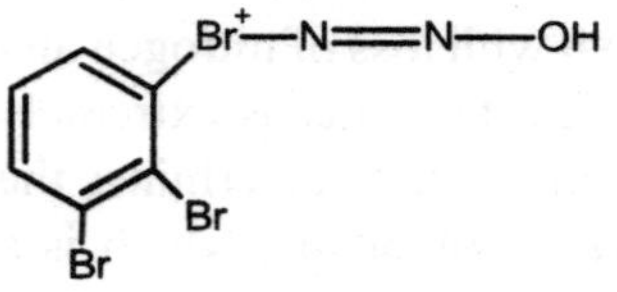

Br$^+$—N═N—OH, Br, Br

(2,3-dibromophenyl)(hydroxydiazenyl)bromonium

and the corresponding nitrosoamine [Br$^+$—NH—N═O, Br, Br] 2, 3 dibromophenyl) (nitrosoamino) bromonium

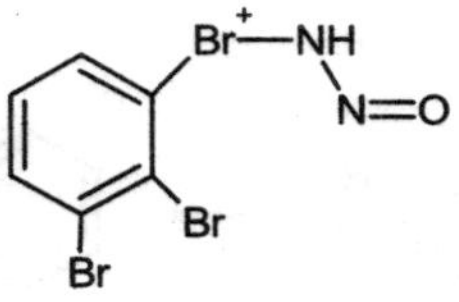

These isomerides recall the somewhat similar aliphatic nitroso compounds

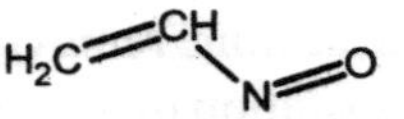

H_2C═CH—N═O Nitrosoethene

H_3C—C≡N^+—OH Ethylidyne(hydroxy)ammonium

and their formation is similar to that of the isomeric nitro – compounds. In chemical behaviour the isomerides differ in accordance with dizohydrates resembles hydroxy acids, where as the nitrosoamine are indifferent pseudo acids.

Decomposition of Diazomium Salts and Compound

Decomposition of diazo compounds includes the hydrolysis of diazo compounds. The compounds during hydrolysis decompose into phenol and nitrogen.

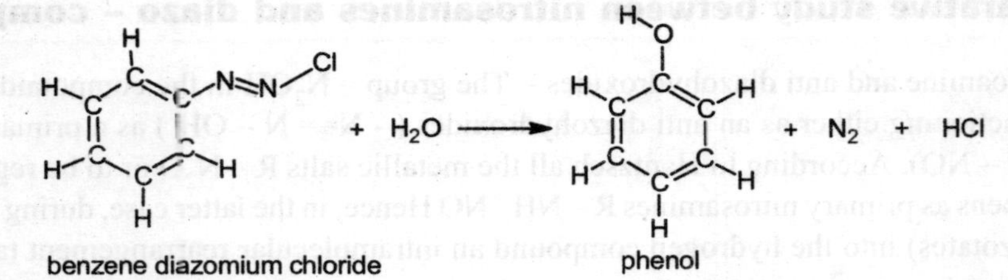

$+ H_2O \longrightarrow$ $+ N_2 + HCl$

benzene diazomium chloride phenol

In presence of alcohol the benzene diazomium chloride is decomposed into hydrocarbon alcoxy derivative and nitrogen

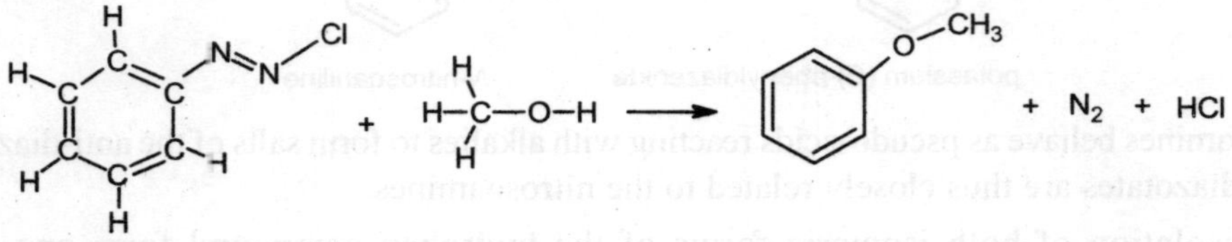

$+ H{-}C(H)(H){-}O{-}H \longrightarrow$ $+ N_2 + HCl$

Benzene diazomium chloride Methanol Methoxybenzene

According to some modern experiment the diazo salt slowly breaks down with loss of nitrogen and liberation of free phenyl or other aryl radical. The compound and its covalent tautomer is extremely reactive and attacks any molecule in its neighbourhood. The part of the change which determines the observed rate of reaction is therefore the slow decomposition of the diazo compounds, which is a unimolecular reaction and largely independent of the medium in which it occurs.

Acetone reacts with diazo compounds as shown below:

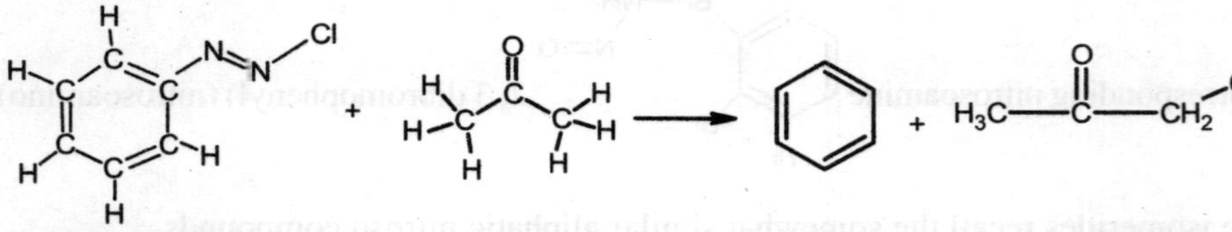

$+ \longrightarrow + H_3C{-}C(=O){-}CH_2Cl$

Benzene diazomium chloride Acetone Benzene 1-chloroacetone

Phenyl radical on being set free immediately attacks the acetone removing hydrogen and producing benzene. The production of 1-chloroacetone indicates the liberation of a neutral active form of chlorine, since the chlorine anion is stable and non reactive.

The chlorine is freely liberated in nascent form, when metals like antimony, bismuth lead and mercury react with the diazo compound.

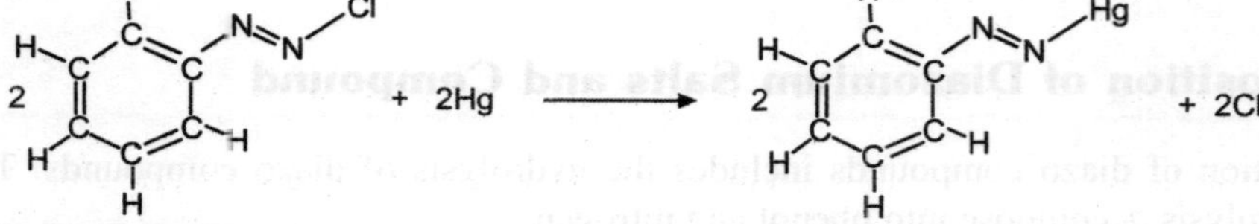

$2 + 2Hg \longrightarrow 2 \; (C_6H_5{-}N{=}N{-}Hg^+) + 2Cl$

Benzene diazomium chloride

When benzene diazomium hydroxide reacts with bromobenzene, bromodiphenyl and free nitrogen is produced.

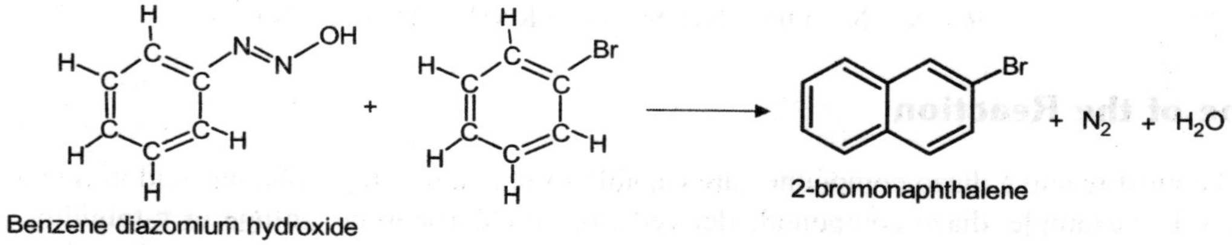

Benzene diazomium hydroxide + → 2-bromonaphthalene + N_2 + H_2O

These nitro phenyls are also produced when nitrobenzene reacts with benzene diazomium hydroxide

benzene diazomium hydroxide + nitrobenzene → 2-nitronaphthalene + N_2 + H_2O

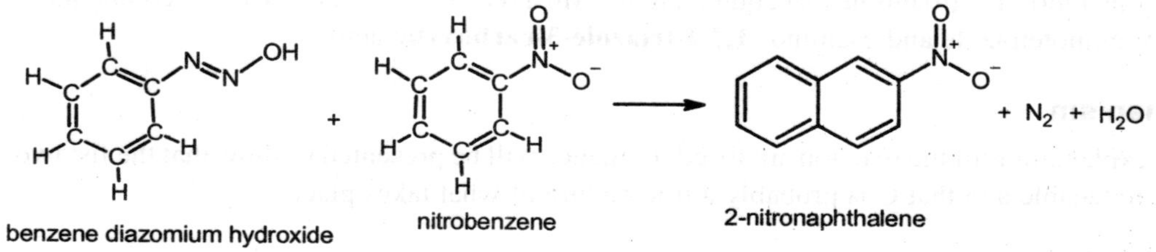

The Photo Sensitivity of Diazo Compounds

When the diazomium compounds are treated with an excess of silver oxide or silver nitrate, a very unstable diazomium hydroxides is produced, the normal, that is the (*syn-*) metallic diazotates on careful addition of acetic acid do not yield the hydrates, but deposit the corresponding diazo hydrides. Hydroxyphenyl diazomium salts, containing the OH group in the ortho and the para position, on treatment with alkalies form internal anhydrides of the diazomium type (1) or quinone type (2)

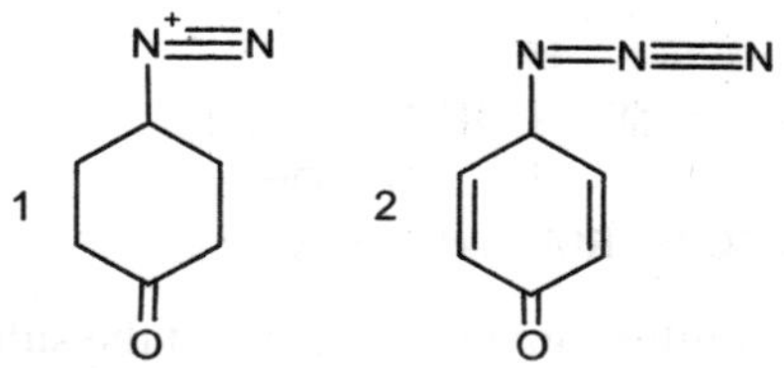

1 2

Hence they are described as quinone diazides

Reaction of Diazo Compounds with Sulphamic Acid

Sulphamic acid has long been used both in industry and in the laboratory to remove excess nitrous acid following diazotization of amines. Obviously, it has been generally assumed that the diazo compounds were unaffected by sulphamic acid under the conditions employed. It has been demonstrated in this Laboratory that a number of diazo compounds do react with sulphamic acid even in strongly acid solution. In fact, in certain cases this reaction is a rapid and quantitative one yielding the products shown in the net equation

$$R—N—N—OH + NH_2SO_1H \rightarrow RNH_2 + H_2SO_4 + N1$$

Scope of the Reaction

Only the most reactive diazo compounds are capable of reacting with sulphamic acid in mineral acid solution. For example, diazo compounds derived from p-chloroaniline, aniline or #-toluidine do not react with sulphamic acid in strong acid solution. 2,5-Dichloroaniline and p-nitroaniline yield diazo compounds which react very slowly with sulphamic acid while the diazo compounds from 2,4 dinitroaniline and 2,6-dichloro-4-nitroaniline react quite rapidly. Most reactive of all are the diazo compounds derived from 5-aminotetrazole and 5-amino- **1,2,4-triazole-3-carboxylic** acid.

Mechanism

Three explanations of the reaction are listed. Evidence will be presented to show that the fist two, A and B, are untenable and that C is probably a true picture of what takes place.

(*a*)

$$R-N=N-OH \xrightleftharpoons{H_2O}$$

$$R-NH_2 + HNO_2 \xrightarrow[RNH_2 + N_2 + H_2SO_1 + H_2O]{NH_2SO_2H} \quad 2$$

(*b*)

$$NH_2SO_2H \xrightarrow{H_2O} NH_2^+ + SO_4H^-$$

$$R-N=N-OH + NH_2 + \longrightarrow [R-N=N-NH_2^+] + H_2O$$

$$\longrightarrow [R-\underset{H}{\overset{H}{N}}-N=N\underset{OH}{H}] \xrightarrow{H_2O} RNH_2^+ + N_2 + H_2O$$

(*c*)

$$[R-N=N-\overset{H}{N}-SO_3H] \xrightleftharpoons{2} [R-\underset{H}{\overset{H}{N}}-N=N-\underset{OH}{SO_2H}] \quad 3$$

$$R-N=N-OH + NH_2SO_2H \quad RNH_2 + N_2 + H_2SO_4$$

Explanation A is incorrect. If equilibrium (1) existed and the sulphamic acid were merely reacting with nitrous acid, it is obvious that the role of sulphamic acid in (2) could be played by urea. However, it was shown by experiment that urea did not react at all to reverse the diazotization reaction. Explanation B breaks down on at least two counts. First of all the hydrolysis of sulphamic acid in cold strongly acid solution is at least ten thousand fold too slow to account for the reaction. The reaction of excess sulphamic acid with the diazo compound derived from 5-amino-l,2,4-triazole-3-carboxylic acid at 0-5ºC is complete within five minutes. However, no barium sulphate precipitate results when an aqueous solution of sulphamic acid, hydrochloric acid, and barium chloride is kept a 5ºC for a week. Secondly, the 5-diazo-1,2,4-triazole-3-carboxylic acid did not react at all with ammonium chloride in dilute hydrochloric acid.

Explanation C is proposed as the correct mechanism for the reaction. Since the first stage of the reaction does not involve hydrolysis of either of the two reactants (mechanisms A or B) it follows that the first step in the reaction directly involves the diazo compound and the sulphamic acid.

Step (1) of mechanism C appears to be the only logical way for the reaction to start if the final products are to include nitrogen and the original amine.

Experimental

5-Diazo-1,2,4-triazole-3-carboxylic Acid. Amino-1,2,4-triazole-3-carboxylic acid (50 g.) was dissolved by warming in water (400 ml.) and concentrated hydrochloric acid (200 ml.). Diazotization was effected at -5° by addition of a slight excess of sodium nitrite solution. The white solid diazo compound was separated by filtration, washed thoroughly with ice water and pressed on the funnel. One gram of this stable press cake was titrated in cold acid solution with 0.05 *N* 2-naphthyl-amine hydrochloride solution to determine its strength. The yield of isolated diazo compound usually amounted to 75% of the theoretical.

Reaction with Sulphamic Acid. The filter cake of the diazo compound (0.05 mole) was slurried in 1 *N* hydrochloric acid solution (200 ml.) and treated with a solution (100 ml.) of sulphamic acid (0.15 mole). Nitrogen was evolved rapidly and the diazo compound was entirely destroyed within five minutes as shown by failure to couple with α-naphthylamine. When acidic solutions of ammonium chloride or urea replaced the sulphamic acid solution, no reaction took place even in several days.

In a second experiment a suspension of the diazo compound was "reversed" to the amine with a slight excess of sulphamic acid, rediazotized, and the newly formed precipitate removed by filtration. This precipitate proved to be identical with the original diazo compound by identical X-ray diffraction patterns.

In a third experiment aliquots of a dilute solution of diazotized 5-amino-l,2,4-triazole-3-carboxylic acid were

(a) titrated with 0.05 *N* 2-naphthylamine hydrochloride and (b) "reversed" to the amine with sulphamic acid, rediazotized, and titrated with 2-naphthylamine. The results of these titrations gave a minimum of 94% yield for the "reversal" and rediazotization.

Substitution reactions of diazonium ions

Diazonium ions are present in solutions such as benzenediazonium chloride solution. They contain an -N2+ group. In the case of benzenediazonium chloride, this is attached to a benzene ring.

Benzenediazonium chloride looks like this:

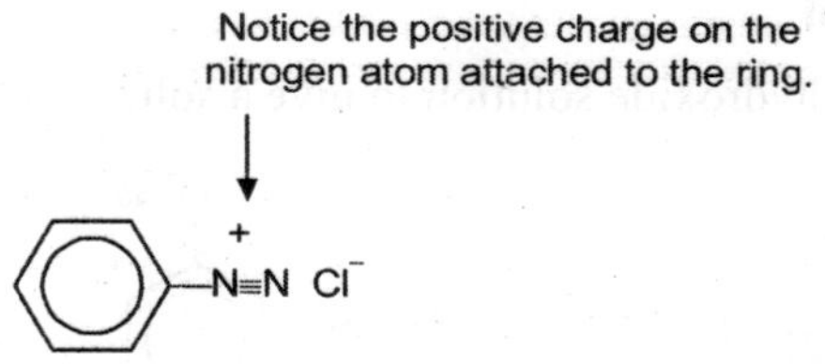

Notice the positive charge on the nitrogen atom attached to the ring.

$C_6H_5-\overset{+}{N}\equiv N\ Cl^-$

Benzenediazonium chloride

In this set of reactions of the diazonium ion, the -N2+ group is replaced by something else. The nitrogen is released as nitrogen gas.

Substitution by an -OH group

To get this reaction, all you need to do is warm the benzenediazonium chloride solution. The diazonium ion reacts with the water in the solution and phenol is formed - either in solution or as a black oily liquid (depending on how much is formed). Nitrogen gas is evolved.

$$C_6H_5-\overset{+}{N}{\equiv}N + H_2O \longrightarrow C_6H_5-OH + N_2 + H^+$$

This is the same reaction that you get if you react phenylamine with nitrous acid in the warm. The diazonium ion is formed first and then immediately reacts with the water in the solution to give phenol.

Note: In this, and all the other reactions of diazonium ions on this page, I am going to write the attached groups at the side of the benzene ring rather than at the top. There's no very sophisticated reason for this! When I write the equations for the coupling reactions further down this page, that's the way I want to draw the structures, because I think it makes them clearer. I am just trying to keep everything else consistent with that.

Substitution by an iodine atom

This is a good example of the use of diazonium salts to substitute things into a benzene ring which are otherwise quite difficult to attach. (That's equally true of the previous reaction, by the way.)

If you add potassium iodide solution to the benzenediazonium chloride solution in the cold, nitrogen gas is given off, and you get oily droplets of iodobenzene formed.

There is a simple reaction between the diazonium ions and the iodide ions from the potassium iodide solution.

$$C_6H_5-\overset{+}{N}{\equiv}N + I^- \longrightarrow C_6H_5-I + N_2$$

Coupling reactions of diazonium ions

In the substitution reactions above, the nitrogen in the diazonium ion is lost. In the rest of the reactions on this page, the nitrogen is retained and used to make a bridge between two benzene rings.

The reaction with phenol

Phenol is dissolved in sodium hydroxide solution to give a solution of sodium phenoxide.

$$C_6H_5OH + NaOH \longrightarrow C_6H_5O^-Na^+ + H_2O$$

Sodium phenoxide

The solution is cooled in ice, and cold benzenediazonium chloride solution is added. There is a reaction between the diazonium ion and the phenoxide ion and a yellow-orange solution or precipitate is formed.

The product is one of the simplest of what are known as azo compounds, in which two benzene rings are linked by a nitrogen bridge.

Note: You might wonder where the hydrogen in the -OH group comes from. It was pushed off that same ring when the nitrogen became attached.

I have made a positive decision not to give names for the products of these coupling reactions. The problem is that there are lots of different variations on the names in common use. Almost every book you look in calls them something different. One web source quoted 45 versions for the compound sometimes called "aniline yellow" (see below). Admittedly, 29 of these were terms like "fast spirit yellow" and "Sudan yellow R", but 16 were variations on chemical names.

Any name which I give is quite likely to conflict with one which you have got from another source - teacher, lecturer or textbook. I'm not going to add to your confusion! Learn the structures, and don't worry too much about the names.

The reaction with naphthalen-2-ol

Naphthalen-2-ol is also known as 2-naphthol or beta-naphthol. It contains an -OH group attached to a naphthalene molecule rather than to a simple benzene ring. Naphthalene has two benzene rings fused together.

The reaction is done under exactly the same conditions as with phenol. The naphthalen-2-ol is dissolved in sodium hydroxide solution to produce an ion just like the phenol one. This solution is cooled and mixed with the benzenediazonium chloride solution.

An intense orange-red precipitate is formed - another azo compound.

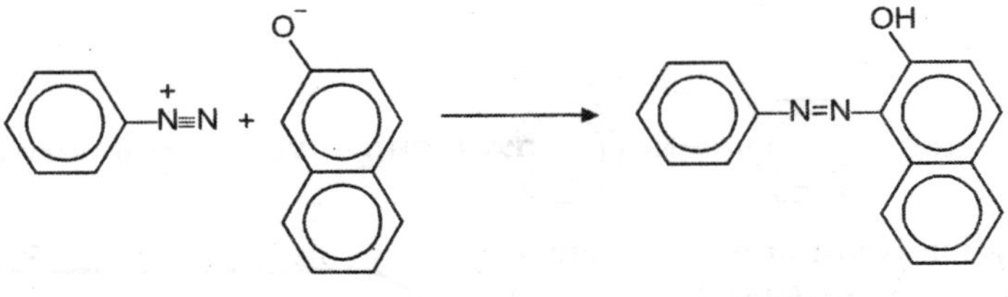

The reaction with phenylamine (aniline)

Some liquid phenylamine is added to a cold solution of benzenediazonium chloride, and the mixture is shaken vigorously. A yellow solid is produced.

$C_6H_5-\overset{+}{N}\equiv N + C_6H_5-NH_2 \longrightarrow C_6H_5-N=N-C_6H_4-NH_2 + H^+$

These strongly coloured azo compounds are frequently used as dyes known as azo dyes. The one made from phenylamine (aniline) is known as "aniline yellow" (amongst many other things - see note above).

Azo compounds account for more than half of modern dyes.

The use of an azo dye as an indicator - methyl orange

Azo compounds contain a highly delocalised system of electrons which takes in both benzene rings and the two nitrogen atoms bridging the rings. The delocalisation can also extend to things attached to the benzene rings as well.

If white light falls on one of these molecules, some wavelengths are absorbed by these delocalised electrons. The colour you see is the result of the non-absorbed wavelengths. The groups which contribute to the delocalisation (and so to the absorption of light) are known as a chromophore.

Note: A Canadian university site describes "chromophore" as "one of those useful but sloppy words whose meaning depends somewhat on the context."

Some sources take the chromophore as being the whole of the two benzene rings and the nitrogen bridge. Others seem to take it as just the -N=N- group.

I'm taking the line that it is the two benzene rings plus the -N=N- group. If your examiners take the more restricted meaning, obviously you should go with what they want. The only way you will find that out is to look at recent exam papers and mark schemes.

Modifying the groups present in the molecule can have an effect on the light absorbed, and so on the colour you see. You can take advantage of this in indicators.

Methyl orange is an azo dye which exists in two forms in equilibrium with each other:

$HO_3S-C_6H_4-N=N-C_6H_4-N(CH_3)_2$ Red from of methyl orange

$\rightleftharpoons$

$H^+ + {}^-O_3S-C_6H_4-N=N-C_6H_4-N(CH_3)_2$ Yellow from of methyl orange

As the hydrogen ion is lost or gained there is a shift in the exact nature of the delocalisation in the molecule, and that causes a shift in the wavelength of light absorbed. Obviously that means that you see a different colour.

we shall read about the construction of methyl orange in the chapter 29 Dyes stain and pigments

When you add acid to methyl orange, the equilibrium is shifted to give more of the red form. Methyl orange is red in acidic solutions (in fact solutions of pH less than 3.1).

If you add an alkali, hydrogen ions are removed and the equilibrium shifts to give more of the yellow form. Methyl orange is yellow at pH's greater than 4.4.

In between, at some point there will be equal amounts of the red and yellow forms and so methyl orange looks orange.

Uses of Dizomium Compounds

Owing to their sensitivity towards light these compounds are used in the manufacture of photographic tracing paper. In the process of manufacturing 1, 2, 4 aminonapthol sulphonic acid (diazotized) and resorcinol is used for the purpose.

Some More Diazo Compounds

HYDRAZINE

Introduction

Hydrazine is the chemical compound with formula N_2H_4. This species is widely used in chemical synthesis. It is perhaps best known as a component in rocket fuel. Hydrazine is also found in tobacco products. It has an ammonia-like odor.

Synthesis

The odour Curtius synthesized free hydrazine for the first time in 1889 via a circuitous route.

Hydrazine is produced in the Olin Raschig process from sodium hypochlorite and ammonia, a process developed in 1907. This method relies on the reaction of chloramine with ammonia. The Olin Raschig process is an industrial process used to produce hydrazine and developed by the Olin Corporation. Sodium hypochlorite solution is mixed with ammonia at 5 °C to form chloramine and sodium hydroxide, which is then rapidly added to anhydrous ammonia under pressure and heated to 130 °C to generate hydrazine, water and sodium chloride (waste product). Ammonia is used in a 33 fold excess.

In the Atofina-PCUK cycle, hydrazine is produced in several steps from acetone, ammonia, and hydrogen peroxide. Acetone and ammonia first react to give the imine followed by oxidation with hydrogen peroxide to the oxaziridine, a three-membered ring containing carbon, oxygen, and nitrogen, followed by ammonolysis to the hydrazone, a process that couples two nitrogen atoms. This hydrazone reacts with one more equivalent of acetone, and the resulting azine is hydrolyzed to give hydrazine, regenerating acetone. Unlike the Raschig process, this process does not produce salt.

Hydrazine can also be produced via the ketazine and the peroxide processes

Physical Properties

General	
Systematic name	Hydrazine
Molecular formula	N_2H_4
Molar mass	32.05 g/mol
Appearance	Colourless liquid
Properties	
Density and phase	1.01 g/mL (liquid)
Solubility in water	infinitely miscible
Other solvents	miscible with polar organic solvents
Melting point	274 K (1 °C)
Boiling point	387 K (114 °C)
Thermochemistry	
$\Delta_f H^0_{gas}$	95.35 kJ/mol
$\Delta_f H^0_{liquid}$	50.63 kJ/mol
$\Delta_f H^0_{solid}$	37.63 kJ/mol
$S^0_{gas, 1 bar}$	238.66 J/mol.K
$S^0_{liquid, 1 bae}$	121.52 J/mol.K

Molecular structure and properties

Conceptually, hydrazine arises via coupling a pair of ammonia molecules by removal of one H per molecule. Each H_2N-N subunit is pyramidal. The N-N distance is 1.45 Å. The molecule adopts a gauche conformation, as seen also in hydrogen peroxide. The rotational barrier is twice that of ethane.

Hydrazine has a liquid range and density like water's. It has basic properties comparable to ammonia but 15-times weaker It can be di-protonated only with difficulty

$$N_2H_4 + H+ \rightarrow [N_2H_5]+ K = 8.5 \times 10^{-7}$$

(for ammonia K = 1.78×10^{-5})

$$[N_2H_5]+ + H+ \rightarrow [N_2H_6]_2+ K = 8.4 \times 10\text{-}16$$

Chemical Properties

Hydrazine is a powerful reducing agent. It is attractive as a reducing agent due to its high hydrogen content, and friendly by-product of nitrogen. It will reduce a number of important metal salts to the element, including silver and nickel.

Producing 148.6 kcal/mol in its oxidation reaction, hydrazine has an impressive affininty for oxygen:

$$N_2H_4 + O_2 = N_2 + 2H_2O$$

It is used in this capacity to remove oxygen from boiler systems, and as an additive to many substances to prevent oxidative deterioration.

Hydrazine is a weak base (Kb = 8.5×10^{-7} at 25 C), more reactive than ammonium sulphide and calcium oxide, but less alkali than sodium hydride, calcium hydride, potasium hydroxide, and sodium hydroxide. Hydrazine is incompatable with acids.

Hydrazine is a reasonable solvent, completely dissociating alkali-metal halides, tetraalkylammonium halides, and many carboylic acids. However, a number of oxides, sulphates, and carbonates are practically insoluble. Since ammonia's proton affinity is so similar, it is usually used rather than hydrazine as a solvent, being safer and more readily available.

Alkali metal hydrazindes, formed by dissolving the metals into hydrazine, are one of the few substances which behave as bases. These tend to be pyrophoric with air, and sodium hydrazide (NaN2H3), for example, explodes violently when heated above 100 C. As a result, alkali metals are incompatable with hydrazine. Alkali metals are common to reduce melting temperatures in some glasses-glass composites, for example.

Platinum-group metal catalysis such as platinum black and Raney nickel are incompatible with hydrazine, causing exothermic decomposition. A hydrazine-oxygen mixture in the presence of platinum ignites at a mere 30 C, while hydrazine-air mixtures on an iron rust surface ignite at 24 C and on a metal burner ignite at 130 C. On glass surfaces, auto-ignition temperature is 270 C. Contact with many porous substances such as rusty surfaces, earth, asbestos, wood, or cloth can cause oxidation or explosion.

Hydrazine spontaneously explodes upon contact with calcium oxide, barium oxide, iron oxides, copper oxide, chromate salts, and many others. Calcium is a possible metal tank alloy ingredient, the oxide of which forms on tank surfaces. Metallic manganese, lead, and copper alloys also spontaneously participate in violent reactions. It is incompatable with ammonia and with organic compounds containing easily reduced functional groups. Hazardous polymerisation does not occur, however.

All flammable liquids, inclusing hydrazine, are incompatable with chemicals such as nitrates, permanganates, chromic acid, hydrogen peroxide, nitric acid, sodium peroxide, halogens, or halogenated chemicals such as phosphorous tetrachloride, oxidizing materials, and highly oxygenated or halogenated solvents.

Hydrazine is sometimes listed as incompatable with metals and combustable materials, as its presence greatly increases burning rates in an oxygenated environment such as Earth.

Dry chemical, alcohol-resistant foams, or CO_2 fire extinguishers all are effective against hydrazine reactions. Water may be ineffective against hydrazine fires, although it's useful in keeping affected materials cool. As the vapour is heavier than air, it may travel along the ground or with ventilation to an ignition source, and "flash back" much like hydrocarbon vapours. Unless fires can be fought from a distance with hose-holders and such, it is best just to let it burn itself out.

Hydrazine is listed among shock-sensitive chemicals, as a chemical prone to rapidly decompose or explode when struck, vibrated, or otherwise agitated. It is flammable in mixtures with air from 4.7% to 100% hydrazine.

The presence of hydrogen increases thermodynamic stability by promoting decomposition into ammonia, rather than to nitrogen and hydrogen. Ammonia dissociation has a reaction energy of -44.8 kcal/mol, rather than -22.8 for decomposition into the gasses.

Butyl rubber and polyvinyl chloride provide good to excellent protection, and should be considered for plumbing materials.

Biological Effects

Hydrazine is a dangerous, although non-cumulative poison, is corrosive to eyes, skin, and mucous membranes, and is a probable human carcinogen. Its reducing effects make it extremely destructive to tissues, and it has a variety of adverse systemic effects. Monomethyl hydrazine has a half-life of 2 days in the environment, and does not bioaccumulate.

Caustic burns to the skin are the immediate result of contact with the liquid. Hydrazine dissolves hair, and causes caustic-like burns on skin.

The vapour is extremely irritating to the eyes and temporary blindness can result, and eye contact with the liquid causes burns and possibly permanent damage. Flushing with large quantities of water for 15 minutes is the recommended treatment.

Vapours cause irritation to the nose, throat, and respiratory tract. High concentrations cause spasm, inflammation, chemical pneumonitis, and pulmonary edema, and coma in humans, being fatal to half of rats exposed for four hours at 570 ppm and half of mice at 252 ppm. Other symptoms of exposure may include a burning sensation, coughing, wheezing, laryngitis, shortness of breath, headache, nausea, and vomiting. Symptoms from chronic, nonacute exposure include lethargy, vomiting, itching and burning sensations, tremors, conjunctivitis, and contact dermititis.

Readily absorbed through the skin, it can inhibit certain enzyme systems causing reduced body metabolism. Other systemic effects may include damage to liver, kidneys, nervous system, and red blood cells. For skin contact with rabbits, its LD-50 is 91 mg/kg, and with guinea pigs it is 190 mg/kg. For acute oral doses with rats, its LD-50 is 60 mg/kg, and with mice it is 59 mg/kg.

Hydrazine has a fishy or ammonia-like odor detectable at a mean of 3.7 ppm. The normal 8-hour OSHA permissible exposure limit (PEL) is 1 ppm, and the immediately dangerous to life and health (IDLH) value is 50 ppm. Hydrazine can be detected in the blood of exposed individuals.

Recommended safe levels for hydrazine's carcinogenic properties are far lower than those of biological effects stemming from chemical properties. EPA Cancer Risk Level (1-in-a-million excess lifetime risk) is 2.0×10^{-73}. The EPA suggests that air containing 0.02 mg/m^3 of hydrazine directly increases cancer risks by 1 part in ten thousand.

The chemical is not considered to have adequate warning properties. Hydrazine detectors, similar to ammonia detectors, are used where the chemical is handled, and detector cards are used which discolour in the presence of hydrazine.

Uses of Hydrazine

Hydrazines are part of many organic syntheses, often those of practical significance in pharmaceuticals, such as antituberculants, as well as in textile dyes and photography.

Hydrazone formation

Illustrative of the condensation of hydrazine with a simple carbonyl is its reaction with acetone to give the azine. This azine reacts further with hydrazine to afford the hydrazone:[4]

$$2\,(CH_3)_2CO + N_2H_4 \rightarrow 2H_2O + [(CH_3)_2C{=}N]_2$$

$$[(CH_3)_2C{=}N]_2 + N_2H_4 \rightarrow 2(CH_3)_2C{=}NNH_2$$

The acetone azine is an intermediate in the Atofina-PCUK synthesis. Direct alkylation of hydrazines with alkyl halides in the presence of base affords alkyl-substituted hydrazines, but the reaction is typically inefficient due to poor control on level of substitution (same as in ordinary amines). The reduction of hydrazones to hydrazines present a clean way to produce 1,1-dialkylated hydrazines.

Wolff-Kishner reduction

Hydrazine is used in the Wolff-Kishner reduction, a reaction that transforms the carbonyl group of a ketone or aldehyde into a methylene (or methyl) group via a hydrazone intermediate. The production of the highly stable dinitrogen from the hydrazine derivative helps to drive the reaction.

Heterocyclic chemistry

Being bifunctional, with two amines, hydrazine is a key building block for the preparation of many heterocyclic compounds via condensation with a range of difunctional electrophiles. With 2,4-pentanedione, it condenses to give the dimethylpyrazole. In the Einhorn-Brunner reaction hydrazines react with amides to give triazoles.

Sulphonation

Being a good nucleophile, N_2H_4 is susceptible to attack by Sulphonyl halides and acyl halides. The tosylhydrazine also forms hydrazones upon treatment with carbonyls.

Deprotection of phthalimide

Hydrazine is used to cleave N-alkylated phthalimide derivatives. This scission reaction allows phthalimide anion to be used as amine precursor.

Reducing agent

Hydrazine is a convenient reductant because the by-products are typically nitrogen gas and water. Thus, it is used as an antioxidant, an oxygen scavenger, and a corrosion inhibitor in water boilers and heating systems. It is also used to reduce metal salts and oxides to the pure metals in electrodeless nickel plating and plutcnium extraction from nuclear reactor waste.

Hydrazinium salts

Hydrazine is converted to solid salts by treatment with mineral acids. A common salt is hydrazine sulphate, $[N_2H_5]HSO_4$, which probably should be called hydrazinium sulphate. The salt of hydrazine and hydrazoic acid N_5H_5 were of scientific interest, because of the high nitrogen content and the explosive properties.

Other industrial uses

Hydrazine is used in many processes including: production of spandex fibers, as a polymerization catalyst, a blowing agent, in fuel cells, solder, fluxes, and photographic developers, as a chain extender in urethane polymerizations, and heat stabilizers. In addition, a semiconductor deposition technique using hydrazine has recently been demonstrated, with possible application to the manufacture of thin-film transistors used in liquid crystal displays. Hydrazine in a 70% hydrazine, 30% water solution is used to power the EPU (emergency power unit) on the F-16 fighter plane.

Rocket fuel

Hydrazine is also used as rocket fuel starting in World War II for the Messerschmitt Me 163, under name B-Stoff (hydrazine hydrate) and in a mixture with methanol (M-Stoff) as C-Stoff).

Hydrazine is also used as a low-power monopropellant for the maneuvering thrusters of spacecraft, and the Space Shuttle's Auxiliary Power Units. In addition, monopropellant hydrazine-fueled rocket engines are often used in terminal descent of spacecraft. A collection of such engines were used in both Viking landers as well as the Phoenix lander scheduled to launch in August 2007.

In all hydrazine monopropellant engines the hydrazine is passed by a catalyst such as iridium metal supported by high-surface area alumina or carbon nanofibers, or more recently molybdenum nitride on alumina, which causes it to decompose into ammonia, nitrogen gas, and hydrogen gas according to the following rections:

$$3\,N_2H_4 \rightarrow 4NH_3 + N_2$$

$$N_2H_4 \rightarrow N_2 + 2H_2$$

$$4NH_3 + N_2H_4 \rightarrow 3N_2 + 8H_2$$

These reactions are extremely exothermic (the catalyst chamber can reach 800°C in a matter of milliseconds), and they produce large volumes of hot gas from a small volume of liquid hydrazine, making it an efficient thruster propellant.

The explosive Astrolite is made by combining hydrazine with ammonium nitrate.

Preparation of different hydrazine derivates

Hydrazine is used for the preparation of different derivatives which is described below:

Derivatives of Hydrazine

PHENYLHYDRAZINE

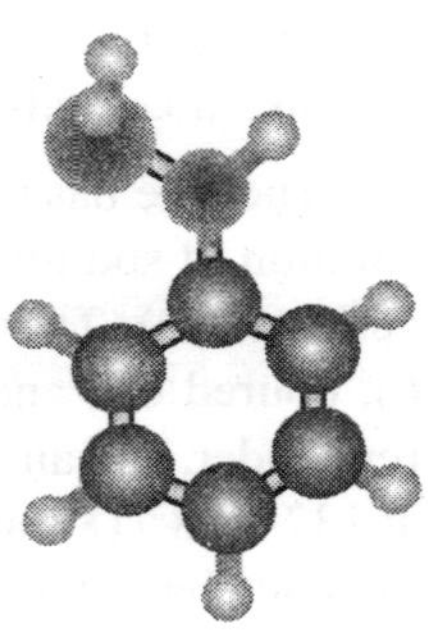

Introduction

Phenylhydrazine was the first hydrazine derivative characterized, reported by **Emil Fischer** in 1875.He prepared it by reduction of a phenyl diazonium salt using sulphite salts. Fischer used phenylhydrazine to characterize sugars via formation of hydrazones with the sugar aldehyde. He also demonstrated in this first paper many of the key properties recognized for hydrazines.

Preparation

In a 1000ml. round-bottomed flask, fitted with a mechanical stirrer, is placed 104.5 cc. of concentrated commercial hydrochloric acid (sp. gr. 1.138). The flask is surrounded with a freezing mixture of ice and salt, and, when the contents are at 0°, stirring is started and 50.0 g. of cracked ice is added, or more ice can be added and the external cooling dispensed with; then 37.2 g. (36.4 cc., 4 moles) of aniline, also cooled to 0°, is run in during five minutes. The mixture is treated with 500 g. more of cracked ice, and a cold solution (0°) of 29.0 g. (4 moles) of technical sodium nitrite dissolved in 60.0 cc. of water is allowed to run in slowly (twenty to thirty minutes) from a separatory funnel, the end of which is drawn to a small tip and reaches nearly to the bottom of the flask. During this addition, the stirrer is operated rather vigorously, and the temperature is kept as near 0° as possible by the frequent addition of cracked ice (about 500gm).

In the meantime, a sodium sulphite solution is prepared by dissolving 89.0 g. (20 moles) of sodium hydroxide, of about 90 per cent purity, in about 100ml. of water and then diluting to 600ml. A few drops of phenolphthalein solution are added and sulfur dioxide passed in, first until an acid reaction is indicated and then for two or three minutes longer. During the addition of the sulfur dioxide, the solution is cooled with running water. On account of the strong alkaline solution, the original colour produced by the phenolphthalein is very faint, but this slowly increases until it becomes deep just before the acid point is reached. It is best to remove a small sample of the liquid from time to time, dilute with three or four volumes of water, and add a drop more of phenolphthalein (Note 1).

The sodium sulphite solution is placed in a 100ml. flask and cooled to about 5°. Approximately 50.0 g. of cracked ice is added, and then, with mechanical stirring, the diazonium salt solution is run in as rapidly as possible (Note 2). The mixture becomes a bright orange-red. The flask is now warmed to about 20° on a steam bath, until the solid sodium sulphite, which has separated while cooling, redissolves. The total amount of liquid is now about 80ml. One-half of this is poured into another 1000ml. flask, and both halves are warmed on the steam bath to 60–70°, until the colour becomes quite dark (thirty to sixty minutes). Sufficient hydrochloric acid (30.0–40.0 cc.) is now added (Note 3) to each flask to make the solutions acid to litmus. The heating is continued and the colour gradually becomes lighter until, after four to six hours, the solutions have become nearly colourless; they may be heated overnight, if desired.

To the hot solutions is now added about one-third of their volume of concentrated hydrochloric acid (2 l. to each portion), and the mixtures are cooled, first in running water, then in a freezing mixture, to 0° (Note 4). The phenylhydrazine hydrochloride precipitates in the form of slightly yellowish or pinkish crystals which may be filtered off and dried (Note 5).

The free base is liberated by adding to the phenylhydrazine hydrochloride 100ml of a 25 per cent solution of sodium hydroxide. The phenylhydrazine separates and is taken up with benzene (two 300-cc. portions). The combined extractions are well dried with 20.0 g. of solid sodium hydroxide (Note 6), poured off, and distilled. Most of the benzene may be distilled under ordinary pressure, and the remainder, and any low-boiling impurities, under diminished pressure. The pure phenylhydrazine boils at 137–138°/18 mm., and is obtained as a pale yellow liquid (Note 7). It can be crystallized on cooling in an ice bath; the crystals melt at 23°. The crude phenylhydrazine from two lots of aniline (74.4 g.) is best distilled at one time and gives 69.5–72.5 g. of pure product (80–84 per cent of the theoretical amount).

Notes

1. If the sodium sulphite solution contains an excess of alkali, a black tar tends to form when the solution is warmed, and very little phenylhydrazine is obtained. Great care must be taken in determining the end point in the neutralization of the sodium hydroxide by the sulfur dioxide.

It is best to use freshly prepared sodium sulphite for the reduction, since the commercial quality is poor and gives a lower yield of phenylhydrazine. A cylinder of liquid sulfur dioxide should, of course, be available.

2. The rapid addition of the diazonium salt solution to the sodium sulphite seems to be advantageous.

3. If the sodium sulphite-diazonium salt mixture is acidified before warming or before becoming dark, the red colour of the solution does not disappear on heating, and the precipitated phenylhydrazine hydrochloride obtained is coloured red.

Most published directions for the preparation of phenylhydrazine specify the use of zinc dust and acetic acid following the reduction with sodium sulphite. No improvement in the quality or quantity of the product was obtained by using zinc and acetic acid.

4. In order to obtain the maximum yield, it is necessary to cool the hydrochloric acid solution of the phenylhydrazine hydrochloride from 20° to 0°, before filtration. From 5 to 10 per cent of product separates between these two temperatures. When this is done, no more phenylhydrazine hydrochloride is obtained by concentration of the mother liquor. An increase in the amount of hydrochloric acid above 200ml. for the precipitation of the hydrochloride produces no increase in yield of product.

5. The phenylhydrazine hydrochloride may be purified by crystallizing from water. A 600-cc. portion of water is used for 100 g. of crude hydrochloride, and the solution boiled a short time with a few grams of animal charcoal. After filtering, 200 cc. of concentrated hydrochloric acid is added, and the mixture cooled to 0°. Pure white crystals in a yield of 85–90 g. are obtained.

6. The benzene solution of phenylhydrazine should be well dried before distilling, since the presence of moisture causes an increased amount of foaming to take place just after the benzene has distilled off.

When the distillation is carried out carefully, practically no phenylhydrazine distills with the benzene or other low-boiling impurities.

Care should be taken that the free base contains no trace of hydrochloride, for this compound catalyzes the decomposition of phenylhydrazine above 100°.

7. Pure phenylhydrazine dissolves in dilute acetic acid to yield a perfectly clear solution.

Care should be taken when working with large quantities of phenylhydrazine, since the product may cause serious injury to the skin. The vapors of phenylhydrazine should not be inhaled.

Discussion

Phenylhydrazine can be prepared by the reduction of benzenediazonium salts by means of a miscellany of reducing agents, of which sodium sulphite1 is preferred. Although this method is given in several laboratory manuals, the results were not found entirely satisfactory. The present directions provide for a lengthy but essential heating of the diazonium-sulphite mixture, omit the useless zinc dust reduction, and supply exact details for preparation on a fairly large laboratory scale. The electrolytic reduction of benzene diazonium salts has been reported to furnish a quantitative yield of phenylhydrazine.

Reaction Equation

$$C_6H_5NH_2\cdot HCl \xrightarrow{NaNO_2,\ HCl} C_6H_5N_2Cl$$

$$C_6H_5N_2Cl \xrightarrow{NaOH,\ SO_2} C_6H_5NHNH_2\cdot HCl$$

$$C_6H_5NHNH_2\cdot HCl \xrightarrow{HCl,\ then\ NaOH} C_6H_5NHNH_2$$

Physical Properties

Boiling point (decomposes):	243.5°C
Melting point:	19.5°C
Relative density (water = 1):	1.1
Solubility in water, g/100 ml at 25°C:	14.5
Vapour pressure, Pa at 20°C:	10
Relative vapour density (air = 1):	3.7
Flash point:	88°C

Auto-ignition temperature:	174°C
Explosive limits, vol% in air:	1.1–1.2
Octanol/water partition coefficient as log Pow:	1.25
Molecular Formula	$C_6H_8N_2$
Formula Weight	108.14112
Composition	C(66.64%) H(7.46%) N(25.90%)
Molar Refractivity	34.76 ± 0.3 cm^3
Molar Volume	96.0 ± 3.0 cm^3
Parachor	255.4 ± 4.0 cm^3
Index of Refraction	1.643 ± 0.02
Surface Tension	49.9 ± 3.0 dyne/cm
Density	1.125 ± 0.06 g/cm^3
Polarizability	13.78 ± 0.5 $10^{-24} cm^3$
Monoisotopic Mass	108.068748 Da
Nominal Mass	108 Da
Average Mass	108.1411 Da

Chemical Properties

Phenylhydrazine forms monoclinic prisms that melt to an oil around room temperature which may turn yellow to dark red upon exposure to air. Phenylhydrazine is miscible with ethanol, diethyl ether, chloroform, and benzene. It is sparingly soluble water.

Phenyl hydrazine is a very strong reducing agent, produces cuprous oxide from Fehling's reagent. While getting involved in the reduction reaction phenylhydrazine gets decomposed into benzene nitrogen and hydrogen.

H_2N–NH (Phenylhydrazine) $\xrightarrow{2(H)}$ Benzene + N_2 + $2H_2$

Phenylhydrazine　　Benzene

When reduced with strong reducing agents like hydrogen sulphide, phenylhydrazine breaks into aniline and ammonia.

$H_2N-NH-C_6H_5 \xrightarrow{2(H)} C_6H_5NH_2 + NH_3$

Phenylhydrazine Aniline

Phenylhydrazine reacts with ethyl aceto-acetate to produce 3-methyl 1-phenyl 5 pyzolone and this on methylation produces antipyrine, which is a drug used as an analgesic.

Phenylhydrazine + Ethyl 3-oxobutanoate → 3-methyl-1-phenyl-5-pyrazolone $\xrightarrow{CH_3OH}$ 1,5-dimethyl-2-phenyl-1,2-dihydro-3 H-pyrazol-3-one antipyrine

Phenylhydrazine is prepared in large scale for the production of antipyrine.

Nitroso and Hydroxylamine Derivatives of Hydrocarbons

Mono – nitroso derivatives are obtained in general, by the reaction of potassium dichromate on aniline

Aniline $\xrightarrow[H_2SO_4]{K_2Cr_2O_7}$ N-hydroxyaniline $\xrightarrow[H_2SO_4]{K_2Cr_2O_7}$ Nitrosobenzene

Generally these compounds are very volatile and exist in different molecular state. The solid aromatic nitroso – compounds are colourless and bimolecular, but in solution or when fused, the great majority of them assume blue or green colour and give molecular weights corresponding to the monomolecular formula C_6H_5 NO. On further oxidation the nitroso – compounds readily pass into nitro compounds and on reduction they yield amino compounds.

The typical aromatic represent to this class, nitrobenzene C_6H_5NO, is obtained by oxidizing β-phenyl hydroxyl amine with potassium dichromate and sulphuric acid.

N-hydroxyaniline $\xrightarrow[H_2SO_4]{K_2Cr_2O_7}$ Nitrosobenzene

It is also produced by the oxidation of aniline by acidified potassium permanganate with sulphuric acid in presence of a little formaldehyde.

Aniline $\xrightarrow[H_2SO_4]{K_2Cr_2O_7 + CH_2O}$ Nitrosobenzene

It crystallizes in colourless volatile needle less (m.p 68) and possesses a powerful characteristic smell. In the molten state or in solution it is emerald green.

It is readily oxidized to nitrobenzene or reduced to aniline. Nitro benzene condenses with aniline to form azo benzene and with β-phenylhydroxylamine to form azoxybenzene.

β-Arylhydroxylamines are prepared by reducing aromatic nitro compounds with neutral reagents such as zinc dust and ammonium chloride solution, or ammonium sulphide.

Nitrobenzene $\xrightarrow[NH_4Cl]{Zn}$ N-hydroxyaniline + Chlorobenzene

They are also obtained by electrochemical reduction. Where acetic acid and sodium acetate solution is used as cathode solution.

Phenylhydroxylamine is obtained as white powder with a melting point of 81°C. The powdered substance produces violent sneezing. Atmospheric oxygen converts it into potassium dichromate it produces nitrobenzene.

It reduces Fehling's reagent and Tollen's reagent even in cold. It produces salts with acids and when warmed with inorganic mineral acids it is converted into p-aminophenol.

N-hydroxyaniline $\xrightarrow{HCl}$ 4-aminophenol

Nitrous acid converts it into a nitroso derivative $C_6H_5N(NO)OH$. This nitroso phenylhydroxylamine is more conveniently obtained by the action of nitric oxide on an ethereal solution of phenyl magnesium bromide.

NO ⟶ N_2O_2 ⟶ O=N—N(—O—Mg—Br)(C_6H_5) ⟶ O=N—N(OH)(C_6H_5)

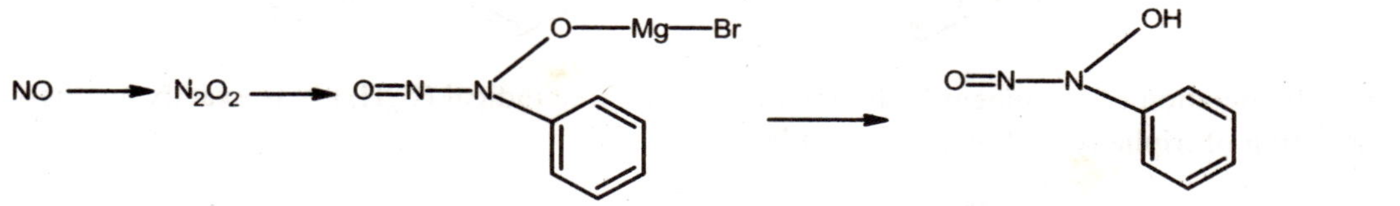

1-hydroxy-2-oxo-1-phenylhydrazine

Diphenyl-hydroxylamine

This compound is prepared by treating nitrosobenzene with phenyl magnesium bromide

$C_6H_5N{=}O$ + C_6H_5Mg—Br $\xrightarrow{+H_2O}$ $(C_6H_5)_2N$—OH + MgBr

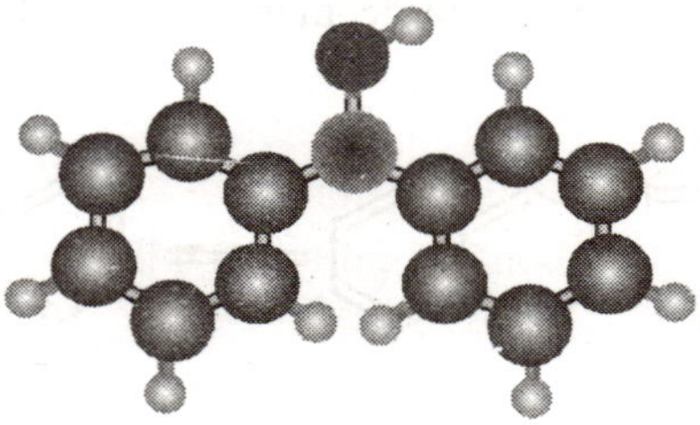

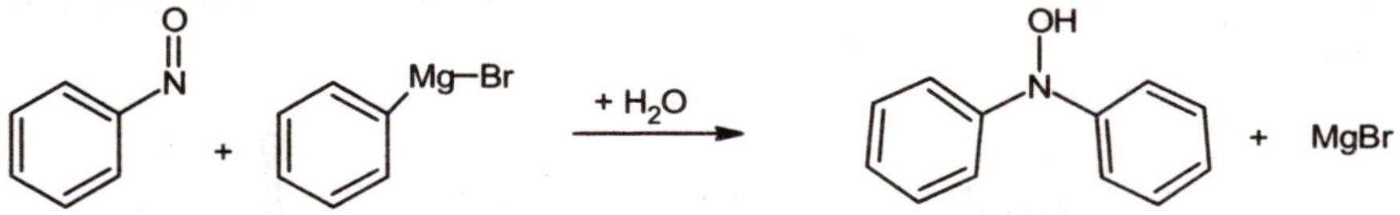

Nitrosobenzene N-hydroxyN-phenylaniline

It is a beautiful looking crystalline compound which decomposes at 60°C and is of interest in connection with the discovery of tetra aryl hydrazines. Nitrogen diphenyl bears the same relation to diphenyl – hydroxyl amine as nitric oxide to nitrous acid.

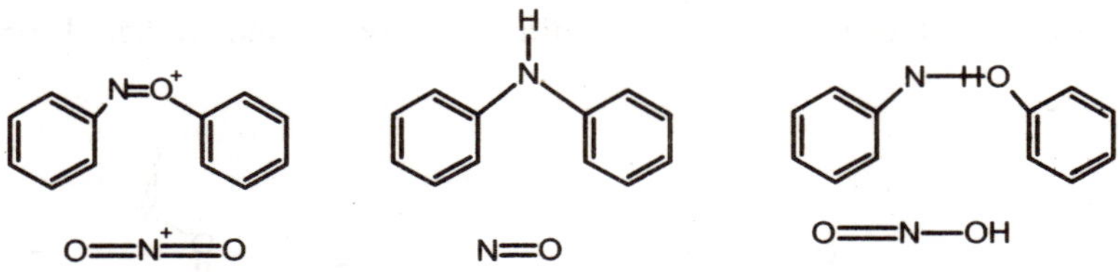

N=O⁺ N—H N—HO

O=N⁺=O N=O O=N—OH

The presence of nitrogen diphenyl in a solution of tetra phenyl hydrazine can be detected by its unsaturated properties.

When for example nitric oxide is passed into a solution of hydrazine in toluene at 90°C diphenyl nitrosoamine is produced by the combination of two divalent nitrogen groups.

NO

The dissociation of tetraphenyl – hydrazine into the free radical $(C_6H_5)_2N$ is exactly analogous to the formation of triphenyl will be discussed later.

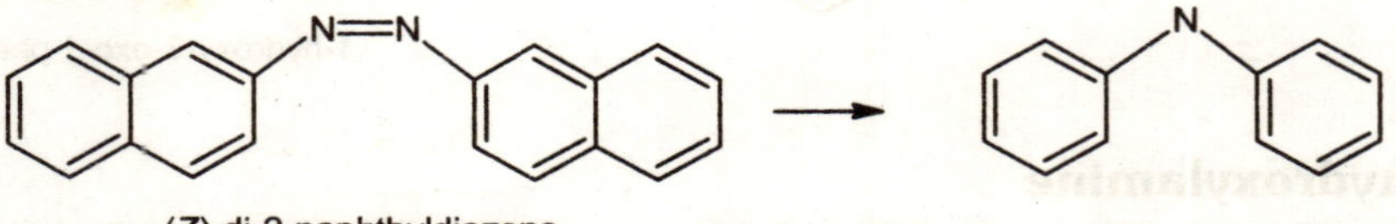

(*Z*)-di-2-naphthyldiazene

Another compound which tends to dissociate into a derivative of divalent nitrogen is hexaphenyl tetrazane. In the solid state this is mono – molecular, but in any aqueous state it largely exists as triphenyl hydrazyl.

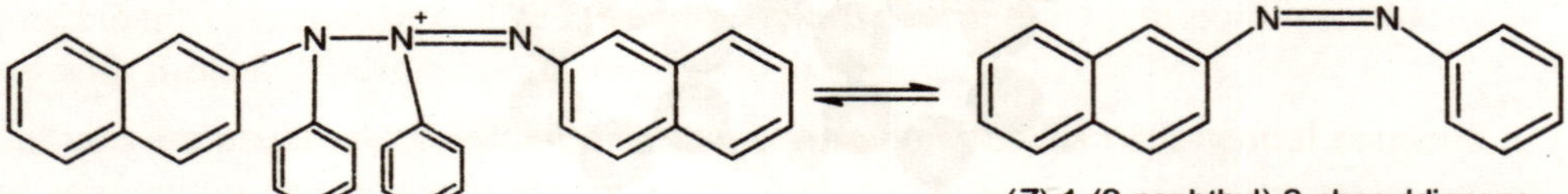

(*Z*)-1-(2-naphthyl)-2-phenyldiazene

Diphenyl – nitric oxide, $(C_6H_5)_2N=O$ is obtained formed diphenyl – hydroxylamine by oxidation with silver oxide.

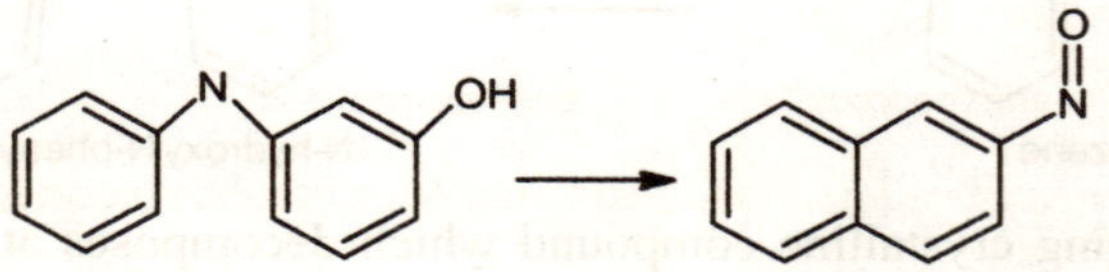

It is similar to nitrogen dioxide form which it is derived by the replacement of an oxygen atom by two benzene resides. It crystallizes in deep red needles, and in many way resembles NO_2 but is of a deeper red. It unites with other radicals such as nitrogen dioxide and triphenyl methyl, with NO_2 it produces,

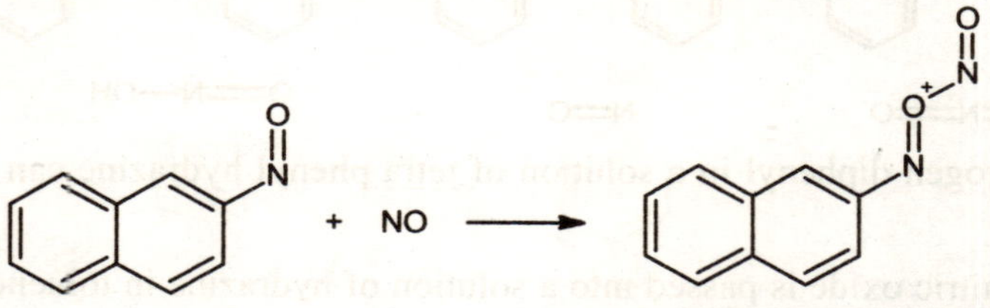

+ NO

2-nitrosonaphthalene

(*E*)-(2-naphthylimino)(nitroso)oxonium

which then undergoes inter molecular rearrangement to give p – nitro – diphenylamine

{phenyl[(phenylimino)oxonio]amino}oxidanide

3-nitro-*N*-phenylaniline

AZOXY COMPOUNDS

Introduction

These compounds are the one which contains oxygen with the azo group

The Azoxy compounds are generally prepared by heating nitro derivatives, with an alcoholic solution of sodium methoxide; sodium amalgam, or magnesium and ammonium chloride solution can also be employed as the reducing agent. They are yellow or red in colour, crystallize well and on further reduction yield azo – hydrazo and amine compounds, with moderately warm concentrated sulphuric acid, they isomerizes to hydroxy – azo – compounds.

AZOXYBENZENE

Introduction

This compound is one of the most important Azoxy compound. Generally prepared by the reaction of nitro benzene and alcoholic solution of sodium methoxide

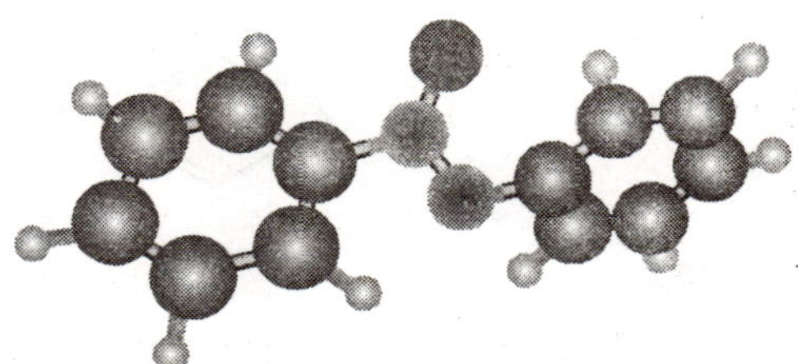

Preparation

Sodium arsenite is prepared by dissolving 22.6 g. (1.1 moles) of powdered arsenious oxide, made into a paste with a little water, in a solution of 27.5 g. (6.9 moles) of sodium hydroxide dissolved in 60.0 cc. of water. This solution, diluted with 600 cc. of water, is poured into a 2-l. three-necked flask provided with a reflux condenser and a mechanical stirrer, and 15.0 g. (12.5 cc., 1.2 moles) of freshly distilled nitrobenzene is added (Note 1).

The mixture is refluxed on an oil bath for eight hours with constant and vigorous stirring (Note 2). After removing the oil bath, the reaction mixture is allowed to cool to about 80°, while stirring is continued, and is then transferred to a separatory funnel previously heated to about the same temperature in an oven (Note 3).

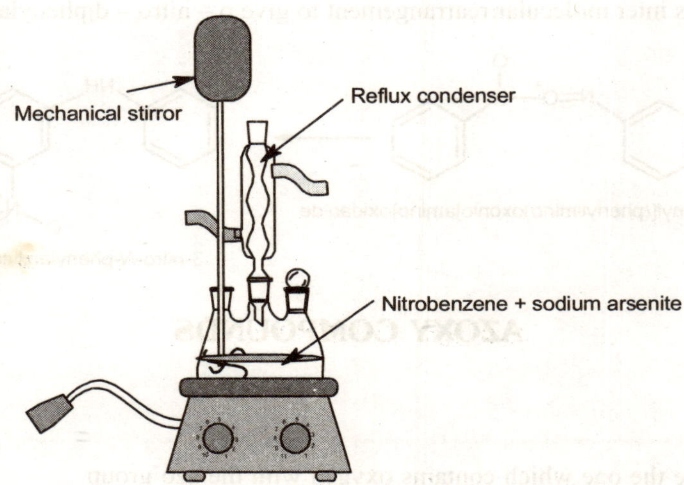

Fig. 30.3. Showing the preparation of Azoxy benzene

The upper layer of oil is separated (Note 4), run at once into an open vessel, and washed with water to which a little hydrochloric acid has been added. Yellow crystals form at once (Note 5), and the yield of azoxybenzene melting at 35.5–36.5° is 10.2 g. (85 per cent of the theoretical amount) (Note 6) and (Note 7).

Reaction equation

$$4\ C_6H_5NO_2 \xrightarrow[18NaOH,\ H_2O,\ \Delta]{3As_2O_3} 2\ C_6H_5{-}N{=}\overset{+}{N}(\rightarrow O){-}C_6H_5$$

Notes

1. The excess of sodium arsenite and the eight-hour period of heating ensure the complete utilization of nitrobenzene. This makes it unnecessary to use steam distillation or other processes for the removal of unreacted nitrobenzene.

Crude nitrobenzene may be used, but a good grade is recommended. When crude nitrobenzene is used, a darker product having a slightly lower melting point is usually obtained.

2. The internal temperature should be about 104°, and the temperature of the bath should not greatly exceed 115°. A smaller flask might be used were it not for the danger of foaming which would result from accidental stopping of the stirrer. If the reaction is interrupted, the oil bath must be removed even though the stirring has been stopped; otherwise, on resumption of stirring, the superheated material may be ejected through the condenser. With these precautions in mind it is unnecessary to have a continuous eight-hour period of heating.

Contrary to the general statement found in the literature, azoxybenzene is somewhat volatile with steam. Therefore, the presence of oil drops in the condenser at the end of the eight-hour period of refluxing is no criterion of unaltered nitrobenzene. Azoxybenzene is easily volatile with steam at 140–150°.

3. In this way the separation is effected at about 60°, and the danger of the solution's cooling to a point where sodium arsenate separates is avoided. Should the sodium arsenate separate because of undue cooling, the mixture is heated again, with stirring, until the arsenate redissolves. On dilution with sufficient water to keep the arsenate in solution, the oil settles to the bottom and does not separate readily from the liquid. Furthermore, the volume of solution is so large with such dilution that it is less easily handled.

4. The solution from which the original oil separated will yield, when diluted with an equal volume of water, a small additional quantity of azoxybenzene. This may be recovered by extraction with benzene, but the quantity of compound so obtained does not justify this extra procedure.

5. The presence of hydrochloric acid accelerates crystallization. Should crystallization be retarded, it is recommended that the oil be seeded with a crystal of azoxybenzene.

6. Recrystallization from 50 cc. of hot 95 per cent alcohol gives 72 g. of azoxybenzene. The recrystallized product melts at the same temperature, 35.5–36.5°, as the crude material but is distinctly lighter in colour.

7. The following alternative procedure for preparing azoxybenzene is convenient.

In a 1-l. three-necked flask fitted with a reflux condenser and an efficient stirrer (fig 103) there are placed 60 g. of sodium hydroxide, 200 cc. of water, and 41 g. (34.2 cc., 0.33 mole) of nitrobenzene. The flask is immersed in a water bath kept at 55–60°, and 45 g. (0.23 mole) of dextrose is introduced in portions, with continuous stirring, in the course of one hour. The temperature of the bath is then raised to 100° and kept there for two hours. The hot mixture is poured into a 2-l. long-necked flask and steam-distilled to removed nitrobenzene and aniline. This requires some twenty minutes, during which time about 2 l. of distillate passes over. When the distillate is clear, the residue is poured into a beaker and cooled well in an ice bath. The azoxybenzene, which solidifies, is collected, the lumps are ground in a mortar, and the product is washed with water and dried. The yield of material melting at 34–35° is 26–27 g. (79–82 per cent of the theoretical amount). Crystallization from 15 cc. of methyl alcohol gives material melting at 35–35.5° with 90 per cent recovery.

Discussion

Azoxybenzene has been prepared by reduction of nitrobenzene with alcoholic potassium hydroxide,1 with sodium amalgam, with hydrogen in the presence of lead oxide, with methyl alcohol and sodium hydroxide, with sodium methoxide and methyl alcohol, with lead suboxide in alkaline suspension, with dextrose in alkaline suspension (see (Note 7), above), and electrolytically; by oxidation of azobenzene with chromic anhydride; by treatment of β-phenylhydroxylamine with alkaline potassium permanganate, with nitrobenzene,10 with mineral acids, and with mercury acetamide; and by oxidation of aniline with hydrogen peroxide, and with acid permanganate solution in the presence of formaldehyde.

Physical Properties

Molecular Formula	= $C_{12}H_{10}N_2O$
Formula Weight	= 198.2206
Composition	= C(72.71%) H(5.08%) N(14.13%) O(8.07%)
Melting Point	= 36ºC
Molar Refractivity	= 60.61 ± 0.5 cm^3
Molar Volume	= 181.5 ± 7.0 cm^3
Parachor	= 467.6 ± 8.0 cm^3
Index of Refraction	= 1.582 ± 0.05
Surface Tension	= 44.0 ± 7.0 dyne/cm
Density	= 1.09 ± 0.1 g/cm^3
Polarizability	= 24.02 ± 0.5 $10^{-24}cm^3$
Monoisotopic Mass	= 198.079313 Da
Nominal Mass	= 198 Da
Average Mass	= 198.2206 Da

Chemical Properties

It forms yellow crystals, when warmed with concentrated sulphuric acid it isomerizes into p-hydroxy-azobenzene.

(phenyl-NNO-azoxy)benzene $\xrightarrow[\text{Conc.}]{H_2SO_4}$ 4-(phenyl-ONN-azoxy)phenol

Azoxybenzene and other azo compounds are believed to have similar structures

e.g.

An unsymmetrical azo compound however, was shown by Angeli to give rise in some instances to two azo-oxy compounds. The symmetrical formula has therefore been abandoned.

The unsymmetrical structure has also been conclusively produced by polarimetric methods.

For the Students who want a Brief Summary

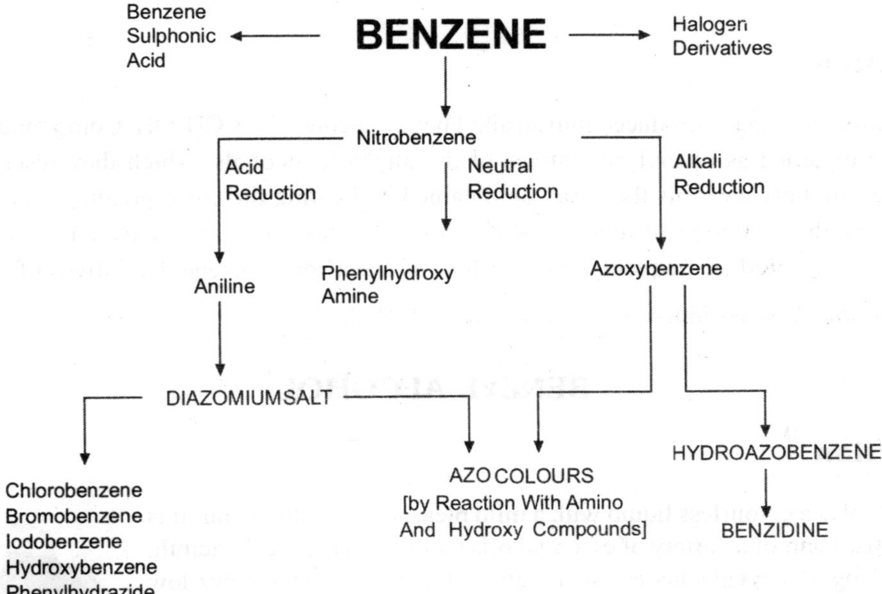

CHAPTER

31

Aromatic Alcohol Aldehydes and Ketones

Introduction

When a hydroxyl group is introduced into an alkyl benzyl alcohol $C_6H_5CH_2OH$. Compounds of this type may also be regarded as phenyl derivatives of the aliphatic alcohols, which they resemble in most repects. Like the fatty alcohols they may be obtained by heating the corresponding chloro-derivative with water, by the reduction of aldehydes ketones and esters. The secondary and tertiary alcohols formed by this last method frequently pass by loss of water into benzene derivatives of olefins.

Now we shall discuss individually about these alcohols.

BENZYL ALCOHOL

Introduction

Benzyl alcohol is a colourless liquid with a mild pleasant aromatic odour. It is a natural constituent of a variety of essential oils including jasmine, hyacinth, and ylang-ylang. Benzyl alcohol is also a useful solvent due to its polarity, low toxicity, and low vapor pressure.

Benzyl alcohol is partially soluble in water (4 g/100 mL) and completely miscible in alcohols and ether.

Preparation

1. From Benzyl chloride: By hydrolysis with alkalies, benzyl chloride is boiled with K_2CO_3 solution until the pungent smell of the chloride vanishes. The product is extracted with ether and subjected to fractional distillation. Benzyl alcohol is collected at 206°C.

$$C_6H_5CH_2-Cl + K_2CO_3 + H_2O \longrightarrow C_6H_5CH_2OH + 2KCl + CO_2$$

2. *From Benzaldehyde:* By the action of potassium hydroxide on benzaldehyde

$$C_6H_5CHO + KOH \longrightarrow C_6H_5CH_2OH + C_6H_5COOK$$

benzaldehyde; phenylmethanol (benzyl alcohol); potassium benzoate

The semi-solid product is dissolved in water and extracted with ether. Ether separates benzyl alcohol from potassium benzoate. Benzyl alcohol is recovered by distillation.

3. *Grignard Reagent:* By the action of dry formaldehyde on Grignard reagent in ethereal solution and subsequent decomposition with HCl

$$C_6H_5MgBr + HCHO \longrightarrow C_6H_5CH_2OMgBr \xrightarrow{HCl} C_6H_5CH_2OH + Mg(Cl)Br$$

Grignard's Reagent; formaldehyde; benzyl alcohol

4. *From Benzoic Acid:* By electrolytic reduction

$$C_6H_5COOH + 4H \longrightarrow C_6H_5CH_2OH + H_2O$$

benzoic acid; phenylmethanol

5. *By the action of Benzyl amine:* By the action of nitrous acid on benzylamine

$$C_6H_5CH_2NH_2 + HNO_2 \longrightarrow C_6H_5CH_2OH + N_2 + H_2O$$

1-phenylmethanamine (Benzylamine); phenylmethanol

Physical Properties

Density	= 1.044 g/cm^3
Melting point	= –15 °C
Boiling point	= 205 °C
Molecular Formula	= C_7H_8O
Formula Weight	= 108.13782
Composition	= C(77.75%) H(7.46%) O(14.80%)
Molar Refractivity	= 32.70 ± 0.3 cm^3
Molar Volume	= 103.2 ± 3.0 cm^3
Parachor	= 260.8 ± 4.0 cm^3
Index of Refraction	= 1.546 ± 0.02
Surface Tension	= 40.7 ± 3.0 dyne/cm
Density	= 1.047 ± 0.06 g/cm^3
Polarizability	= 12.96 ± 0.5 10^{-24}cm^3
Monoisotopic Mass	= 108.057515 Da
Nominal Mass	= 108 Da
Average Mass	= 108.1378 Da

Chemical Properties

In chemical properties it closely relates aliphatic alcohols.

1. With Sodium: It forms Sodium benzoate

2 $C_6H_5CH_2OH$ + 2Na ⟶ 2 $C_6H_5CH_2ONa$ + H_2

benzyl alcohol — sodium benzoate

2. With acids: It forms esters

2 $C_6H_5CH_2OH$ + $HO{-}C(=O){-}CH_3$ ⟶ $C_6H_5CH_2O{-}C(=O){-}CH_3$ + H_2O

benzyl alcohol — benzyl acetate

3. With Acid Anhydrides or Acid Chlorides:

benzyl alcohol + acetic anhydride → benzyl acetate + H_2O

benzyl alcohol + acetyl chloride → benzyl acetate + HCl

4. With dehydrating agents: It forms ether

benzyl alcohol $\xrightarrow{-H_2O}$ dibenzyl ether

5. With phosphorous pentachloride: It produces benzyl Chloride

benzyl alcohol + PCl_5 → (chloromethyl) benzene + $POCl_3$ + HCl

6. On oxidation: It gives either benzaldehyde or benzoic acid.

benzyl alcohol $\xrightarrow[-H_2O]{[O]}$ benzaldehyde $\xrightarrow[-H_2O]{[O]}$ benzoic acid

7. On reduction: It produces toluene with hydrogen iodide

benzyl alcohol + 2HI → toluene + H_2O + I_2

8. Nitration Sulphonation and Halogenation

(i) Action of Ammonia: Benzyl amine is produced ammonia is passed through benzyl alcohol in presence of aluminium chloride in ether.

$$C_6H_5CH_2OH + NH_3 \xrightarrow[\text{in ether}]{AlCl_3} C_6H_5CH_2NH_2 + H_2O$$

benzyl alcohol → 1-phenylmethanamine benzylamine

(ii) Action of Conc Nitric Acid: Conc. Nitric Acid oxidizes benzyl alcohol to benzoic acid

$$C_6H_5CH_2OH \xrightarrow{HNO_3} C_6H_5COOH + NO_2$$

benzyl alcohol → benzoic acid

(iii) Action of Conc. Sulphuric acid: It Dehydrates and produces dibenzyl ether.

Uses

Benzyl alcohol is used as a general solvent for inks, paints, lacquers, and epoxy resin coatings. It is also a precursor to a variety of esters, used in the soap, perfume, and flavour industries, as well as for pharmaceuticals. It exhibits bacteriostatic and antipruritic properties. It is also used as a photographic developer.

Illustrative Organic Synthetic Uses

In organic synthesis, benzyl esters are popular protecting groups because they can be removed by mild hydrogenolysis.

BnOH reacts with acrylonitrile to give N-benzylacrylamide. This is an example of a Ritter reaction:

$$C_6H_5CH_2OH + NCCHCH_2 \rightarrow C_6H_5CH_2N(H)C(O)CHCH_2$$

Nanotechnological Uses

Benzyl alcohol has been used as a dielectric solvent for the dielectrophoretic reconfiguration of nanowires.

Health and Safety

Benzyl alcohol is used as a bacteriostatic preservative in parenteral (IV) medications. Benzyl alcohol is also known for its toxic effects including respiratory failure, vasodilation, hypotension, convulsions, and paralysis. Sixteen Neonatal deaths have been associated with the use of benzyl alcohol as a preservative in saline flush solutions. Preservative free solutions are now being used for the infant population.

β-PHENYL ETHYL ALCOHOL

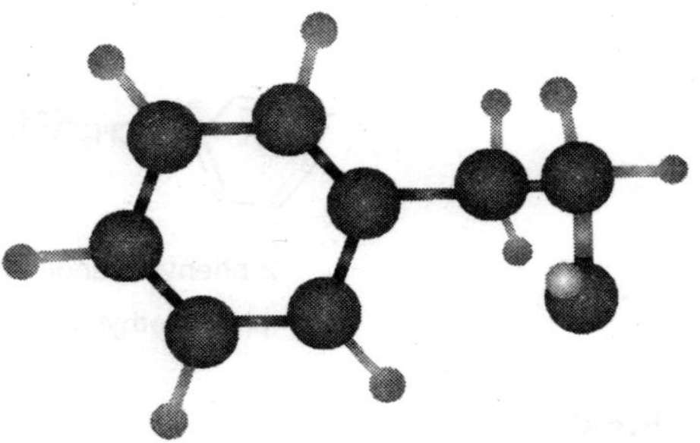

Introduction

This alcohol occurs in rose oil neroli and geramium oil as their flavoring agent this synthetic product is used in perfume making

Preparation

It is prepared by reacting excess benzene with ethyl oxide in the presence of a catalyst called aluminium tri chloride.

$$\text{benzene} + \underset{\text{ethylene oxide}}{H_2C\overset{O}{—}CH_2} \xrightarrow[10\text{ - }12\ °C]{AlCl_3} C_6H_5–CH_2–CH_2–OH$$

benzene + ethylene oxide → 2-phenylethanol (β-phenyl ethyl alcohol)

Physical Properties

Melting Point	= –27°C
Boilign point	= 218°C
Molecular Formula	= $C_8H_{10}O$
Formula Weight	= 122.1644
Composition	= C(78.65%) H(8.25%) O(13.10%)
Molar Refractivity	= 37.33 ± 0.3 cm^3
Molar Volume	= 119.7 ± 3.0 cm^3
Parachor	= 300.6 ± 4.0 cm^3
Index of Refraction	= 1.535 ± 0.02
Surface Tension	= 39.6 ± 3.0 dyne/cm
Density	= 1.020 ± 0.06 g/cm^3
Polarizability	= 14.80 ± 0.5 $10^{-24}cm^3$
Monoisotopic Mass	= 122.073165 Da
Nominal Mass	= 122 Da
Average Mass	= 122.1644 Da

Chemical Properties

With hot aqueous sodium hydroxide this alcohol gets dehydrated to styrene

2-phenylethanol
β - phenyl ethyl alcohol

styrene

Uses

Phenylethyl is a structural skeleton found in sympathomimetic amines that have important physiological functions within the body as neurotransmitters in the central nervous system and hormones in the blood circulation. Phenyl ethyl alcohol might be an useful intermediate to prepare nervous system drugs. Phenyl ethyl alcohol is used as an biocide (antimicrobial, antiseptic, or disinfectant) and as an aromatic essence (particularly rose and jasmine, lily) and as a preservative in pharmaceuticals and perfumery.

Coniferyl alcohol is a colourless crystalline solid. It is a phytochemical, one of the monolignols. It is synthetized via the phenylpropanoid biochemical pathway. When polymerized, coniferyl alcohol forms lignin or lignans.

Coniferyl alcohol is an intermediate in biosynthesis of eugenol and of stilbene and coumarin.Gum benzoin contains significant amount of coniferyl alcohol and its esters.

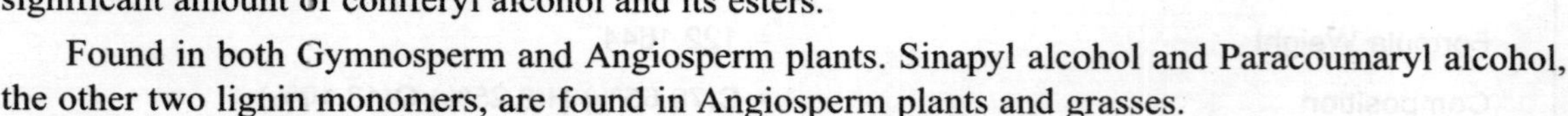

Found in both Gymnosperm and Angiosperm plants. Sinapyl alcohol and Paracoumaryl alcohol, the other two lignin monomers, are found in Angiosperm plants and grasses.

Paracoumaryl alcohol, also called ***p*-coumaryl alcohol, 4-coumaryl alcohol, 4-hydroxycinnamyl alcohol**, or **4-(3-hydroxy-1-propenyl)phenol**, is a phytochemical, one of the monolignols. It is synthetized via the phenylpropanoid biochemical pathway. When polymerized, *p*-coumaryl alcohol forms lignin or lignans.

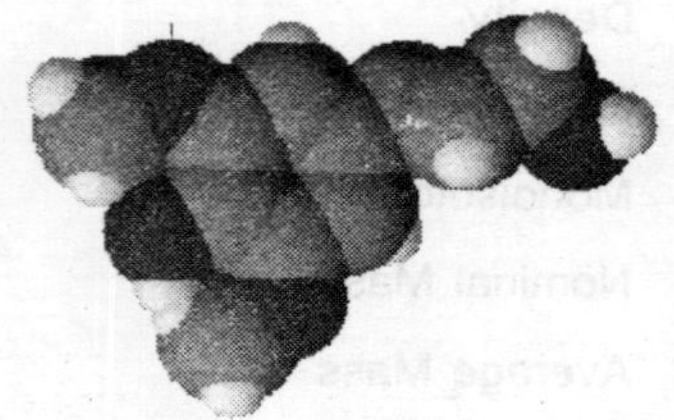

Sinapyl alcohol, also called **sinapoyl alcohol** or **4-(3-hydroxy-1-propenyl)-2,6-dimethoxy-phenol**, is a phytochemical, one of the monolignols. It is synthetized via the phenylpropanoid biochemical pathway. When polymerized, sinapyl alcohol forms lignin or lignans.

Sinapyl alcohol is also a precursor for biosynthesis of stilbene and coumari.

Aldehydes

Aromatic aldehydes have the same typical – CHO group attached directly to the benzene ring. Among these benzaldehyde is the most important.

BENZALDEHYDE

Introduction

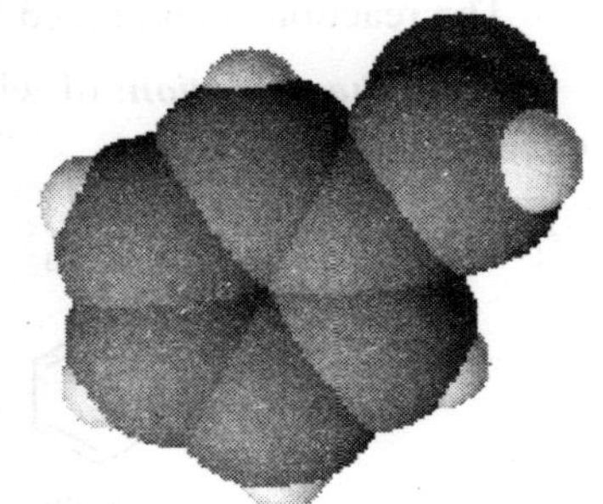

Benzaldehyde was obtained in early time from a crystalline glucoside amylgdalin, which is present bitter almond. When almonds are crushed in water the enzyme emulsin (also present in almonds) hydrolysis the glucoside into benzaldehyde

$$\underset{\text{Amylgdalin}}{C_{20}H_{27}NO_{11}} + 2H_2O \rightarrow \underset{\text{Benzaldehyde}}{C_6H_5CHO} + \underset{\text{Glucose}}{2C_6H_{12}O_6}$$

Preparation

Benzaldehyde can be prepared by the following methods:

1. ***By oxidation of benzyl alcohol***: with copper nitrate or lead nitrate, in current of carbon dioxide

$$C_6H_5CH_2OH + [O] \longrightarrow C_6H_5CHO + H_2O$$

phenylmethanol benzyl alcohol → benzaldehyde

2. ***By the distillation of calcium benzoate***: with calcium formate

$$(C_6H_5COO)_2Ca + (HCOO)_2Ca \longrightarrow 2C_6H_5CHO + 2CaCO_3$$

calcium benzoate + calcium formate → benzaldehyde

3. ***Rosenmound's Reaction***: By the reduction of Benzyl Chloride

$$C_6H_5COCl + 2H \xrightarrow[\text{BaSO}_4]{\text{Platinized}} C_6H_5CHO + HCl$$

benzoyl chloride → benzaldehyde

4. By the hydrolysis of Benzal chloride: This is the manufacturing process

(dichoromethyl)benzene + H_2O —(Milk of lime)→ benzaldehyde + 2HCl

The reaction is employed for the manufacture of benzaldehyde.

5. By the oxidation: of with manganese dioxide and sulphuric acid in the presence of a catalyst, copper sulphate.

toluene + 2O —($MnO_2 + H_2SO_4$, ($CuSO_4$))→ benzaldehyde + H_2O

Benzaldehyde obtained by this method is very pure and specially used in perfumery.

6. By Etard's Reaction: This reaction is used in the laboratory to prepare benzaldehyde. Toluene in carbon disulphide is oxidized by treating it with chromyl Chloride (CrO_2Cl_2) and decomposing resulting addition compound with water.

$$3CrO_2Cl_2 + H_2O \rightarrow 2CrCl_3 + H_2CrO_4 + 3O$$

toluene —(2O)→ benzaldehyde + H_2O

This is a laboratory process must be tried by students.

7. Friedel Crafts Reactions: benzene is treated with a mixture of carbon monoxide and dry HCl gas in the presence of anhydrous aluminium trichloride.

benzene + HCl + CO —($AlCl_3$)→ benzaldehyde + HCl

This reaction is in fact the modified form of Friedel crafts reaction.

Physical Properties

Density and phase	1.0415 g/ml, liquid
Solubility in water	Slightly soluble (0.6 wt at 20 °C)
Solubility in ethanol	Miscible
Melting point	–26 °C
Boiling point	178.1 °C
Viscosity	1.4 cP at 25 °C
Standard enthalpyof formation DfH°liquid	–86.8 kJ/mol
Standard enthalpyof combustion DcH°liquid	–3525.1 kJ/mol

Chemical Properties

It shows most of the reactions of aliphatic aldehydes but it does not under go polymerization and condensation. It also shows some characteristics reactions of benzene nucleus.

Resemblance with Aliphatic Aldehyde

1. Oxidation: It gives benzoic acid

$$C_6H_5CHO + [O] \longrightarrow C_6H_5COOH$$

benzaldehyde benzoic acid

Benzaldehyde is therefore, reducing agent and gives positive reaction in silver mirror test.*

$$C_6H_5CHO + Ag_2O \longrightarrow C_6H_5COOH + 2Ag$$

benzaldehyde benzoic acid

2. Reduction: It gives benzyl alcohol

$$C_6H_5CHO + 2H \longrightarrow C_6H_5CH_2OH$$

benzaldehyde phenylmethanol benzyl alcohol

*benzaldehyde does not reduce Fehling's Reagent it is the difference betwen reducing sugar and benzaldehyde.

3. With Sodium Bisulphite: It gives Bisulphite compound.

benzaldehyde + $NaHCO_3$ → sodium hydroxy(phenyl)methanesulfinate

The bisulphite compounds are crystalline and can be decomposed by warming with dilute sulphuric acid. The reaction is therefore used for the identification and purification of benzaldehyde.

sodium hydroxy(phenyl)methanesulfinate + H_2SO_4 → benzaldehyde + $NaHCO_3$ + H_2O + SO_2

4. With Hydrogen Cyanide: It forms cyanohydrin

benzaldehyde + HCN → isocyano(phenyl)methanol

5. With Hydrazine: It gives Oxime

benzaldehyde + H_2N-OH → benzaldehyde oxime

6. With Benzyl amine: It gives phenyl hydrazone

benzaldehyde + benzylamine → benzaldehyde phenylhydrazone phenylhydrazone + H_2O

7. ***With Phosphorous Pentachloride***: It gives Benzal chloride

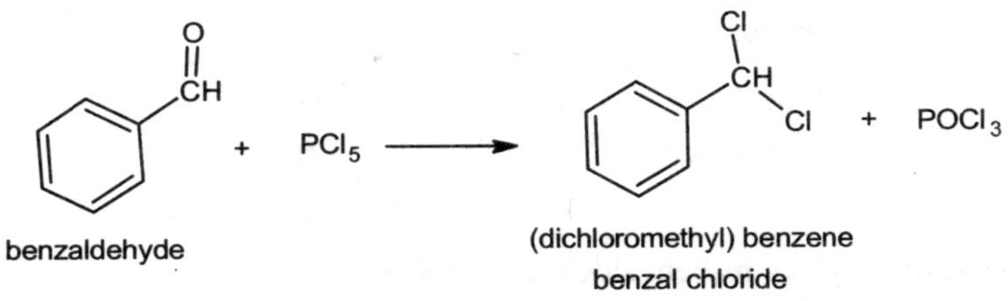

$$C_6H_5CHO + PCl_5 \longrightarrow C_6H_5CHCl_2 + POCl_3$$

benzaldehyde (dichloromethyl) benzene benzal chloride

8. ***Schiff's Reaction***: It restores the original pink colour of the reagent decoloured by sulphur dioxide.

Distinction from Aliphatic aldehydes

9. Benzaldehyde does not reduce Fehling's Reagent

10. ***With Ammonia***: It gives a complex compound known as hydrobenzamide

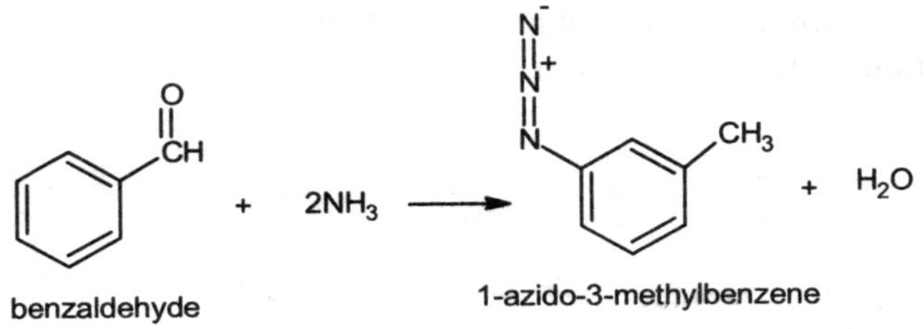

$$C_6H_5CHO + 2NH_3 \longrightarrow CH_3C_6H_4N_3 + H_2O$$

benzaldehyde 1-azido-3-methylbenzene

11. ***With Conc. Sodium Hydroxide:*** It produces Sodium benzoate and benzyl alcohol

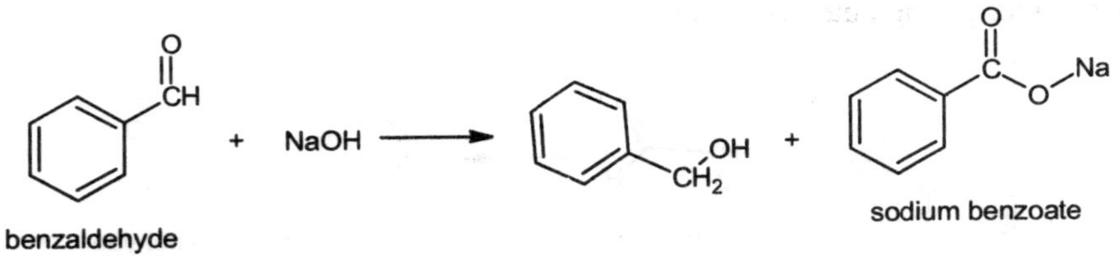

$$C_6H_5CHO + NaOH \longrightarrow C_6H_5CH_2OH + C_6H_5COONa$$

benzaldehyde sodium benzoate

12. ***Chlorination:*** It forms Benzoyl chloride

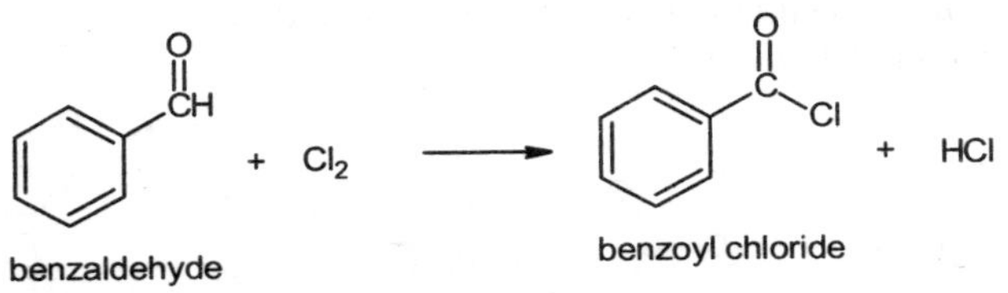

$$C_6H_5CHO + Cl_2 \longrightarrow C_6H_5COCl + HCl$$

benzaldehyde benzoyl chloride

Chlorine does not attack the benzene nucleus. The reaction is used for the preparation od benzoyl chloride.

benzaldehyde + nitric acid → 3-nitrobenzaldehyde + H_2O

14. Sulphonation: It gives m-benzaldehyde sulphonic acid

benzaldehyde + sulfuric acid → 3-formylbenzenesulfonic acid + H_2O

15. Benzoin Condensation: On warming with alcoholic solution of potassium cyanide two molecules of benzaldehyde combines to form Benzoin.

benzaldehyde $\xrightarrow[CH_3CH_2OH]{KCN}$ benzoin

16. The Claisen Reaction: Benzaldehyde condenses with acetaldehyde, in the presence of a little sodium hydroxide, giving cinnamic aldehydes

benzaldehyde + acetaldehyde → (2*Z*)-3-phenylacrylaldehyde + H_2O

17. Perkin reaction: Benzaldehyde condenses with sodium acetate to form sodium cinnamate in the presence of acetic anhydride

benzaldehyde + sodium acetate $\xrightarrow{\text{acetic anhydride}}$ sodium (2Z)-3-phenylacrylate (Sodium cinnamate) + H_2O

18. Condensation with Aniline: It gives benzalidine aniline also known as Schiff's Base or anil

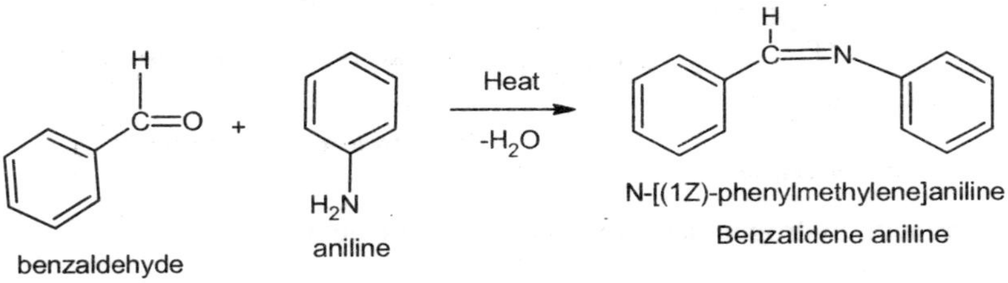

benzaldehyde + aniline —Heat, $-H_2O$→ N-[(1Z)-phenylmethylene]aniline

Benzalidene aniline

Uses

While it is commonly employed as a commercial food flavourant (almond flavour) or industrial solvent, benzaldehyde is used chiefly in the synthesis of other organic compounds, ranging from pharmaceuticals to plastic additives. It is also an important intermediate for the processing of perfume and flavouring compounds and in the preparation of certain aniline dyes.

Glaciologists LaChapelle and Stillman reported in 1966 that benzaldeyde and N-heptaldehyde inhibit the recrystallization of snow and therefore the formation of depth hoar. This treatment may prevent avalanches caused by unstable depth hoar layers. However, the chemicals are not in widespread use because they damage vegetation and contaminate water supplies

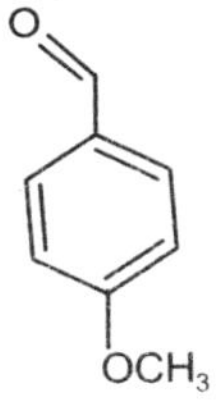

Anisaldehyde, or anisic aldehyde or 4-methoxybenzaldehyde, is an organic compound that consists of a benzene ring substituted with an aldehyde and a methoxy group. It is a clear colorless liquid with a strong aroma. Anisaldehyde is found in anise, from which it gets its name. It is similar in structure to vanillin. Anisaldehyde is used as an intermediate in the synthesis of other compounds important in pharmaceuticals and perfumery. A solution of anisaldehyde in acid and ethanol is frequently used to stain Thin Layer Chromatography plates.

OCH_3

Salicylaldehyde, or 2-hydroxybenzaldehyde, is the chemical compound with the formula C_6H_4CHO-2-OH. This colourless oily liquid has a bitter almond odor. Salicylaldehyde is a key precursor to a variety chelating agents, some of which are commercially important. It can be prepared from phenol and chloroform by heating with sodium hydroxide in a Reimer-Tiemann reaction.

Salicylaldehyde is a common highly functionalized arene that has often been exploited as a precursor to still other chemicals, which are shown in Fig. 31.1.

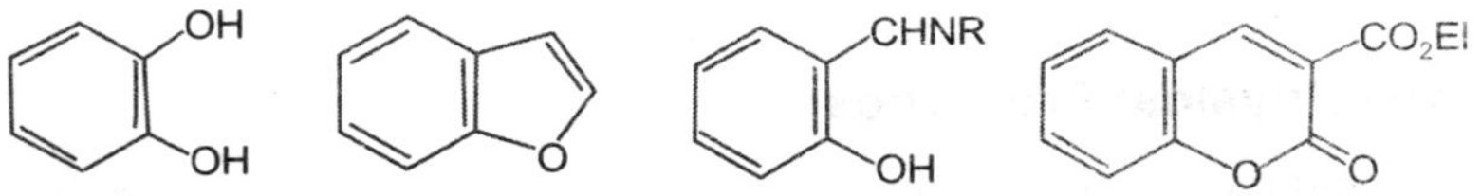

OH, OH; O; CHNR, OH; CO_2Et, O, O

Fig. 31.1: Illustrative compounds derived from salicylaldehyde. From the left: catechol, benzofuran, a salicylaldehydimine (R = alkyl or aryl), 3-carbethoxycoumarin.

- Hydroxylamine gives salicylaldoxime.

$$\text{salicylaldehyde} + H_2N{-}OH \xrightarrow{H^+} H{-}\underset{OH}{\overset{H}{C}}{-}NH{-}OH + H_2O$$

salicylaldehyde (hydroxyamino)methanol

- Oxidation with hydrogen peroxide gives salicylic acid

$$\text{salicylaldehyde} + H_2O_2 \longrightarrow \text{salicylic acid}$$

salicylaldehyde salicylic acid

- Condensation with acetic anhydride gives a derivative of the heterocycle coumarin.

$$\text{CHO, OH-benzene} + 2\,(CH_3CO)_2O \xrightarrow{CH_3COONa} \text{coumarin} + CH_3COOH$$

- etherification with chloroacetic acid followed by cyclisation gives the heterocycle benzofuran.

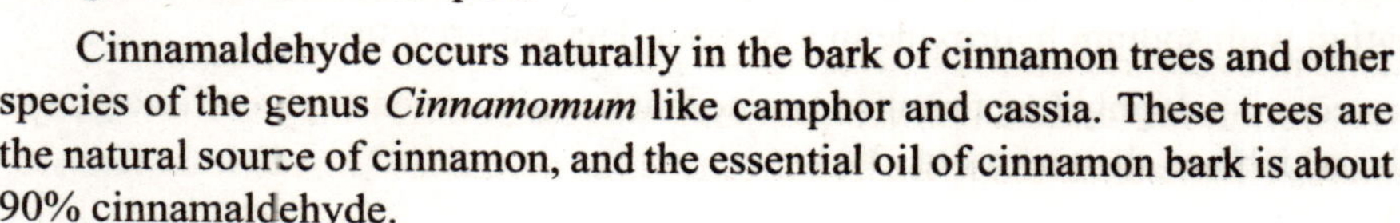

salicylaldehyde chloroacetic acid 1-benzofuran

Cinnamic aldehyde or **cinnamaldehyde** (more precisely ***trans*-cinnamaldehyde**, the only naturally-occurring form) is the chemical compound that gives cinnamon its spice.

Cinnamaldehyde occurs naturally in the bark of cinnamon trees and other species of the genus *Cinnamomum* like camphor and cassia. These trees are the natural source of cinnamon, and the essential oil of cinnamon bark is about 90% cinnamaldehyde.

Structure and Physical Properties

A yellow oily liquid more viscous than water, cinnamaldehyde smells strongly of cinnamon.

Concentrated cinnamaldehyde is a skin irritant, and the chemical is toxic in large doses, but no agencies suspect the compound is a carcinogen or poses a long-term health hazard. Most cinnamaldehyde is excreted in urine as cinnamic acid, an oxidized form of cinnamaldehyde.

Both aromatic and an aldehyde, cinnamaldehyde has a mono-substituted benzene ring. A conjugated double bond (alkene) makes geometry of the compound planar. Cinnamaldehyde is technically *trans*-cinnamaldehyde because the terminal carbonyl is on the opposite side of the aromatic ring over the rigid double bond.

A technique to synthesize cinnamaldehyde was published in 1884. Several methods of synthesis are now known, but cinnamaldehyde is most economically obtained from the steam distillation of the oil of cinnamon bark. The compound can be prepared from related compounds like cinnamyl alcohol, (the alcohol form of cinnamaldehyde), but the first synthesis from unrelated compounds was the aldol condensation of benzaldehyde and acetaldehyde.

Applications

The most obvious application for cinnamaldehyde is as a flavoring. 50 grams can be purchased for around $20.00, and concentrations required for flavoring food items like chewing gum, ice cream, candy, and beverages range from 9 to 4900 ppm (that is, less than 0.5%).

Cinnamaldehyde is used in some perfumes of natural, sweet, or fruity scents. Almond, apricot, butterscotch, and other aromas may partially employ the compound for their pleasant smells.

Cinnamaldehyde is also used as a fungicide. Proven effective on over 40 different crops, cinnamaldehyde is typically applied to the root systems of plants. Its low toxicity and well-known properties make it ideal for agriculture. To a lesser extent, cinnamaldehyde is an effective insecticide, and its scent is also known to repel animals like cats and dogs.

Cinnamaldehyde is also known as a corrosion inhibitor for steel and other ferrous alloys in corrosive fluids. It can be used in combination with additional components such as dispersing agents, solvents and other surfactants.

Ketones

There are two types of aromatic ketones:

(a) True Aromatic Ketones: In which both the radicles are aromatic, such, such as Benzophenone, Benzoin and Diketone, benzil.

BENZOPHENONE

Introduction

Benzophenone is the aromatic ketone with the formula $(C_6H_5)_2CO$, generally abbreviated Ph_2CO. Substituted benzophenones such as oxybenzone and dioxybenzone are used in sunscreen.

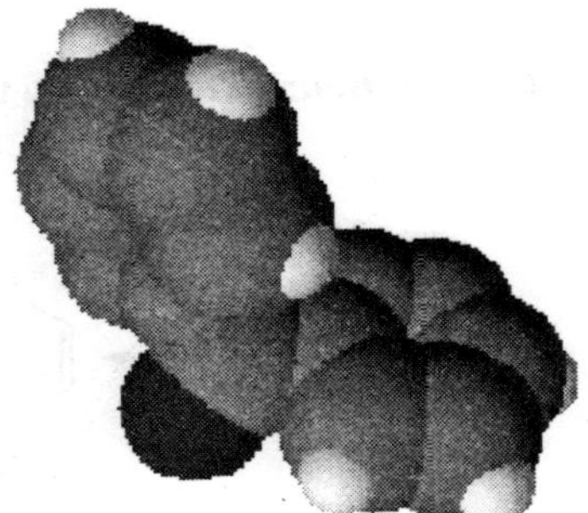

Preparation

Benzophenone can be prepared by the reaction of benzene with carbon tetrachloride followed by hydrolysis of the resulting diphenyldichloromethane, or by Friedel-Crafts acylation of benzene with benzoyl chloride in the presence of a Lewis acid (e.g. aluminum chloride) catalyst. Benzophenone can also be prepared by following methods

1. From Benzene: By the action of carbonyl chloride on benzene

benzene + phosgene → benzophenone

2. From Grignard reagent: When Grignard Reagent reacts with one of the C = O group benzophenone is produced.

benzene + dioxomethane → Grignard reagent, C_6H_5Mg → H_2O → benzophenone + 2MgOH

3. By oxidation of Diphenyl Methane: When diphenyl methane is oxidized by chromic acid benzophenone is produced.

1,1'-methylenedibenzene —Cr_2O_3→ benzophenone + H_2CrO_4

4. From Benzaldehyde: By the oxidation of benzaldehyde

benzaldehyde —oxidation→ benzoic acid —calcium hydroxide→ calcium benzoate —distillation→ benzophenone

Physical Properties

Density and phase	= 1.11 g/cm^3, solid
Solubility in water	= insoluble
Other solvents	= benzene, THF, ethanol, propylene glycol
Melting point	= 47.9 °C
Boiling point	= 305.4 °C
Molecular Formula	= $C_{13}H_{10}O$
Formula Weight	= 182.2179
Composition	= C(85.69%) H(5.53%) O(8.78%)
Molar Refractivity	= 56.04 ± 0.3 cm^3
Molar Volume	= 167.5 ± 3.0 cm^3
Parachor	= 426.9 ± 4.0 cm^3
Index of Refraction	= 1.583 ± 0.02
Surface Tension	= 42.1 ± 3.0 dyne/cm
Density	= 1.087 ± 0.06 g/cm^3
Polarizability	= 22.21 ± 0.5 10^{-24}cm^3
Monoisotopic Mass	= 182.073165 Da
Nominal Mass	= 182 Da

Chemical Properties

Benzophenone is soluble in 80% ethanol and in soluble in water.

Benzophenone is reduced into secondary alcohol, benzydrol, the presence of zinc dust and sodium hydroxide

O=C (benzophenone) —Zn / NaOH→ HO—HC (diphenylmethanol)

benzophenone

diphenylmethanol
benzydrol

But benzophenone reacts with hydrogen iodide produces hydrocarbon called diphenyl methane.

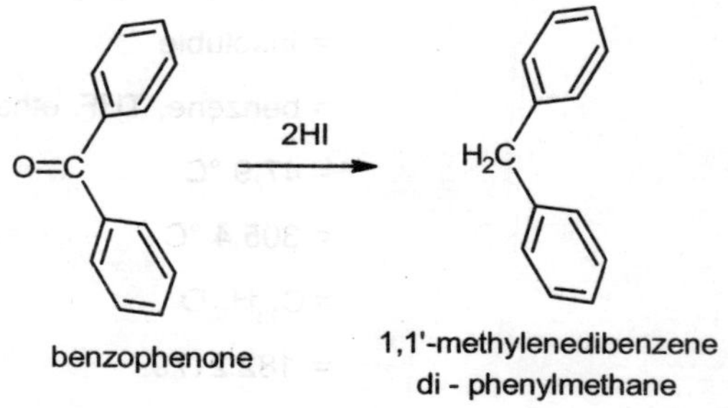

2HI

benzophenone

1,1'-methylenedibenzene
di - phenylmethane

With phenyl magnesium bromide, it forms triphenylcarbinol

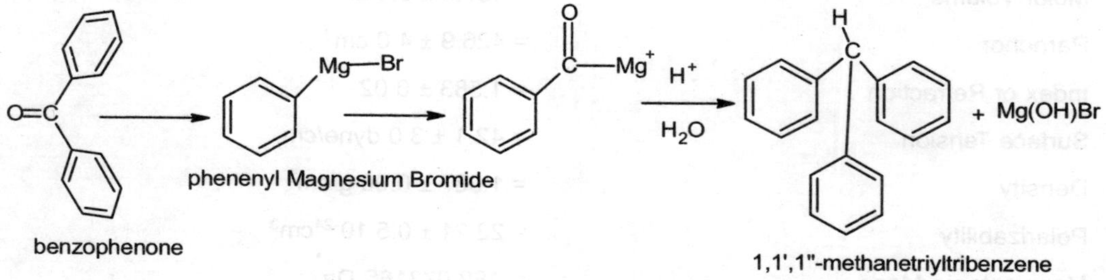

phenenyl Magnesium Bromide

benzophenone

H^+ / H_2O + Mg(OH)Br

1,1',1"-methanetriyltribenzene

Uses

Benzophenone is used as a constituent of synthetic perfumes and as a starting material for the manufacture of dyes, pesticides and drugs (especially anxiolytic and hypnotic drugs). Benzophenone is used as a photoinitiator of UV-curing applications in inks, adhesive and coatings, optical fiber as well as in printed circuit boards. Photoinitiators are compounds that break down into free radicals upon exposure to ultraviolet radiation. Photoinitiators undergo a unimolecular bond cleavage upon irradiation to yield free radicals (benzoin esters; benzil ketals; alpha-dialkoxy acetophenones; alpha-hydroxy alkylphenones; alpha-amino alkylphosphine; acylphosphine oxides). Another type of photoinitiators undergo a bimolecular reaction where the excited state of the photoinitiator interacts with a second molecule (a coinitiator) to generate free radicals (benzophenones, amines; thioxanthones, titanocenes). Ultraviolet radiation has more energy than visible light, and thus degrade the physical properties such as the appearance of organic substances and plastics. Benzophenones can act as optical filters or deactivate substrate molecules that have been excited by light for the protection polymers and organic substances. They, cosmetic grades, are used as sunscreen agents to reduce skin damage by blocking UV-A, B.

BENZOIN

Introduction

Benzoin or 2-Hydroxy-2-phenylacetophenone or 2-Hydroxy-1,2-Diphenylethanone or desyl alcohol or bitter almond oil camphor is an organic compound consisting of an ethylene bridge flanked by phenyl groups and with a hydroxyl and a ketone functional group. It comes as off-white crystals, with a light camphor odour.

Preparation

Benzoin is synthesized from benzaldehyde in the benzoin condensation.

The **Benzoin condensation** is a condensation reaction between two aromatic aldehydes, especially benzaldehyde that is catalyzed by a cyanide. The reaction product is an aromatic acyloin with benzoin as the parent compound. The reaction mechanism for this organic reaction was already proposed in 1903 by A. J. Lapworth

Reaction Mechanism

In the first step in this reaction the cyanide ion (as sodium cyanide) reacts with the aldehyde in a nucleophilic addition. Umpolung reverses the polarity of the carbonyl group and the rearranged intermediate adds to the second carbonyl group in a second nucleophilic addition. Proton transfer and elimination of the cyanide ion affords the benzoin. This is a reversible reaction.

The cyanide ion is a very specific catalyst and serves three different purposes in the course of the reaction. It acts as a nucleophile, it facilitates proton abstraction in the umpolung by its inductive effect and it is also the leaving group in the final step. The benzoin condensation is in effect a dimerization and not a condensation because a small molecule like water is not released in this reaction. For this reason the reaction is also called a **benzoin addition**. Both aldehydes have a different purpose. One aldehyde donates a proton and one aldehyde accepts a proton. 4-dimethylaminobenzaldehyde is an efficient proton donor while benzaldehyde is both a proton and a donor. In this way it is possible to synthesise asymmetric benzoins.

Physical Properties

Molar mass	= 212.25 g/mol
Density	=1.31 g/cm^3
Melting point	=137 °C
Boiling point	= 344 °C
Molecular Formula	= $C_{14}H_{12}O_2$
Formula Weight	= 212.24388
Composition	= C(79.22%) H(5.70%) O(15.08%)
Molar Refractivity	= 62.30 ± 0.3 cm^3
Molar Volume	= 179.9 ± 3.0 cm^3
Parachor	= 479.9 ± 4.0 cm^3
Index of Refraction	= 1.609 ± 0.02
Surface Tension	= 50.6 ± 3.0 dyne/cm
Density	= 1.179 ± 0.06 g/cm^3
Polarizability	= 24.70 ± 0.5 10^{-24}cm^3
Monoisotopic Mass	= 212.08373 Da
Nominal Mass	= 212 Da
Average Mass	= 212.2439 Da

Uses

Its main uses are:

- catalyst in photopolymerization and a photoinitiator
- raw material for benzil by organic oxidation with nitric acid or oxone. In one study, this reaction is carried out with atmospheric oxygen and basic alumina in dichloromethane.

BENZIL

Benzil is the organic compound with the formula $(C_6H_5CO)_2$, generally abbreviated $(PhCO)_2$. Benzil is also employed as a photoinitiator for free-radical curing of polymer networks. Ultraviolet radiation decomposes benzil, creating free-radical species which propagate throughout polymer material, creating cross-links between individual polymer chains.

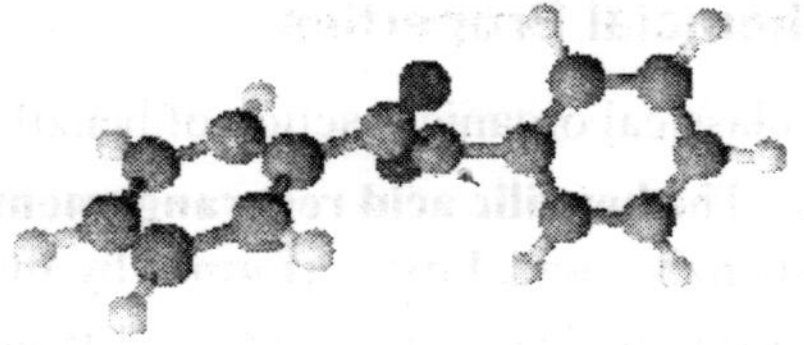

Benzil can be prepared in the laboratory by the benzoin condensation of benzaldehyde followed by oxidation of benzoin using copper sulfate.

benzaldehyde —(KCN, C_2H_5OH)→ benzoin —([O], $CuSO_4$)→ benzil

BENZOIN CONDENSATION — BENZIL FORMATION

Physical Properties

Molecular Formula	= $C_{14}H_{10}O_2$
Formula Weight	= 210.228
Composition	= C(79.98%) H(4.79%) O(15.22%)
Molar Refractivity	= 61.24 ± 0.3 cm^3
Melting point	= 94-95 °C
Boiling point	= 346-348 °C
Molar Volume	= 180.3 ± 3.0 cm^3
Parachor	= 473.1 ± 4.0 cm^3
Index of Refraction	= 1.594 ± 0.02
Surface Tension	= 47.3 ± 3.0 dyne/cm
Density	= 1.165 ± 0.06 g/cm^3
Polarizability	= 24.28 ± 0.5 $10^{-24}cm^3$
Monoisotopic Mass	= 210.06808 Da
Nominal Mass	= 210 Da
Average Mass	= 210.228 Da
Density and phase	= 1.23 g/cm^3, solid
Solubility in water	= insoluble
Solubility in ethanol	= soluble

Chemcial Properties

A classical organic reaction of benzil is the Benzilic acid rearrangement to benzilic acid.

The **benzilic acid rearrangement** is the rearrangement reaction of benzil with potassium hydroxide to benzilic acid. First performed by Justus Liebig in 1838 this reaction type is displayed by 1,2-diketones in general. The reaction product is an α-hydroxy-carboxylic acid.

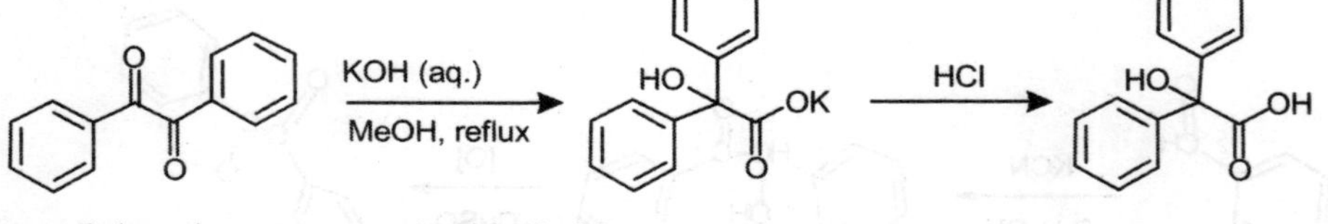

KOH (aq.) MeOH, reflux → HO ... OK → HCl → HO ... OH

Certain acyloins also rearrange in this fashion.

This diketone reaction is related to other rearrangements: the corresponding keto-aldehyde (one alkyl group replaced by hydrogen) rearranges in a Cannizzaro reaction, the corresponding 1,2-diol reacts in a pinacol rearrangement.

Reaction mechanism

The reaction is a representative of 1,2-rearrangements. These rearrangements usually have migrating carbocations but this reaction is unusual because it involves a migrating carbanion. The long established reaction mechanism updated with in silico data [2] is outlined in *scheme 2*.

A hydroxide anion attacks one of the ketone groups in **1** in a nucleophilic addition to the hydroxyl anion **2**. The next step requires a bond rotation to conformer **3** which places the migrating group R in position for attack on the second carbonyl group in a concerted step with reversion of the hydroxyl group back to the carbonyl group. This sequence resembles a nucleophilic acyl substitution. Calculations show that when R is methyl the charge build-up on this group in the transition state can be as high as 0.22 and that the methyl group is positioned between the central carbon carbon at a separation of 209 pm.

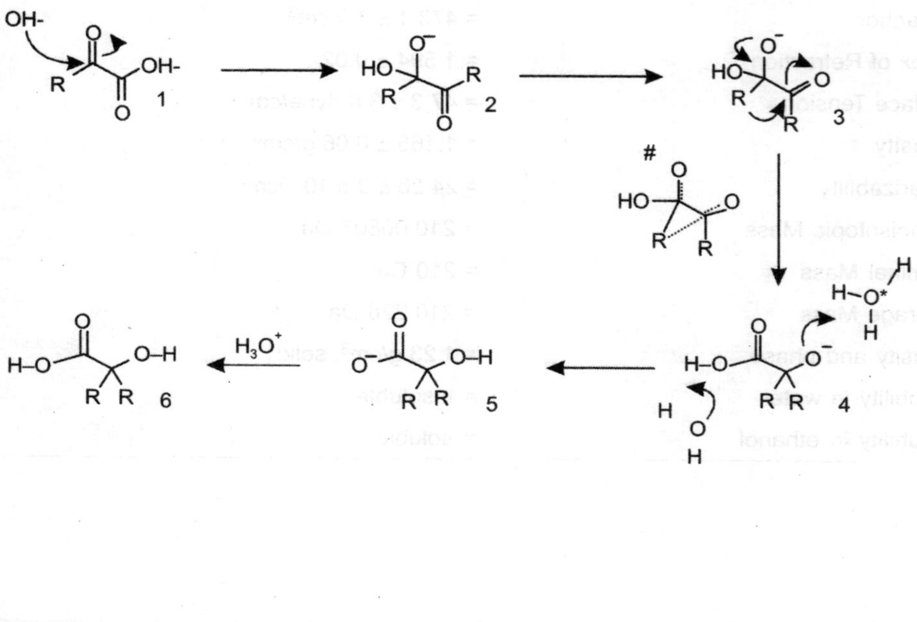

OH- R O OH- O 1 → HO O⁻ R R O 2 → HO O⁻ R O R 3 → # HO O O R R → H-O O O⁻ R R 4 H-O*-H H O H → O⁻ O O-H R R 5 → H_3O^+ → H-O O O-H R R 6

The carboxylic acid in intermediate **4** is less basic than the hydroxyl anion and therefore proton transfer takes place to intermediate **5** which can be protonated in acidic workup to the final α-hydroxy-carboxylic acid **6**. Calculations show that an accurate description of the reaction sequence is possible with the participation of 4 water molecules taking responsibility for the stabilization of charge buildup. They also provide a shuttle for the efficient transfer of one proton in the formation of intermediate 5.

From a molecular orbital point of view this rearrangement may at a first glance not obvious. Contrary to a carbocationic rearrangement as in the Wagner-Meerwein rearrangement in which the empty carbocationic orbital interacts positively and symmetry allowed with the filled pi orbital HOMO of the central C-C bond (situation **A** in *scheme 3*), a filled carbanionic orbital should not be able to escape a symmetry forbidden MO overlap with the LUMO which is the empty antibonding pi orbital having one node (situation **B**).

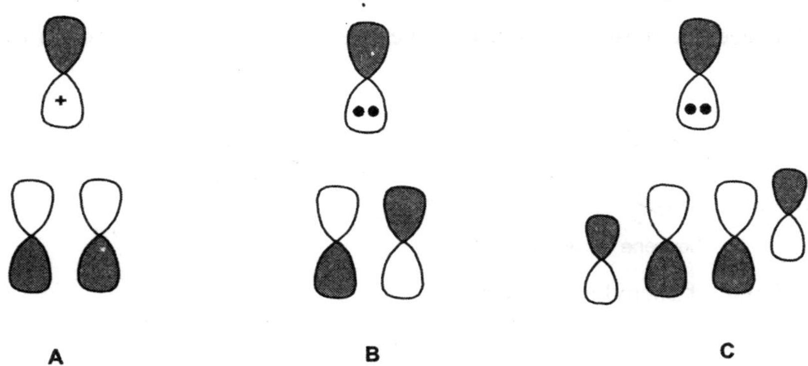

In reality a 1,2-diketone LUMO is a 4 electron system without any nodes in the central C-C bond and a symmetry allowed transition is possible (Situation **C**). In other words the transition states of both a carbocationic rearrangement and the benzilic rearrangement obey the Woodward-Hoffmann rules because the involves respectively 2 electrons and 6 electrons (n=0 and 1 in the 4n+2 Hückel's rule).

Uses

This diketone is used as an intermediate in organic synthesis. Also used in organic synthetic intermediate, used as pesticide, also used as photosensitizer of UV resin.

ACETOPHENONE

Introduction

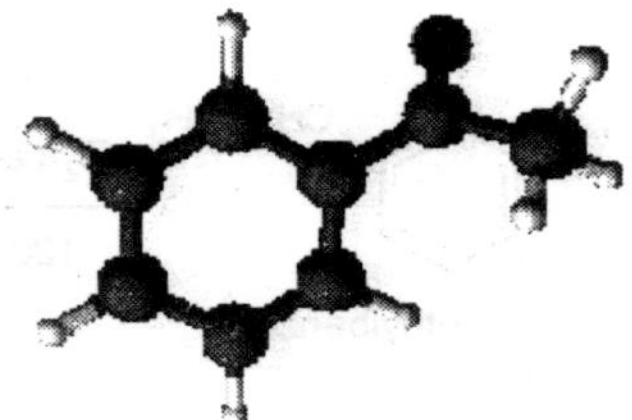

Acetophenone is a crystalline ketone that is used as a solvent for cellulose ethers and esters in the manufacture of alcohol-soluble resins.

Preparation

Acetophenone is prepared by the following methods.

1. By distilling a mixture of calcium acetate and calcium benzoate.

calcium benzoate + calcium acetate → acetophenone

2. By adding acetyl chloride to benzene in the presence of aluminium trichloride

benzene + acetyl chloride $\xrightarrow{AlCl_3}$ acetophenone

3. The oxidation of phenyl methyl carbinol.

phenyl methyl carbinol $\xrightarrow{O}$ acetophenone

4. Industrially Acetophenone is prepared by the action of ethyl benzene and aqueous manganese acetate (15%) is fed from top at 126°C and 2 atmosphere pressure, while air is blown at the bottom. The liquid is pumped to a similar second reactor to meet more air. Total time of contact is about 1½ hours and conversion is about 26%. Benzoic acid, a co-product, is removed with aqueous sodium hydroxide, the upper layer on distillation yields about 68% Acetophenone and 32% α-phenylethyl alcohol. The mixture is catalytically dehydrogenated to Acetophenone at 200°C in a stream of nitrogen.

ethylbenzene $\xrightarrow[120\ °C]{O_2/catalyst}$ acetophenone + 1-phenylethanol $\xrightarrow[200\ °C/N_2]{catalyst}$ acetophenone + H_2

5. From Cumene: About 0.8% Acetophenone is produced during manufacture of phenol and of phenol are thus made the amount of Acetophenone recovered is appreciable. After distilling of phenol are separated by fractionation or with a solvent for e.g. butyl acetate which only dissolved phenol.

Physcial Properties

Melting Point	= 19 -20°C
Boiling point	= 198 -204°C
Specific Gravity	= 1.03
Molecular Formula	= C_8H_8O
Formula Weight	= 120.14852
Composition	= C(79.97%) H(6.71%) O(13.32%)
Molar Refractivity	= 36.28 ± 0.3 cm^3
Molar Volume	= 120.9 ± 3.0 cm^3
Parachor	= 292.4 ± 4.0 cm^3
Index of Refraction	= 1.511 ± 0.02
Surface Tension	= 34.1 ± 3.0 dyne/cm
Density	= 0.993 0.06 h/cm^3
Polarizability	= 14.38 0.5 10^{-24} cm^3
Monoisotopic Mass	= 120.05751 Da
Nominal Mass	= 120 Da
Average Mass	= 120.1485 Da
Log P	= 1.66 ± 0.22

Chemical Properties

It closely relates with aliphatic ketones.

1. Oxidation: It gives benzoic acid

$$C_6H_5COCH_3 \longrightarrow C_6H_5COOH + CO_2 + H_2O$$

1-phenylethanone benzoic acid

2. Reduction: It gives Phenyl methyl carbinol

1-phenylethanone + 2H → benzoic acid + CO_2 + H_2O

This is a distinction from the aliphatic compound.

3. With sodium Bisulphite: It does not react with sodium bisulphite.

4. With Hydrogen Cyanide: It forms cyanohydrin.

acetophenone + HCN → Cyanohydrin acetophenone

5. With Hydrazine: It gives oxime

acetophenone + HCN → acetophenone oxime + H_2O

6. With benzylamine: It forms hydrozone

acetophenone + $NH-NH_2$ → acetophenone phenyl hydrazone + H_2O

7. With Phosphorous Pentachloride: It forms Acetophenone dichloride

acetophenone + PCl_5 (phosphorous pentachloride) → (1,1-dichloroethyl) benzene acetophenone dichloride + $POCl_3$ (phosphorous oxychloride)

8. Nitration and Sulphonation: Acetophenone may be converted into Nitro and sulphuric, derivatives by the displacement of nulceur hydrogen.

9. Meta derivatives are formed

10. Halogenation: On warming with halogen it under goes substitution in side chain

$$C_6H_5-CO-CH_3 + Br_2 \longrightarrow C_6H_5-CO-CH_2-Br + HBr$$

acetophenone 2-bromo-1-phenylethanone

Uses

It is used as a polymerization catalyst for the manufacture of olefins. It is used as an intermediate for pharmaceuticals, agrochemicals and other organic compounds. It also has been used as a drug to induce sleep. Its is used in tear gas (especially as the form of chloro acetophenone) and warfare. It is used as an solvent for plastics, resins, cellulose ethers, and esters. The dimer (dipnone) is used as a plasticizer. Actophenone and its derivatives, having additionally substituted saturated alkyls, oxygenated alkyl groups, thio groups, additional aromatic groups, unsaturated aliphatic side chains, and other functional groups, are ingredients of flavour and fragrance for in soaps, detergents, cosmetics, and perfumes as well as in foods, beverages, and tobacco.

CHAPTER

32

Aromatic Acids

Introduction

In this chapter we are going to discuss about the aromatic acids, their physical and chemical properties and uses and their derivatives.

Monocarboxylic Acids

BENZOIC ACID

Introduction

It occurs in gum Benzoin, Peru balsam and many other resins – from which in can be obtained but sublimation. It is also found in the urine of the horse or ox as Hippuric acid or benzoyl glycine.

Benzoic acid can be recovered from Hippuric acid by boiling it with hydrochloric acid and then cooling the solution, when benzoic acid crystals separates out.

$$\text{hippuric acid} + HCl + H_2O \longrightarrow \text{benzoic acid} + HCl\,H_2C(NH_2){-}COOH$$

hippuric acid — benzoic acid — aminoacetic acid glycine hydrochloride

Preparation

Historical Preparations

The first industrial process involved the reaction of benzotrichloride (trichloromethyl benzene) with calcium hydroxide in water, using iron or iron salts as catalyst. The resulting calcium benzoate is converted to benzoic acid with hydrochloric acid. The product contains significant amounts of chlorinated benzoic acid derivatives. For this reason, benzoic acid for human consumption was obtained by dry distillation of gum benzoin. Even after the discovery of other synthesis methods, it was forbidden to use benzoic acid of other source than gum benzoin.

1. By oxidation : By the oxidation of benzyl alcohol or benzaldehyde with potassium permanganate

$$\text{benzyl alcohol} \xrightarrow{2O} \text{benzaldehyde} + H_2O \longrightarrow \text{benzoic acid}$$

2. By the hydrolysis of Benzo nitrile: With dilute sulphuric acid

$$\text{isocyanobenzene (benzonitrile)} \xrightarrow[2H_2O]{H_2SO_4} \text{benzoic acid}$$

3. Grignard Reaction: By treating magnesium phenyl bromide with carbon dioxide and subsequent hydrolysis benzoic acid

4. By treating benzene with Carbonyl Chloride, in the presence of aluminium trichloride and hydrolyzing the product (Friedel Craft's reaction)

benzene + phosgene $\xrightarrow{AlCl_3}$ benzoyl chloride + HCl → benzoic acid

5. Cannizaro's Reaction: By treating benzaldehyde with potassium hydroxide.

benzaldehyde + K-OH (potassium hydroxide) → benzyl alcohol + potassium benzoate $\xrightarrow{HCl}$ benzoic acid + K—Cl (potassium chloride)

In this reaction one molecule of benzaldehyde is reduced to benzyl alcohol and the other molecule is oxidized to potassium salt of benzoic acid or potassium benzoate.

6. By fusing sodium Benzene Sulphonate: With sodium formate and subsequently decomposing it with HCl

sodium benzene sulphonate + sodium formate $\xrightarrow{fused}$ sodium benzoate + $NaHSO_3$; $\xrightarrow{HCl}$ benzoic acid

7. By Nitration of Benzalmide: When nitrous acid reacts with benzamide benzoic acid is produced.

$$C_6H_5CONH_2 + HNO_2 \longrightarrow C_6H_5COOH + N_2 + H_2O$$

benzamide benzoic acid

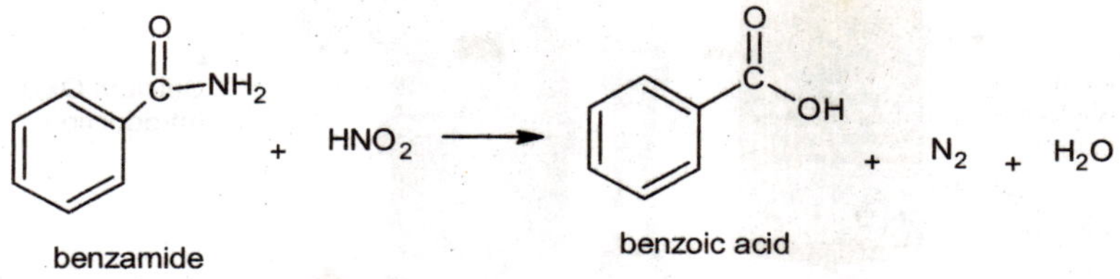

8. Laboratory Method: By the oxidation of Benzyl Chloride, with alkaline potassium permanganate.

(a) $$2KMnO_4 + 2KOH \longrightarrow 2K_2MnO_4 + H_2O + O$$

(b) $$C_6H_5CH_2{-}Cl + H_2O \longrightarrow C_6H_5CH_2OH + HCl$$

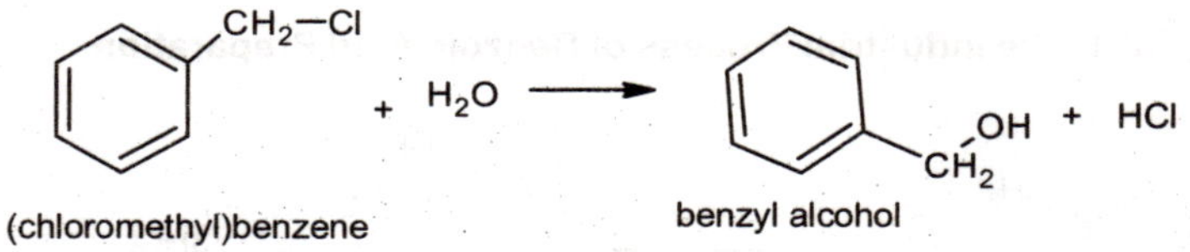

(chloromethyl)benzene benzyl alcohol

(c) $$C_6H_5CH_2OH + 2[O] \longrightarrow C_6H_5COOH + H_2O$$

benzyl alcohol benzoic acid

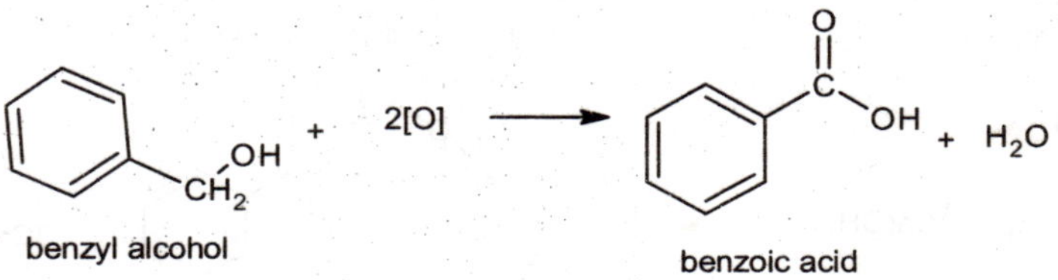

9. By the oxidation of Toluene: Toluene is oxidized with dilute nitric acid or potassium dichromate benzoic acid is produced

$$C_6H_5CH_3 \xrightarrow[K_2Cr_2O_7/HNO_3]{2[O]} C_6H_5COOH + H_2O$$

toluene benzoic acid

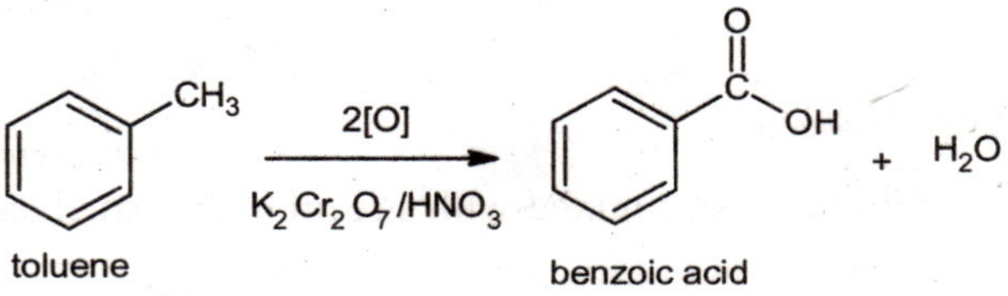

10. Industrial Preparation By the hydrolysis of benzo trichloride with milk of lime. Benzoic acid is manufactured from toluene. Toluene is first chlorinated at its boiling point when benzoic chloride is produced. This on hydrolysis with milk of lime gives calcium benzoate, which on further treatment with acid gives a precipitate of benzoic acid. The acid thus obtained is purified by sublimation or by crystallization.

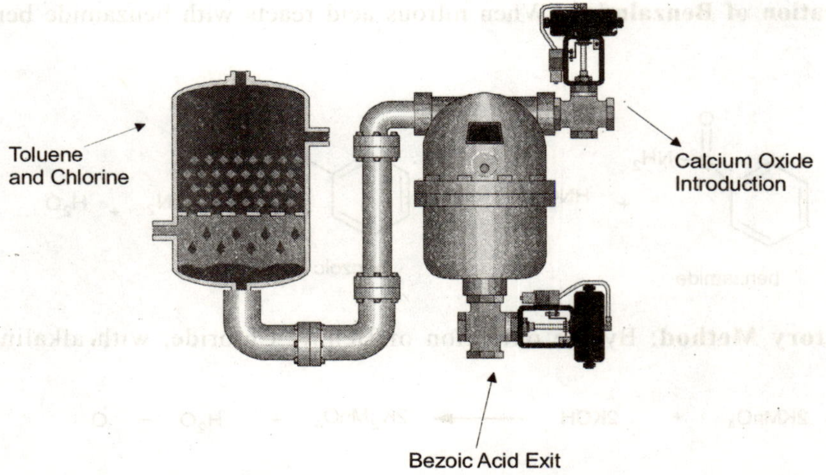

Fig. 32.1. The Industrial Process of Benzoic Acid Preparation

$$\text{toluene} + 3Cl_2 \longrightarrow \text{(trichloromethyl)benzene} + 3HCl$$

$$\text{(trichloromethyl)benzene} + 4Ca(OH)_2 \longrightarrow \text{calcium benzoate} + 3CaCl_2 + H_2O$$

$$\text{calcium benzoate} + HCl \longrightarrow \text{benzoic acid}$$

11. By passing vapours of Phthalic acid mixed with stream over heated catalyst

$$\text{phthalic acid} \xrightarrow[\Delta]{CaO} \text{benzoic acid} + CO_2$$

12. By reacting Cyclopentane with peroxybenzoic acid

cyclopentane + benzenecarboperoxoic acid → 6-oxabicyclo[3.1.0]hexane + benzoic acid

13. Styrene is allowed to react with hydrogen peroxide and carbonic acid, 1 penphenylethane 1, 2 diol is produced and this when oxidized with potassium permanganate benzoic acid is obtained

styrene $\xrightarrow[H_2CO_3]{H_2O_2}$ 1-phenylethane-1,2-diol $\xrightarrow{KMnO_4}$ benzoic acid

Physical Properties

Melting point	= 122.4°C
Boiling Point	= 249°C
Molecular Formula	= $C_7H_6O_2$
Formula Weight	= 122.12134
Composition	= C(68.85%) H(4.95%) O(26.20%)
Molar Refractivity	= 33.18 ± 0.3 cm^3
Molar Volume	= 101.9 ± 3.0 cm^3
Parachor	= 269.4 ± 4.0 cm^3
Index of Refraction	= 1.564 ± 0.02
Surface Tension	= 48.7 ± 3.0 dyne/cm
Density	= 1.197 ± 0.06 g/cm^3
Polarizability	= 13.15 ± 0.5 $10^{-24}cm^3$
Monoisotopic Mass	= 122.036779 Da
Nominal Mass	= 122 Da
Average Mass	= 122.1213 Da
Log P	= 1.89 ± 0.21

Chemical Properties

Benzoic acid is an aromatic mono – carboxylic acid. It combines in itself the reactions of aliphatic monocarboxylic acid and the benzene ring.

The important reactions are:

1. Reactions due to Carboxyl group

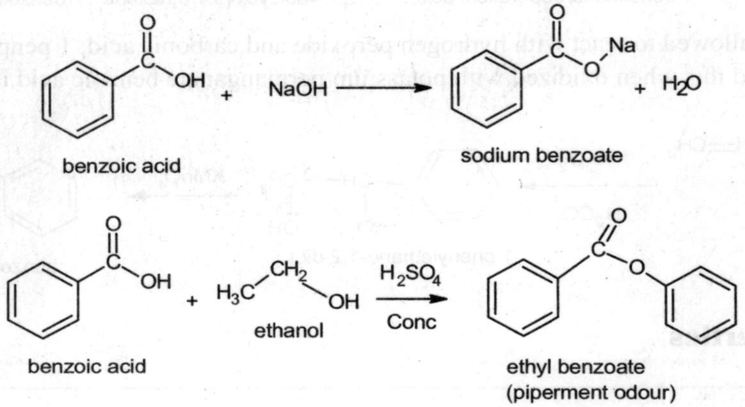

benzoic acid + NaOH → sodium benzoate + H_2O

benzoic acid + ethanol (H_3C-CH_2-OH) $\xrightarrow[\text{Conc}]{H_2SO_4}$ ethyl benzoate (piperment odour)

2. With Phosphorous pentachloride: It gives benzoyl chloride

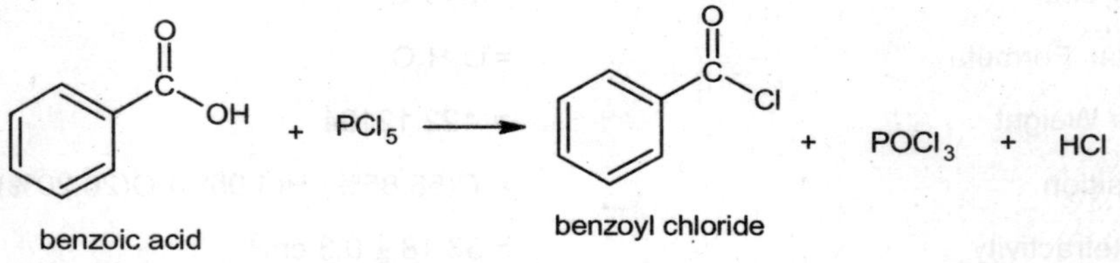

benzoic acid + PCl_5 → benzoyl chloride + $POCl_3$ + HCl

3. With Calcium Oxide: It produces benzene

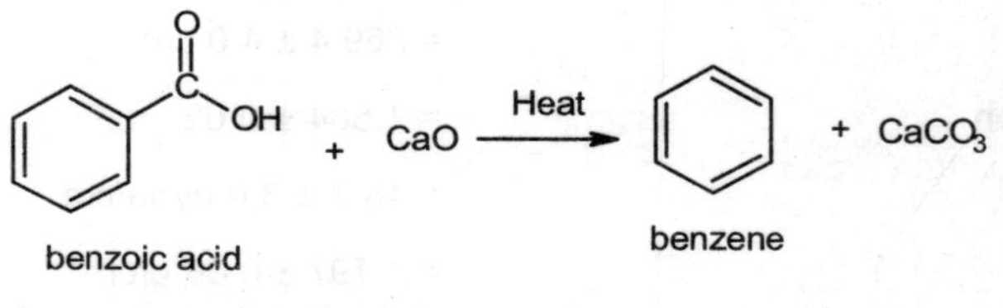

benzoic acid + CaO $\xrightarrow{\text{Heat}}$ benzene + $CaCO_3$

4. With Ammonia: It produces benzamide

benzoic acid + NH_3 → ammonium benzoate $\xrightarrow{\text{distil}}$ Benzamide

5. On dehydration with acetic anhydride: It produces Benzoic anhydride

benzoic acid + acetic anhydride → Benzoic anhydride

Reactions due yo Benzene Nucleus

6. On Nitration it gives m-Nitrobenzoic acid:

benzoic acid + nitric acid → 3-nitrobenzoic acid

7. On sulphonation, It gives m-Sulphobenzoic acid

benzoic acid + sulphuric acid → 3-sulphobenzoic acid

8. On Halogenation It gives m-Chlorobenzoic acid

benzoic acid + Cl_2 → 3-chlorobenzoic acid + HCl

9. Benzoic acid is allowed to react with methanol

benzoic acid + methanol $\overset{H^+}{\rightleftharpoons}$ methyl benzoate + H_2O

10. Benzoic is allowed reacting with thionyl chloride

benzoic acid + thionyl chloride → benzoyl chloride

11. Nitration of Benzoic acid and then reacting with Phosphorous Pentachloride

benzoic acid + nitric acid (two molecules) → 3,5-dinitrobenzoic acid + phosphorous pentachloride → 3,5 Dinitrobenzoyl Chloride

12. Reaction of Benzoic Acid with Isobutyl Alcohol

benzoic acid + isobutyl alcohol $\overset{H^+}{\rightleftharpoons}$ isobutyl benzoate

Derivatives

BENZOYL CHLORIDE

Introduction

Benzoyl chloride, also known as benzenecarbonyl chloride, is a colourless, fuming liquid, C_6H_5COCl, with an irritating odour.

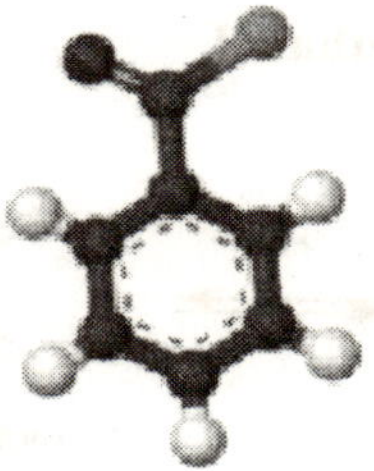

Preparation

1. By the action of phosphorous Pentachloride on benzoic acid:

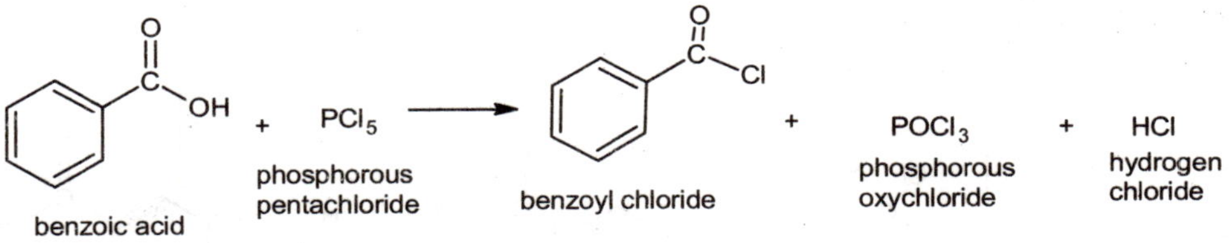

benzoic acid + PCl_5 (phosphorous pentachloride) → benzoyl chloride + $POCl_3$ (phosphorous oxychloride) + HCl (hydrogen chloride)

2. By heating Thionyl Chloride with Benzoic acid:

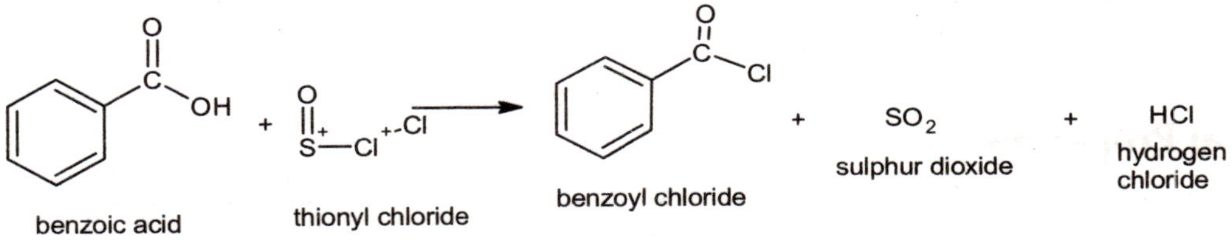

benzoic acid + thionyl chloride → benzoyl chloride + SO_2 (sulphur dioxide) + HCl (hydrogen chloride)

3. By Friedel Crafts Reaction:

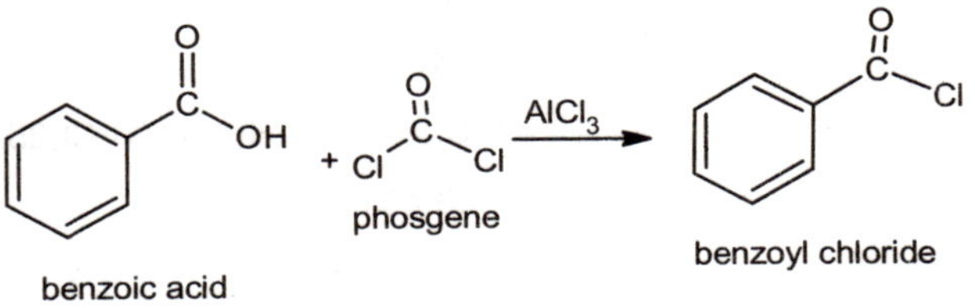

benzoic acid + phosgene $\xrightarrow{AlCl_3}$ benzoyl chloride

4. By the action of Chlorine on benzaldehyde

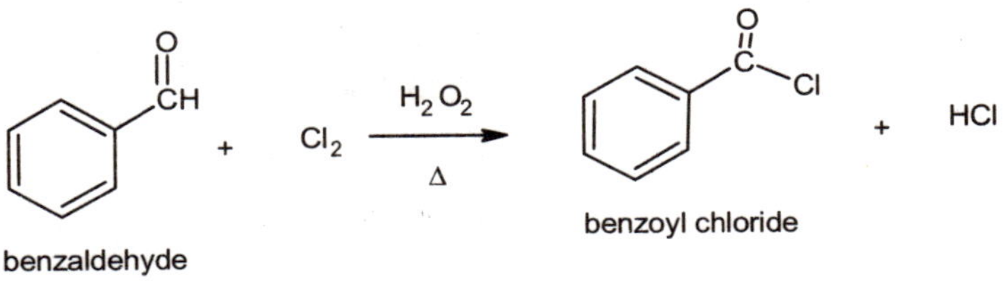

benzaldehyde + Cl_2 $\xrightarrow[\Delta]{H_2O_2}$ benzoyl chloride + HCl

5. From Toluene: Toluene is allowed to react with chlorine then benzo trichloride is produced, sodium carbonate is allowed to react with, producing sodium benzoate. With the action of acid with sodium benzoate benzoic acid is produced. This benzo-trichloride reacts with benzoic acid producing benzoyl chloride

benzo trichloride
(trichloromethyl)benzene

benzoyl chloride

toluene → Cl_2 → (trichloromethyl)benzene benzo trichloride → $NaCO_3$(aq) → sodium benzoate → H^+, H_2O → benzoic acid

Physical Properties

Melting Point	= –1°C
Boiling Point	=197.2°C
Molecular Formula	= C_7H_5ClO
Formula Weight	= 140.567
Composition	= C(59.81%) H(3.59%) Cl(25.22%) O(11.38%)
Molar Refractivity	= 36.49 cm^3
Molar Volume	= 115.8 cm^3
Parachor	= 291.1 4.0 cm^3
Index of Refraction	= 1.542 0.02
Surface Tension	= 39.8 3.0 dyne/cm
Density	= 1.213 0.06 g/cm^3
Polarizability	= 14.46 0.5 $10^{-24}cm^3$
Monoisotopic Mass	= 140.002892 Da
Nominal Mass	= 140 Da
Average Mass	= 140.567 Da
Log P	= 2.21
Flash Point	= 72°C

Chemical Properties

Benzoyl Chloride reacts chemically and some important reactions are being demonstrated below:

1. With water: It produces benzoic acid

benzoyl chloride + H_2O → benzoic acid

2. With Ammonia: It produces benzamide

benzoyl chloride + NH_3 → benzamide + HCl (hydrogen chloride)

3. With Ethyl Alcohol: It produces Ethyl acetate

benzoyl chloride + ethanol → phenyl benzoate / ethyl benzoate + H_2O

4. With Benzene: It produces Benzophenone; this is Friedel Craft's reaction

benzoyl chloride + benzene $\xrightarrow{AlCl_3}$ benzophenone + HCl (hydrogen chloride)

5. On Reduction: It gives Benzaldehyde; this is Rosenmound's Reaction

benzoyl chloride + 2H $\xrightarrow[(BaSO_4)]{Pt}$ benzaldehyde + HCl (hydrogen chloride)

6. When treated with hydroxyl compounds: (phenol or amino compounds (aniline) in the presence of caustic alkalies the corresponding benzoyl derivatives are formed (Schötten Baumann's Reaction)

(*a*) benzoyl chloride + phenol $\xrightarrow{NaOH}$ phenyl benzoate + HCl

This process is known as Benzolation and is used for the detection of –OH and – NH_2 group in organic compounds.

(*b*) benzoyl chloride + phenol $\xrightarrow{NaOH}$ N-phenylbenzamide + HCl

7. With Sodium Benzoate: It gives benzoic anhydride

benzoyl chloride + sodium benzoate $\longrightarrow$ Benzoic anhydride

BENZOYL PEROXIDE

Introduction

Benzoyl peroxide is a chemical in the organic peroxide family. It consists of two benzoyl groups (benzaldehyde with the H of CHO removed) joined by a peroxide group.

Trademarks for benzoyl peroxide include OXY, NeoBenz Micro, Basiron, Brevoxyl, Stioxyl, Panoxyl and Benzac.

Preparation

It is prepared by the reaction of benzoyl chloride and sodium peroxide in ice cold water

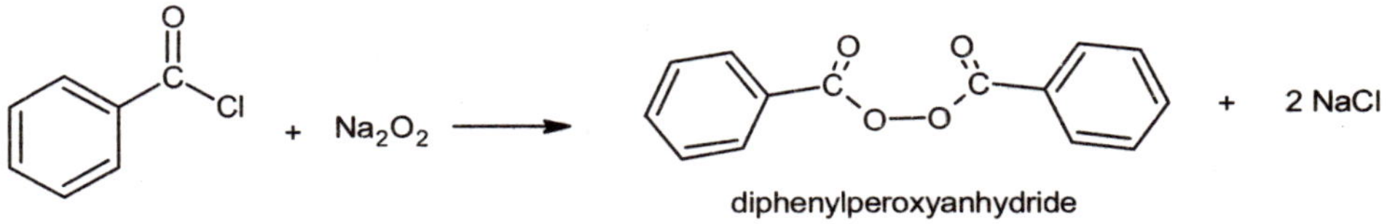

$$C_6H_5COCl + Na_2O_2 \longrightarrow C_6H_5CO{-}O{-}O{-}COC_6H_5 + 2\,NaCl$$

benzoyl chloride

diphenylperoxyanhydride
benzoyl peroxide

Uses

Benzoic acid and its salts are used as a food preservative, represented by the E-numbers E210, E211, E212, and E213. Benzoic acid inhibits the growth of mold, yeast[4] and some bacteria. It is either added directly or it is created from reactions with its sodium, potassium or calcium salt. The mechanism starts with the absorption of benzoic acid in to the cell. If the intracellular pH changes to 5 or lower the anaerobic fermentation of glucose through phosphofructokinase is decreased by 95%. The effectivity of benzoic acid and benzoate is thus dependent on the pH of the food. Acidic food and beverage like fruit juice (citric acid), sparkling drinks (carbon dioxide), soft drinks (phosphoric acid), pickles (vinegar) or other acidified food are preserved with benzoic acid and benzoates.

Concern has been expressed that benzoic acid and its salts may react with ascorbic acid (vitamin C) in some soft drinks, forming small quantities of benzene.

1-Naphthaleneacetic acid (2-(1-naphthyl)acetic acid, NAA) is a white powder, with the molecular formula $C_{12}H_{10}O_2$. It melts at 135 °C in atmospheric pressure.

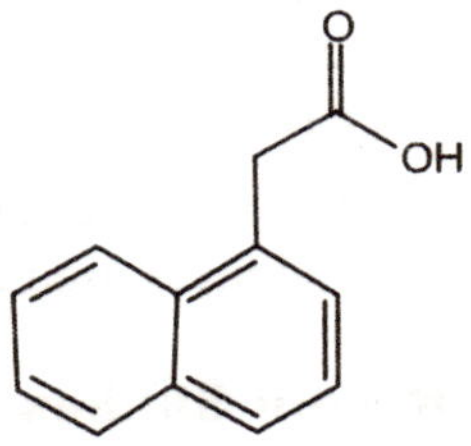

NAA is a plant hormone in the auxin family and is an ingredient in many commercial plant rooting horticultural products; it is a rooting agent and used for the vegetative propagation of plants from stem and leaf cutting.

4-Hydroxybenzoic acid, or *p*-hydroxybenzoic acid, is a phenolic derivative of benzoic acid. It a white crystalline solid that is slightly soluble in water and chloroform, but well soluble in alcohols, ether, and acetone.

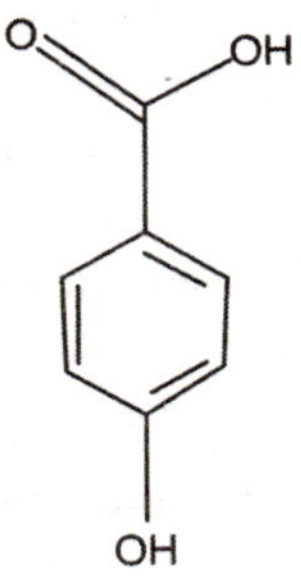

4-Hydroxybenzoic acid is primarily known as the basis for the preparation of its esters, known as parabens, which are used as preservatives in cosmetics.

Chlorogenic acid, an ester of caffeic acid and quinic acid, is a major phenolic compound in coffee, isolated from the leaves and fruits of dicotyledonous plants. This compound, long known as an antioxidant, also slows the release of glucose into the bloodstream after a meal.

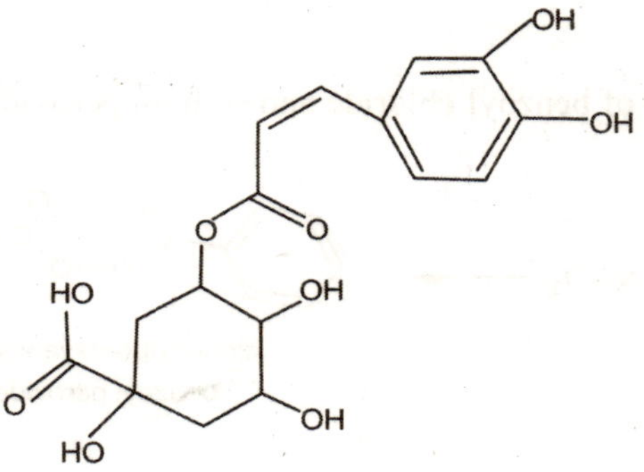

Chemical Properties

Structurally, chlorogenic acid is the ester of caffeic acid with the 3-hydroxyl group of quinic acid.

Biological importance

This acid is an important factor in plant metabolism. It is also an antioxidant and an inhibitor of the tumor promoting activity of phorbol esters; at concentrations as high as 100 µg/L, does not inhibit the 5-lipoxygenase activity of ionophore-stimulated human polymorphonuclear leukocytes.

Chlorogenic acid and caffeic acid are antioxidants in vitro and might therefore contribute to the prevention of cardiovascular disease.

Pharmaceutical and Industrial Applications

This substance is claimed to have anti-viral and anti-bacterial effects with relatively low toxicity and side effects, alongside properties that do not lead anti-microbial resistance. Potential uses are suggested in pharmaceuticals, foodstuffs, feed additives, and cosmetics.

Chlorogenic acid is marketed under the tradename Svetol in Norway and the United Kingdom as a food additive used in coffee, chewing gum, and mints to promote weight reduction.

Recent Studies

Chlorogenic acid has been proven in animal studies in vitro to inhibit the hydrolysis of the glucose-6-phosphate enzyme in an irreversible fashion. This mechanism allows chlorogenic acid to reduce hepatic glycogenolysis (transformation of glycogen into glucose) and to reduce the absorption of new glucose. In addition, in vivo studies on animal subjects have demonstrated that the administration of chlorogenic acid lessens the hyperglycemic peak resulting from the glycogenolysis brought about by the administering of glucagon, a hyperglycemiant hormone. The studies also confirmed a reduction in blood glucose levels and an increase in the intrahepatic concentrations of glucose-6-phosphate and of glycogen.

Melanic acid is the common name chosen by William Prout for 2,5-dihydroxyphenylacetic acid. The compound is also known as homogentisic acid.

Biological Relevance

Melanic acid is a by-product in the poor metabolism associated with alkaptonuria.

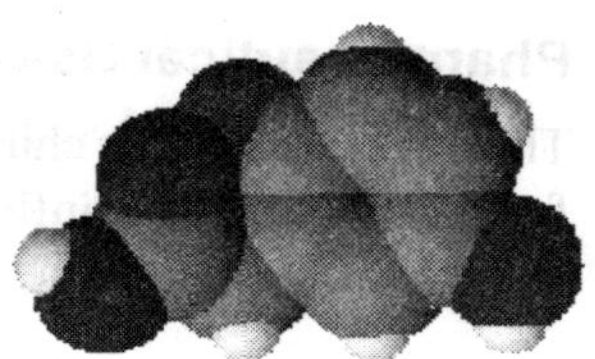

QUINIC ACID

History

This substance was isolated for the first time in the 1800s by French pharmacist Nicolas Vauquelin and further reactions from this acid to synthetize other compounds were studied by German chemist E. Lautemann in 1863.

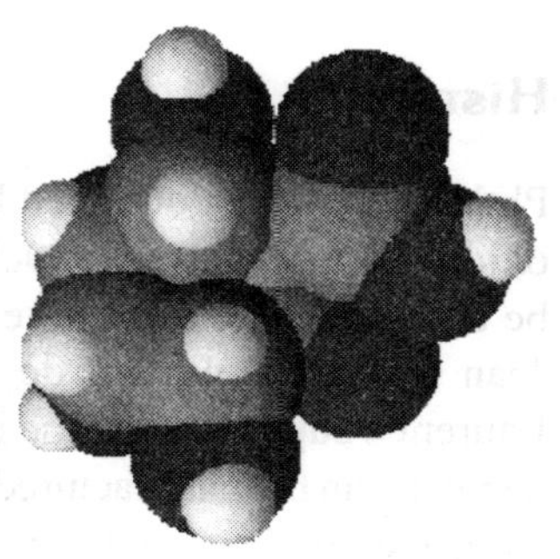

Quinic acid, $C_7H_{12}O_6$ is a crystalline acid obtained from cinchona bark, coffee beans, and other plant products and made synthetically by hydrolysis of chlorogenic acid. Quinic acid is also implicated in the perceived acidity of coffee.

Physical Properties

Melting Point	= 168°C
Molecular Formula	= $C_7H_{12}O_6$
Formula Weight	= 192.16658
Composition	= C(43.75%) H(6.29%) O(49.95%)
Molar Refractivity	= 40.04 ± 0.3 cm^3
Molar Volume	= 105.0 ± 3.0 cm^3
Parachor	= 361.1 ± 6.0 cm^3
Index of Refraction	= 1.687 ± 0.02
Surface Tension	= 139.5 ± 3.0 dyne/cm
Density	= 1.828 ± 0.06 g/cm^3
Polarizability	= 15.87 ± 0.5 $10^{-24}cm^3$
Monoisotopic Mass	= 192.063388 Da
Nominal Mass	= 192 Da
Average Mass	= 192.1666 Da
Log P	= – 2.01

Industrial Applications

Quinic acid is used as an astringent.

Pharmaceutical Uses

This acid is a versatile chiral starting material for the synthesis of new pharmaceuticals. A new medicament for the treatment of influenza A and B strains called Tamiflu has been successfully developed and launched into the market recently.

Dicarboxylic Acids

PHTHALIC ACID

History

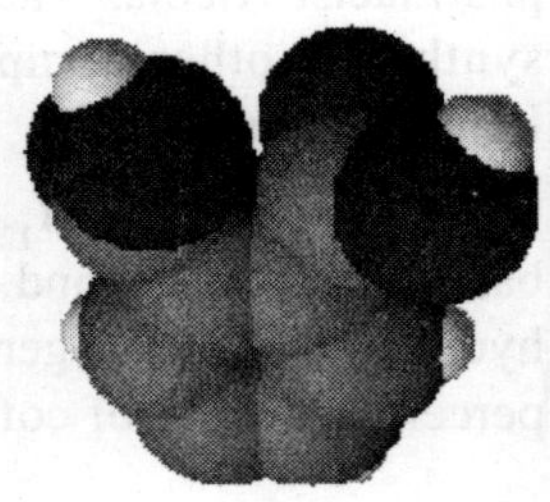

Phthalic acid was obtained by French chemist Auguste Laurent in 1836 by oxidizing naphthalene tetrachloride, and, believing the resulting substance to be a naphthalene derivative, he named it naphthalenic acid. Swiss chemist Jean Charles Galissard de Marignac determined its formula and showed Laurent's supposition to be incorrect, upon which Laurent gave it its present name. It can be manufactured by oxidizing naphthalene tetrachloride (prepared from naphthalene, potassium, chlorate and hydrochloric acid) with nitric acid, or, better, by oxidizing the hydrocarbon with fuming sulfuric acid, using mercury or mercury(II) sulphate as a catalyst.

Preparation

In laboratory Phthalic acid is prepared by the oxidation of o – xylene with nitric acid or potassium permanganate

o-xylene $\xrightarrow[(-\,H_2O)]{3O}$ 2-methylbenzoic acid $\xrightarrow[(-\,H_2O)]{3O}$ phthalic acid

Other Process

Generally phthalic acid is available in the form of anhydride, this is prepared by vapour phase oxidation of naphthalene in the presence of vanadium pentoxide at a high temperature.

naphthalene + O_2 $\xrightarrow[350\text{ - }450\ °C,\ 0.4\text{ - }4.0\ sec]{V_2O_5}$ 1,4-naphthoquinone $\xrightarrow[-\,2H_2O]{-\,2CO_2}$ phthalic anhydride

Naphthalene is melted and pumped to a vaporizer, where preheated primary air is bubbled through the molten material. Secondary air is subsequently added in a mixing section to bring the air-naphthalene ratio to approximately 25: 1 by weight.

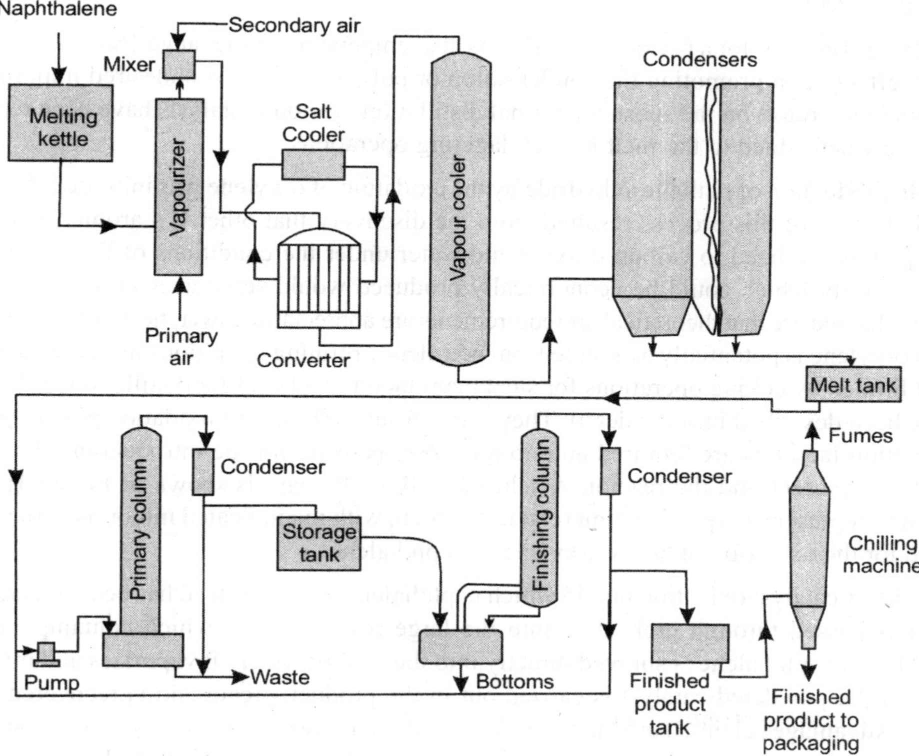

Fig. 32.2. The Industrial Preparation of Phthalic Anhydride

This vapour mixture is then fed to the catalytic reactors or converters, which typically consist of a series of small-diameter tubes packed with the granular, supported vana-dium pentoxide catalyst. The heat of reaction for the formation of anhydride is 3330 cal./kg. of naphthalene, but a certain amount of complete combustion occurs and the actual heat liberated in practice is 5500~6000 cal./kg. of naphthalene feed. This heat is partially removed by a fluid heat-exchange medium, which contacts the exterior of the catalyst tubes. Both boiling mercury and molten nitrate-nitrite salt mixtures are currently in use as cooling media (see *Heal-transfer media)*. An appreciable part of the generated heat is carried out of the reactor section as sensible heat of the gaseous reaction products. From the converter, the hot gases pass through a vapor cooler, where their temperature is reduced to just above the dew point (125-130°C.). The heat-transfer medium used in reducing the temperature is water, and the heat is utilized to generate low-pressure steam. Until very recently, established practice was to send these cooled vapours to large air-cooled condensers, commonly known as "hay barns", where the crude phthalic anhydride crystallized on the walls in the form of long needles or "hay". More recent installations tend to utilize either fin-tube condensers, cyclones, or water scrubbers rather than the traditional "hay barns".

The crude phthalic anhydride is withdrawn from the primary recovery units, melted, and customarily purified by vacuum fractional distillation in a packed column, using pressures which vary considerably. In general, two distillations of the crude product are required, and some producers also use chemical purification procedures.

"Lagering" (heating for a period of 2-10 hours at a temperature in the range 180-280°C.) is reported to be quite effective in promoting the condensation or polymerization of undesired impurities, so that they remain in the retort on the subsequent final distillation. Various catalysts have been reported to be of assistance when added to the melt in such lagering operations.

In 1946 production of phthalic anhydride by the oxidation of o-xylene was initi-ated. The successful commercialization of this process resulted from the discovery that other C's aromatics and paraffins were completely oxidized to carbon dioxide and water under the conditions of the reaction. Thus o-xylene of a purity which could be economically produced would serve as a satisfactory feed. Other advantages claimed are that theoretical air requirements are appreciably lower, heat generation is similarly lower and o-xylene is potentially available from petroleum-refining operations in any desired quantities and is not limited by coking operations for steel manufacture and coal-tar distillation. The process and plant have been described in some detail. They differ from naphthalene-oxidation practice primarily in that no melting facilities are required and no provision is made for the introduction of secondary air between the vaporizer and the reactor. A schematic flow diagram is shown in Figure 1; this can be considered to represent a typical commercial unit which, with the indicated minor modifications, could be utilized for the oxidation of either o-xylene or naphthalene.

A fluidized-catalyst oxidation unit in which naphthalene is used as feed has been operated for some years. Air is passed through preheaters into the large reactor vessel which contains the powdered catalyst. Melted naphthalene is sprayed directly into the reactor, where it vaporizes and mixes with the air. Some of the powdered catalyst is carried out in the product stream and is recovered in cyclones and filters. Advantages claimed are higher purity of product, lower air ratio, less explosion hazard, less labor, lower operating temperatures, and no necessity for housing the unit. At least one engineering contractor offered in 1952 to construct large commercial units for the fluidized-catalyst type of plant, and it may be expected that this type of reactor will shortly be adapted for operation with o-xylene feed.

Physical Properties

Property	Phthalic Acid	Anhydride
Molecular Formula	$C_8H_6O_4$	$C_8H_4O_3$
Formula Weight	166.13084	148.11556
Melting Point	210°C (decomp)	128°C
Molar Refractivity	40.11 cm^3	35.68 cm^3
Molar Volume	114.4 cm^3	102.5 cm^3
Parachor	331.5 cm^3	283.1 cm^3
Index of Refraction	1.617	1.612

Property	Phthalic Acid	Anhydride
Surface Tension	70.3 dyne/cm	58.0 dyne/cm
Density	1.451 g/cm^3	1.444 g/cm^3
Polarizability	15.90 10–24 cm^3	1.14 10-24 cm^3
Monoisotopic Mass	166.026609 Da	148.016044 Da
Nominal Mass	166 Da	148 Da
Average Mass	166.1308 Da	148.156 Da

Chemical Properties

Phthalic anhydride forms white flakes and sublimes readily; it is practically insoluble in cold water, but produces phthalic acid on addition to hot water.

(1) Proximity of carboxyl groups enables phthalic anhydride to condense with benzene in presence of AlCl3 and yield o – benzolybenzoic acid which, by intramolecular cyclisation, forms a new fused ring 9, 10 Anthraquinone.

phthalic anhydride + benzene $\xrightarrow{AlCl_3}$ 2-benzoylbenzoic acid $\xrightarrow[D]{H_2SO_4}$ anthracen-9(10H)-one 9, 10 anthraquinone

(2) If ammonia under pressure is allowed to pass over molten phthalic anhydride at 170°C - 240°C phthalimide is produced

phthalic anhydride + NH_3 → phthalamic acid $\xrightarrow{ammonia}$ phthalonitrile

phthalic anhydride + NH_3 → phthalimide $\xrightarrow{Conc\ ammonia}$ phthalamide

(3) Phthalic anhydride when heated and chloroform is added on its molten state produces 3-chloro-1,5-dihydro-2,4-benzodioxepine

phthalic anhydride + chloroform —D→ 3-chloro-1,5-dihydro-2,4-benzodioxepine + Cl_2

(4) Phthalic anhydride (one molecule) reacts with phenol (two molecule) to produce phenolphthalein

phthalic anhydride + 2 phenol → phenolphthalein

For details see. A chemical analyzer's guide page 282.

(5) Phenol also reacts with resorcinol producing flurescein indicator in the presence of anhydrous zinc chloride.

phthalic anhydride + resorcinol —$ZnCl_2$→ Flurescein Indicator resorcephthalein

(6) Phthalic acid when reacts with calcium oxide produces benzene

phthalic anhydride + CaO → benzene + $2CaCO_3$

(7) Phthalic acid when reacts with phosphorous pentachloride, produces phthaloyl dichloride

phthalic anhydride + PCl_5 → phthaloyl dichloride + 2HCl

(8) Phthalic anhydride reacts with iso - butyl alcohol producing *sec*-butyl benzoate

phthalic anhydride + $H_3C—CH_2—CH(OH)—CH_3$ (butan-2-ol) → *sec*-butyl benzoate

Isomers of Phthalic Acid

ISOPHTHALIC ACID

Isophthalic acid (m-benzenedicarboxylic acid), formula weight 166.13, is a white crystalline solid (colourless needles). It is not yet available as an industrial chemical, but by 1952 at least three or possibly four major firms had progressed sufficiently far with development programs to be offering research quantities to prospective users and were known to be considering its production in1953. Semicommercial production of the dimethyl ester of isophthalic acid was announced by one manufacturer early in 1953. In general, the properties of the acid and its derivatives are more closely related to those of the para isomer, terephthalic acid, than to those of the ortho isomer.

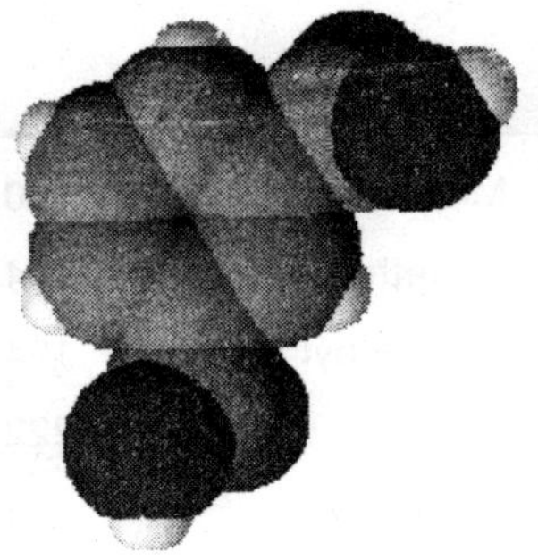

Reactions

Isophthalic acid exhibits the usual reactions of a dibasic carboxylic acid, limited by its low solubility and high melting point. Substitution of the ring positions also occurs, as would be expected.

Esterification Isophthalic acid esterification proceeds at a good rate with mono- hydric alcohols in the presence of conventional catalysts, such as sulfuric acid and *p*- , toluenesulphonic acid. The overall reaction rates for the esterification are greater than 'or the corresponding ortho or para isomer. At a high temperature, for example at the boiling point of 2-ethyl-1-hexanol (183.5°C.), a catalyst is not essential. Isophthalic acid dissolves readily in isophthalate diesters at 200-250°C to form half-esters which do not decompose even during prolonged heating. Advantage may be taken of this ester-interchange reaction to prepare additional quantities of diesters of low boiling alcohols. The reaction of the diester and the acid is carried out at 250°C., for example, and then the required amount of alcohol is added slowly at the same temperature. Benzene may be used for the azeotropic separation of water when necessary. The reaction of isophthalic acid with diols is rapid at 200°C., and an esterification catalyst is not required. To obtain high polymers, however, an ester-interchange catalyst such as litharge should be added.

Dehydration. A linear intermolecular anhydride is formed by treating the acid with acetic anhydride

Formation of Acid Halides. Isophthaloyl halides may be prepared by the conventional methods with phosphorus oxyhalide or pentahalide, or with acetyl halide. The acid halides are frequently used in the preparation of derivatives of isophthalic acid when increased reactivity is desired.

Nylon Salt Formation. Isophthalic acid, like adiphic acid, forms a salt with hexamethylenediamine which is soluble in water.

Hydrogenation. Dihydro- and tetrahydroisophthalic acids are formed by hydrogenation of isophthalic acid in the presence of sodium amalgam. By catalytic hydrogenation under severe conditions, the hexahydro derivative is obtained

Halogenation. Chlorine, bromine, and iodine react with isophthalic acid under suitable conditions to yield the various tetrahalo derivatives.

Table showing the properties of Isothalic Acid Esters

Ester	Formula weight	M.p., °C.	B.p., °C	n_D	d_{20}^{20}
Monmethyl	180.2	193°	—	—	—
Dimethyl	194.2	67–68	12412	—	—
Monoethyl	194.2	115–117	—	—	
Diethyl	222.2	11.5	302	—	1.1389
Di-n-butyl	278.3	—	$191–193_3$	1.4938^{20}	1.045
Diisooctyl	390.6	—	$127–138_{0.4}$	1.4893^{25}	0.987
Di(2-ethylhexyl)	390.6	—	$223–225_1$	1.4859^{20}	0.980
Diallyl	246.3	—	$150–152_{0.9}$	1.5248^{20}	1.114
Diphenyl	318.3	137–138	—	—	—
m-C_6H_4 $(COOCH_2CH_2NR)_2$, where R is:					
$(CH_3)_2$	308.4	—	$216–217_{10}$	—	—

Ester	Formula weight	M.p., °C.	B.p., °C	n_D	d^{20}_{20}
$(CH_3)_2.HCl$	381.3	237	—	—	—
$(CH_3)_3I$	592.3	246–247	—	—	—
$(CH_3)_2C_2H_5I$	620.3	215	—	—	—
$(C_2H_5)_2$	364.5	—	$200_{0.5}$ (approx.)	—	—
$(C_2H_5)_2.HCl$	437.4	198	—	—	—
$(C_2H_5)_2I$	676.4	200	—	—	—

TEREPHTHALIC ACID

Terephthalic acid (p-benzenedicarboxylic acid), , formula weight 166.13, forms colourless needles or an amorphous solid. It has only recently become available as an industrial chemical

Physical and Chemical Properties

Some of the more important properties of terephthalic acid include: m.p.300°C. (sublimises), 425°C. (sealed tube); density 1.510. It is soluble in hot concentrated sulfuric acid, pyridine, and dimethylformamide (DMF). It is slightly soluble in hot ethyl alcohol and insoluble in ether, acetone, chloroform, water, and acetic acid.

Reactions

Terephthalic acid exhibits the characteristic reactions of a dicarboxylic acid, as limited by its low solubility and high melting point. Ring substitution reactions also proceed as expected.

Esterification. Terephthalic acid reacts with the simple aliphatic alcohols in the presence of esterification catalysts, such as concentrated sulfuric acid. A typical one-step reaction in which the diester is obtained from the free acid is the preparation of di-n-butyl terephthalate. An excess of I-butanol and 5% of 95% sulfuric acid (based on terephthalic acid) are refluxed for 6-8 hours so as to distill out the butanol-water azeotropie mixture at 92.5°C. When the reaction mixture clears and the vapor temperature rises to 117°C., the esterification is considered complete, the excess alcohol is removed by distillation, and the product is washed with water and distilled under vacuum. The yield is reported to be approximately 90% (based on terephthalic acid). The cloudy appearance of the reaction mixture is due to the presence of the acid as a solid phase and its disappearance constitutes an additional means of checking the esterification end point in reactions in which the free acid is employed.

In the preparation of some esters, the simple conventional methods cannot be employed, for example, when the water-alcohol azeotropic is so high boiling that degradation of the ester occurs, or when the boiling point is too low and the reaction velocity is accordingly reduced below tolerable values. In such cases, the use of terminating azeotrope-forming assistants, or the use of indirect means such as the acid chloride, may be required.

Dehydration. A linear, intermolecular anhydride is formed by treating the acid with acetic anhydride.

acetic anhydride

teraphthalic anhydride

Formation of Acid Halides. Terephthaloyl halides are obtained by reaction with phosphorus pentahalide and phosphorus oxyhalide or with acetyl halide. The acid halides, because of their increased reactivity, are frequently used in the preparation of various other terephthalic acid derivatives.

terephthalic acid

teraphthalic chloride

Salt Formation. A large number of metal salts of terephthalic acid have been prepared because of their potential usefulness as catalysts, pigments, driers, alkyd resin modifiers, etc. The ammonium, alkali metal, and magnesium salts are water soluble. The salts of other common metals, such as calcium, barium, lead, aluminum, zinc, copper, cobalt, manganese, and ferric iron, are not water-soluble. The copper salt is light blue in colour, the cobalt is purple, and the ferric iron is rust-coloured; the others mentioned are all colourless.

Nitrile Formation. Terephthalonitrile, p-$C_6H_4(CN)_2$, is obtained by reaction of the acid with benzenesulphonamide. It is believed that the diamine could be obtained by hydrogenation of the nitrile.

Hydrogenation. Dihydro- and tetrahydroterephthalic acids are formed by hydrogenation of terephthalic acid in the presence of sodium amalgam. By catalytic hydrogenation under more severe conditions, the hexahydro derivative may be obtained.

Halogenation. Chlorine, bromine, and iodine react with terephthalic acid in fuming sulfuric acid solution. In addition to the tetrahalogenated acid, some hexahalogenated benzene also is produced.

Nitration. Strong nitrating agents convert terephthalic acid to the mononitro derivative, which can be reduced to the corresponding amino acid. The amino acid, by diazotization and hydrolysis, yields hydroxyterephthalic acid

Sulphonation. Fuming sulfuric acid at 200-260°C. Converts terephthalic acid to sulphoterephthalic acid. This reaction is of interest since the sulphonic and group imparts water solubility to the acid and to many of its otherwise insoluble derivatives.

DERIVATIVES

Numerous derivatives of terephthalic acid have been reported. The most important of these are the esters, particularly the dimethyl ester.

Dimethyl terephthalate, $C_6H_4(COOCH_3)_2$, formula weight 194.18, is available as a white, flaked solid. Important physical properties are given in Table below. It is obtained by essentially conventional esterification methods.

Physical Property of Teraphthalic Acid

M.p., °C	140.6	
B.p., °C	288	
d^{150}_{25}	1.07	
Bulk density, grams/cu.cm.	0.59	
Av. sp. heat, eal./(gram)(°C):		
Solid (29–141 °C)	0.326	
Liquid (141–210 °C)	0.464	
Heat of fusion, cal./gram	38	
Het of vaporization, cal./gram	84.9	
Heat of combustion, cal./gram	5,733	
Flash point (open cup), °C	146	
Acidity, m.e./gram	0.005	
Saponification equivalent	97.1–97.2	
Purity, %	99.8+	
Toluene-insoluble. wt. %	0.01 max.	
Solubility, grams/100 ml. solvent at 25 °C., in:		
Water	insol.	
Methyl alcohol	0.86	
Ethyl alcohol	0.82	
Chloroform	26.1	
Carbon tetrachloride	2.65	
Dioxane	7.33	
Solubility, wt. %, in:	At 10°C	At 80°C
Benzene	3.0	35
Toluene	2.2	29
Xylene	1.3	22

The preparation of dimethyl terephthalate offers one of the easiest a Yenues of purification of the acid: in fact, its use instead of the acid in the production of many of the derivatives of terephthalic acid may be advantageous because of its lower melting point, greater solubility in a number of solvents, and generally greater ease of handling. The higher-molecular-,,-eight aliphatic esters and the high-polymer polyester compositions, which represent the only commercially significant end-use items at this time (1952), can all be readily prepared from dimethyl terephthalate. For these reasons there appears to be an excellent chance that this ester rather than the free acid may become the predominant article of commerce as the demand increases.

Physical Properties of Terephthalic Acid Esters

Properties	Diethyl	Di-npropyl	Di-n-butyl	Di(2-ethylhexyl)	Diisooctyl	Diallyl
Appearance	White crystalline solid	Light yellow mobile liquid	Colorless to light green mobile liquid	Colorless to light yellow liquid	Colorless to pale yellow liquid	Light yellow mobile liquid
Formula weight	222.23	250.28	278.34	390.54	390.54	246.25
Freezing point, °C	43–44	25	16.6–18.1	–63.5	Forms amorphous solid	–10
Boiling range, °C	$140–142_3$	$164–166_4$	181–1892	$186–192_{0.1}$	$189–190_{0.05}$	$140–143_1$
Density	1.1022^{45}_{25}	1.0715^{28}_{25}	1.0452^{25}_{25}	0.9823^{25}_{25}	0.9809^{25}_{25}	1.1380^{45}_{25}
Refractive index at 25°C	—	—	1.4960	1.4882	1.4889	1.5244
Viscosity at 25°C, centipoises	—	12.94	15.33	61.2	65.05	12.46
Flash point (open cup), °C	154	171	182	232	218	171
Sapnification equivalent	113	125	139–140	195–197	194–196	124

Polyethylene terephthalate (**PET**, **PETE** or the obsolete PETP or PET-P) is a thermoplastic polymer resin of the polyester family and is used in synthetic fibers; beverage, food and other liquid containers; thermoforming applications; and engineering resins often in combination with glass fiber. It is one of the most important raw materials used in manmade fibers.

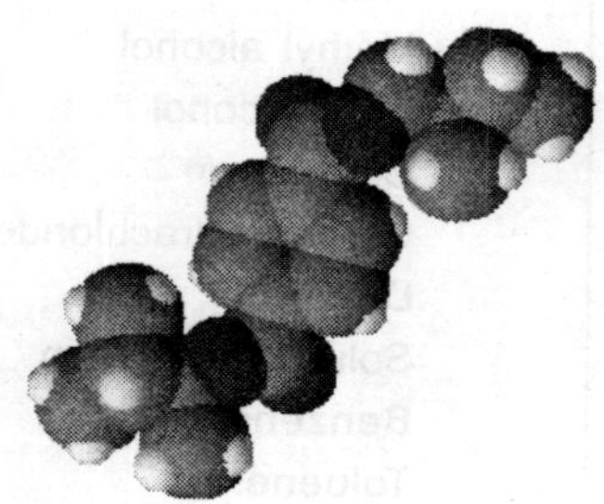

Depending on its processing and thermal history, it may exist both as an amorphous (transparent) and as a semi-crystalline (opaque and white) material. Its monomer can be synthesized by the esterification reaction between terephthalic acid and ethylene glycol with water as a byproduct, or the transesterification reaction between ethylene glycol and dimethyl terephthalate with methanol as a byproduct. Polymerization is through a polycondensation reaction of the monomers (done immediately after esterification/transesterification) with ethylene glycol as the byproduct (the ethylene glycol is recycled in production).

The majority of the world's PET production is for synthetic fibers (in excess of 60%) with bottle production accounting for around 30% of global demand. In discussing textile applications, PET is generally referred to as simply "polyester" while "PET" is used most often to refer to packaging applications.

It is manufactured under trade names **Arnite**, **Impet** and **Rynite**, **Ertalyte**, **Hostaphan**, **Melinex** and **Mylar** films, and **Dacron**, **Diolen**, **Terylene** & **Trevira** fibers.

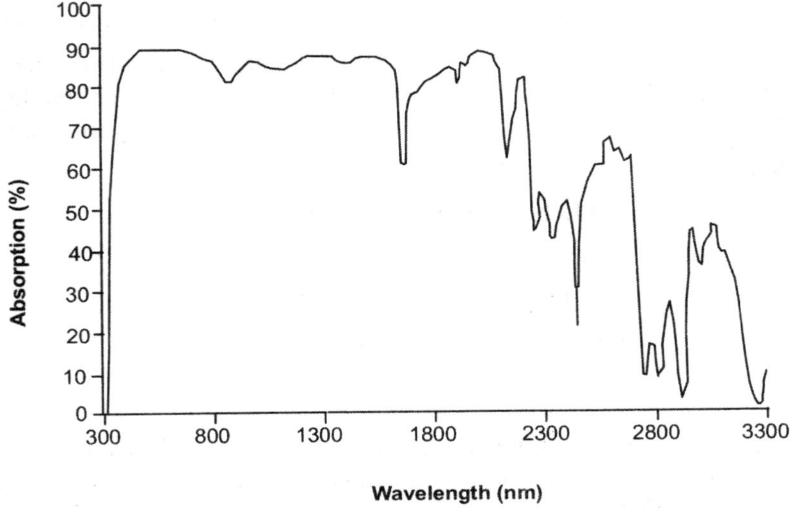

Fig. 32.3. Light absorption spectrum of polyethylene terephthalate measured using a light spectrophotometer (Cary 5000)

PET can be semi-rigid to rigid, depending on its thickness, and is very lightweight. It makes a good gas and fair moisture barrier, as well as a good barrier to alcohol (requires additional "Barrier" treatment) and solvents. It is strong and impact-resistant. It is naturally colourless and transparent.

When produced as a thin film (often known by the tradename Mylar), PET is often coated with aluminium to reduce its permeability, and to make it reflective and opaque. PET bottles are excellent barrier materials and are widely used for soft drinks, (see carbonation). PET or Dacron is also used as a thermal insulation layer on the outside of the International Space Station as seen in an episode of Modern Marvels "Sub Zero". For certain specialty bottles, PET sandwiches an additional polyvinyl alcohol to further reduce its oxygen permeability.

When filled with glass particles or fibers, it becomes significantly stiffer and more durable. This glass-filled plastic, in a semi-crystalline formulation, is sold under the tradename Rynite, Arnite, Hostadur & Crastin.

While all thermoplastics are technically recyclable, PET bottle recycling is more practical than many other plastic applications. The primary reason is that plastic carbonated soft drink bottles and water bottles are almost exclusively PET which makes them more easily identifiable in a recycle stream. PET has a resin identification code of 1. PET, as with many plastics, is also an excellent candidate for thermal recycling (incineration) as it is composed of carbon, hydrogen and oxygen with only trace amounts of catalyst elements (no sulfur) and has the energy content of soft coal.

PET was patented in 1941 by the Calico Printers' Association of Manchester. The PET bottle was patented in 1973.

Intrinsic Viscosity

One of the most important characteristics of PET is referred to as I.V.(intrinsic viscosity).

The I.V. of the material, measured in dl/g (deciliters/gram) is dependent upon the length of its polymer chains. The longer the chains, the stiffer the material, and therefore the higher the I.V. The average chain length of a particular batch of resin can be controlled during polymerization.

An I.V. of about:

0.60 — Would be appropriate for fiber

0.65 — Film

0.76–0.84 — Bottles

0.85 — Tire cord

Drying

PET is hygroscopic, meaning that it naturally absorbs water from its surroundings. However, when this 'damp' PET is then heated a chemical reaction known as hydrolysis takes place between the water and the PET which reduces its molecular weight (IV) and its physical properties. This means that before the resin can be processed in a molding machine, as much moisture as possible must be removed from the resin. This is achieved through the use of a desiccant or dryers before the PET is fed in the processing equipment.

Inside the dryer, hot dry air is pumped into the bottom of the hopper containing the resin so that it flows up through the pellets removing moisture on its way. The hot wet air leaves the top of the hopper and is first run through an after-cooler, because it is easier to remove moisture from cold air than hot air. The resulting cool wet air is then passed through a desiccant bed. Finally the cool dry air leaving the desiccant bed is re-heated in a process heater and sent back through the same processes in a closed loop. Typically residual moisture levels in the resin must be less than 40 ppm before processing. Dryer residence time should not be shorter than about four hours. This is because drying the material in less than 4 hours would require a temperature above 160°C, at which level hydrolysis would begin inside the pellets before they could be dried out.

Copolymers

In addition to pure (homopolymer) PET, PET modified by copolymerization is also available.

In some cases, the modified properties of copolymer are more desirable for a particular application. For example, cyclohexane dimethanol (CHDM) can be added to the polymer backbone in place of ethylene glycol. Since this building block is much larger (6 additional carbon atoms) than the ethylene glycol unit it replaces, it does not fit in with the neighboring chains the way an ethylene glycol unit would. This interferes with crystallization and lowers the polymer's melting temperature. Such PET is generally known as PETG (EastmanChemical and SKchemicals are the only two manufacturers).

Replacing terephthalic acid (right) with isophthalic acid (center) creates a kink in the PET chain, interfering with crystallization and lowering the polymer's melting point.

Another common modifier is isophthalic acid, replacing some of the 1,4- (*para-*) linked terephthalate units. The 1,2- (*ortho-*) or 1,3- (*meta-*) linkage produces an angle in the chain, which also disturbs crystallinity.

Such copolymers are advantageous for certain molding applications, such as thermoforming, which is used to make tray or blister packages from PET sheet (sometimes called APET, for "amorphous PET"). On the other hand, crystallization is important in other applications where mechanical and dimensional stability are important, such as seat belts. For PET bottles, the use of small amounts of CHDM or other comonomers can be useful: if only small amounts of comonomers are used, crystallization is slowed but not prevented entirely. As a result, bottles are obtainable via stretch blow molding ("SBM"), which are both clear and crystalline enough to be an adequate barrier to aromas and even gasses, such as the carbon dioxide in carbonated beverages.

Crystals

Crystallization occurs when polymer chains fold up on themselves in a repeating, symmetrical pattern. Long polymer chains tend to become entangled on themselves, which prevents full crystallization in all but the most carefully controlled circumstances. PET is no exception to this rule; 60% crystallization is the upper limit for commercial products, with the exception of polyester fibers.

PET in its natural state is a crystalline resin. Clear products can be produced by rapidly cooling molten polymer to form an amorphous solid. Like glass, amorphous PET forms when its molecules are not given enough time to arrange themselves in an orderly fashion as the melt is cooled. At room temperature the molecules are frozen in place, but if enough heat energy is put back into them, they begin to move again, allowing crystals to nucleate and grow. This procedure is known as solid-state crystallization.

Like most materials, PET tends to produce many small crystallites when crystallized from an amorphous solid, rather than forming one large single crystal. Light tends to scatter as it crosses the boundaries between crystallites and the amorphous regions between them. This scattering means that crystalline PET is opaque and white in most cases. Fiber drawing is among the few industrial processes that produces a nearly single-crystal product.

Degradation

PET is subject to various types of degradations during processing. The main degradations that can occur are hydrolytic, thermal and probably most important thermal oxidation. When PET degrades, several things happen: discoloration, chain scissions resulting in reduced molecular weight, formation of acetaldehyde and cross-links ("gel" or "fish-eye" formation). Discoloration is due to the formation of various cromophoric systems following prolonged thermal treatment at elevated temperatures. This becomes a problem when the optical requirements of the polymer are very high eg in packaging applications. Acetaldehyde is normally a colorless gas with a fruity smell. It forms naturally in fruit, but it can cause an off-taste in bottled water. Acetaldehyde forms in PET through the "abuse" of the material. High temperatures (PET decomposes above 300 °C or 572 °F), high pressures, extruder speeds (excessive shear flow raises temperature) and long barrel residence times all contribute to the production of acetaldehyde. When acetaldehyde is produced, some of it remains dissolved in the walls of a container and then diffuses into the product stored inside, altering the taste and aroma. This is not such a problem for non-consumables such as shampoo, for fruit juices, which already contain acetaldehyde or for strong-tasting drinks, such as soft drinks. For bottled water, low acetaldehyde content is quite important, because if nothing masks the aroma, even extremely low concentrations (10-20 ppb) of acetaldehyde can produce an off-taste. The thermal and thermooxidative degradation results in poor procesibility characteristics and performance of the material.

One way to alleviate this is to use a copolymer. Comonomers such as CHDM or isophthalic acid lower the melting temperature and reduces the degree of crystallinity of PET (especially important when the material is used for bottle manufacturing). Thus the resin can be plastically formed at lower temperatures and/or with lower force. This helps to prevent degradation, reducing the acetaldehyde content of the finished product to an acceptable (that is, unnoticeable) level. See copolymers, above. Other ways to improve the stability of the polymer is by using stabilizers, mainly antioxidants such as phosphites. Recently, molecular level stabilization of the material using nanostructured chemicals has also been considered.

Antimony

Antimony trioxide (Sb_2O_3) is a catalyst that is often used in the production of PET. It remains in the material and can thus in principle migrate out into food and drinks. Although antimony trioxide is of low toxicity, its presence is still of concern. The Swiss Federal Office of Public Health investigated the amount of antimony migration, comparing waters bottled in PET and glass: the antimony concentrations of the water in PET bottles was higher, but still well below the allowed maximal concentrations. (report available in German and French only) The Swiss Federal Office of Public Health concluded that the health risk of these low concentrations is negligible (1% of the "tolerable daily intake" determined by the WHO).

Re-crystallization

PET can be used to explore the crystallization of amorphous solids. The resin identification code can be used to verify the type of plastic is made of: many plastic beverage bottles have the letters PET or PETE and a code of 1 on the bottom, near the center. When a flame is held several inches below the bottle and slowly brought closer, part of the material will visibly change. This happens because high temperatures melt the PET. This releases the tension that was frozen in during the blow molding process and the polymer chains will shift to a more relaxed and disordered state, which results in shrinkage of the softened area. Because of the decreased order of the polymer chains, there are now fewer crystal nuclei. Consequently, when the crystallites re-form upon cooling they grow larger than the original crystallites in the bottle wall. Because the new crystallites are larger than the wave length of light, they will now cause light to scatter, giving the material an opaque white appearance.

Processing Equipment

There are two basic molding methods, one-step and two-step. In two-step molding, two separate machines are used. The first machine injection molds the preform. The preform looks like a test tube. The bottle-cap threads are already molded into place, and the body of the tube is significantly thicker, as it will be inflated into its final shape in the second step using stretch-blow molding. In the second process, the preforms are heated rapidly and then inflated against a two-part mold to form them into the final shape of the bottle. Preforms (uninflated bottles) are now also used as containers for candy.

In one-step machines, the entire process from raw material to finished container is conducted within one machine, making it especially suitable for molding non-standard shapes (custom molding), including jars, flat oval, flask shapes etc. Its greatest merit is the reduction in space, product handling and energy, and far higher visual quality than can be achieved by the two-step system.

Phenolic Acid

SALICYLIC ACID

Introduction

The name derives from the Latin word for the willow tree (*Salix*), from whose bark it can be obtained.

Salicylic acid is the chemical compound with the formula $C_6H_4(OH)CO_2H$, where the OH group is adjacent to the carboxyl group. This colourless crystalline organic acid is widely used in organic synthesis and functions as a plant hormone. It is probably best known as a compound that is chemically similar but not identical to the active component of aspirin.

Preparation

Methyl salicylate has a fragrant, minty smell that has made it a favourite flavouring in candies. It is also the major constituent of oil of wintergreen, making up over 90% of the essential oil from the wintergreen plant. However, most methyl salicylate used in foods is made synthetically, a cheaper process than its extraction from wintergreen leaves or sweet birch bark. Note that it is a different ester derivative of salicylic acid.

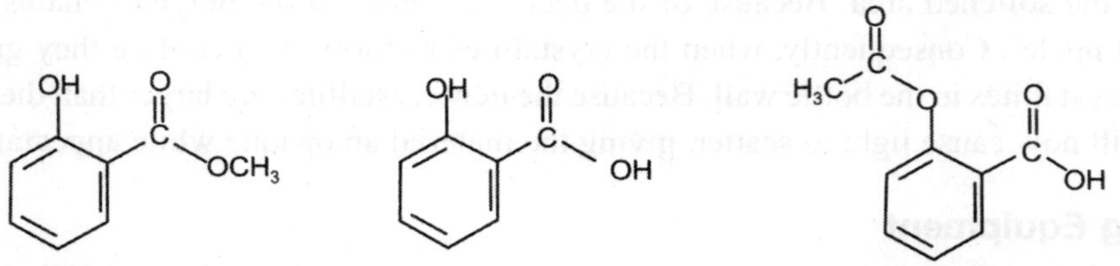

methyl salicylate salicylic acid acetyl salicylic acid

In any synthetic procedure, a chemist must use a series of techniques that can be classified as:

(1) carrying out the reaction,

(2) isolating the product (perhaps using extraction, crystallization, distillation, etc.),

(3) purifying the product (again, perhaps using crystallization, distillation, chromatography, etc.),

(4) and characterizing the product and estimating its purity (perhaps using melting point, spectroscopy, various types of chromatography, etc.).

Steps (2) and (3), the isolation and purification of the reaction product, are often called the "work-up" of the reaction mixture. The types of techniques that are used for each of these steps depend on the energy requirements of the reaction and on the physical and chemical properties of both the starting materials and the products.

For each reaction, the chemist also computes a theoretical and % yield. To do this, you must know the stoichiometry of the reaction, and thus the mechanistic details are often useful to know. With this information, you can tell for example if a substance added to the reaction mixture is a catalyst or if it is a reagent that gets used up in the process. Of the materials that are consumed, the one that is used up 1st is called the limiting reagent. Once the ID of the limiting reagent is known, that can be used to determine the theoretical amount of the product that you hope for. With this in hand, the percent yield can be determined. The overall reaction that you will carry out is shown here, in two steps. The process is an example of ester hydrolysis under strongly basic conditions. We will discuss the mechanistic detail in recitation.

Laboratory Preparation of Salicylic Acid

Procedure

Step 1 :

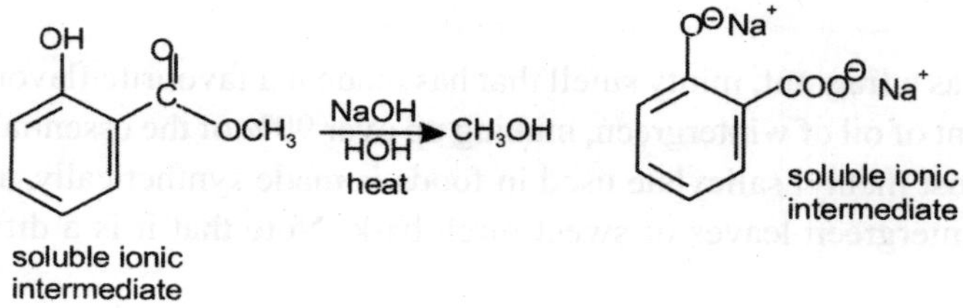

$\xrightarrow[\text{HOH, heat}]{\text{NaOH}}$ CH_3OH + $COO^{\ominus}Na^+$

soluble ionic intermediate

soluble ionic intermediate

Step 2:

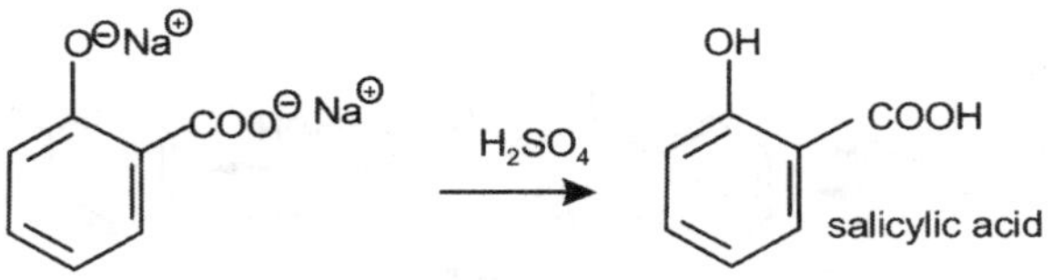

$O^{\ominus}Na^{\oplus}$ $COO^{\ominus}Na^{\oplus}$ $\xrightarrow{H_2SO_4}$ OH COOH

salicylic acid

To carry out the reaction: Clamp a 500-mL round-bottomed flask, to a ring stand or vertical support. Place 1.2 g of sodium hydroxide and 7 mL of water in the flask; swirl the mixture until the solid dissolves. Add 0.5 mL methyl salicylate. A white solid will quickly form. Add one or two boiling stones to the reaction mixture to prevent bumping of the solution when it is heated and place a heating mantle under the flask. Note that neither of our heating mantle sizes is exactly right for a 500 ml round bottom. Attach a water-cooled reflux condenser to the round-bottomed flask. Heat the reaction mixture to its boiling point, and allow it to reflux for 15 min. The solid that formed initially will dissolve as the mixture is warmed. After the reflux period, remove the heating mantle and let the mixture cool to room temperature. Placing a beaker of cold tap water under the flask speeds the cooling process. Carefully add 3M sulfuric acid solution in approximately 1-mL increments until a heavy white precipitate of salicylic acid forms and *remains* when the mixture is well stirred. You will need approximately 4 or 5 mL of the sulfuric acid solution. To isolate the product: After you have added just enough 3M sulfuric acid to give a heavy white precipitate, add 0.5 mL more acid to ensure complete precipitation of the salicylic acid. Cool the mixture in an ice-water bath to 5 °C. Collect the precipitated crude product by vacuum filtration, using a Buchner funnel. Rinse the solid product well with ice-cold water, in the Buchner funnel, with the suction flowing.

To purify the product: Recrystallize the product from water in a 25 or 50 mL Erlenmeyer flask. Follow the usual procedures for recrystallization

To characterize the product and estimate its purity: After the recrystallized product has dried completely, measure its mass and calculate its theoretical and % yield. Measure the melting point range of the product and compare it to the literature value.

Cleanup. Neutralize any excess acid in the filtrate of the synthesis using aqueous sodium carbonate and flush it down the drain with lots of water. The recrystallization filtrate can be flushed directly down the drain with lots of water.

Salicylic acid can also be obtained by the following reactions:

1. By the oxidation of salicyl alcohol

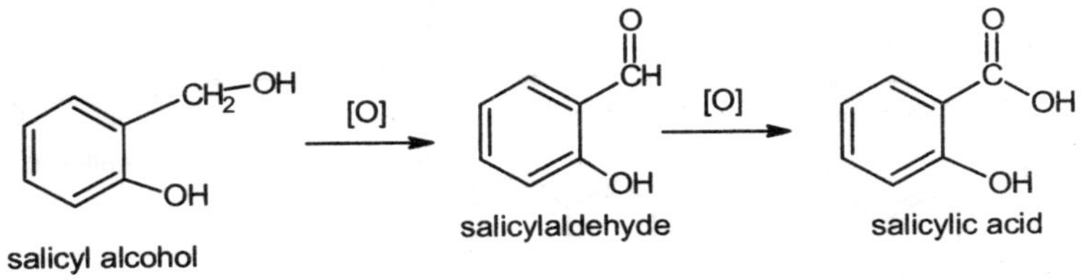

CH_2–OH, OH $\xrightarrow{[O]}$ CH=O, OH $\xrightarrow{[O]}$ C(=O)OH, OH

salicyl alcohol — salicylaldehyde — salicylic acid

2. By Fusing o – Sulphobenzoic acid with sodium hydroxide

2-sulfinobenzoic acid → (2NaOH) → disodium 2-sulfonatobenzoate → (NaOH (fused)) → sodium benzoate → (HCl) → salicylic acid

3. From Anthranilic acid through diazotization

2-aminobenzoic acid → (HCl, $NaNO_2$) → 2-(2-carboxyphenyl)-1-chloroniodiazenium → (H_2O, Δ) → salicylic acid

5. From Benzene **Benzene reacts with methyl chloride in presence of aluminium trichloride**

benzene → (CH_3Cl, ($AlCl_3$)) → toluene → ((H_2SO_4), (HNO_3)) → 1-methyl-2-nitrobenzene → (6H) → o-toluidine → ($NaNO_2$, HCl (cold)) → [(E)-(2-methylphenyl) diazenyl]chloronium → o-cresol → (oxidation) → salicylic acid

Manufacture

This Kolbe's reaction dry sodium phenate is heated with dry carbon dioxide in an auto clave at 140°C under pressure.

p-cresol → ($C_6H_5SO_2Cl$, NaOH aq) → 4-methylphenyl phenyl sulfite → ([O], $KMnO_4$) → 4-[(phenoxysulfinyl)oxy]benzoic acid → (H^+/H_2O) → salicylic acid

The reaction product is then activated with hydrochloric acid, and thus salicylic acid is produced. This acid is finally purified by recrystallization from hot water or alcohol.

7. p – cresol when reacts with benzene sulphonyl chloride producing salicylic acid.

$C_6H_5SO_2Cl$ / NaOH aq; [O] / $KMnO_4$; H^+/H_2O

4-methylphenyl phenyl sulfite

4-[(phenoxysulfinyl)oxy]benzoic acid

salicylic acid

Physical Properties

Molecular Formula	$= C_7H_6O_3$
Formula Weight	= 138.12074
Composition	= C(60.87%) H(4.38%) O(34.75%)
Melting Point	= 160 °C
Boiling Point	= 211 °C
Molar Refractivity	= 35.06 cm^3
Molar Volume	= 100.3 cm^3
Parachor	= 284.4 cm^3
Index of Refraction	= 1.615
Surface Tension	= 64.4 dyne/cm
Density	= 1.375 g/cm^3
Polarizability	= 13.90 cm^3
Monoisotopic Mass	= 138.031694 Da
Nominal Mass	= 138 Da
Average Mass	= 138.1207 Da
pKa	= 2.97
Log P	= 2.06

Chemical Properties

Salicylic acid reacts like an aromatic acid and also as a phenol.

1. With Sodium bi carbonate, it forms monosodium salicylate

salicylic acid + sodium hydrogen carbonate → sodium salicylate + CO_2 + H_2O

2. With Sodium Hydroxide it produces di sodium salicylate

salicylic acid + sodium hydroxide → disodium 2-oxidobenzoate (disodium salicylate) + $2H_2O$

3. With Ethyl alcohol it produces ethyl salicylate

salicylic acid + ethanol → ethyl salicylate

4. With Methyl alcohol it produces methyl salicylate, which smells as oil of wintergreen used in pain relieving ointments.

salicylic acid + methanol → methyl salicylate (oil of wintergreen)

With acetyl chloride it produces acetyl salicylic acid or aspirin with is also used as pain killer

For detail preparation see **A chemical analyser's guide**

OH + $H_3C-C(=O)-Cl$ ⟶ aspirin

salicylic acid — acetyl chloride — acetyl salicylic acid aspirin

6. With phosphorous pentachloride it produces o – Chlorobenzyl chloride

salicylic acid + pentachlorophosphorane ⟶ 2-chlorobenzoyl chloride + $2POCl_3$ + 2HCl

7. Action of Heat

(*a*) On heating strongly, it decomposes into phenol and carbon dioxide

salicylic acid —heat strongly⟶ phenol + CO_2

The above decomposition can be brought about more by heating with lime

8. With ferric chloride it produces a violet colour (Uffelman's reaction)

salicylic acid + $FeCl_3$ ⟶ voilet (coloured compound) + Cl_2

9. With Ammonia and Calcium chloride a white precipitate of calcium salicylate

salicylic acid + $CaCl_2$ + $2NH_3$ ⟶

10. It can be halogenated, nitrated and sulphonated in the conventional method

Uses

Medicinal and Cosmetic Uses

Also known as 2-hydroxybenzoic acid (one of several beta hydroxy acids (compare to AHA), salicylic acid is the key additive in many skin-care products for the treatment of acne, psoriasis, callouses, corns, keratosis pilaris and warts. It treats acne by causing skin cells to slough off more readily, preventing pores from clogging up. This effect on skin cells also makes salicylic acid an active ingredient in several shampoos meant to treat dandruff. Use of straight salicylic solution may cause hyperpigmentation on unpretreated skin for those with darker skin types (Fitzpatrick phototypes IV, V, VI), as well as with the lack of use of a broad spectrum sunblock.

Salicylic acid is also used as an active ingredient in gels which remove verrucas*.

Other Uses

- Salicylic acid is toxic if ingested in large quantities, but in small quantities is used as a food preservative and antiseptic in toothpaste. For some people with salicylate sensitivity even these small doses can be harmful.
- Sodium Salicylate is a useful phosphor in the vacuum ultraviolet with nearly flat quantum efficiency for wavelengths between 100 to 1000 Angstroms. It flouresces in the blue at 4200 Angstroms. It is easily prepared on a clean surface by creating a supersatured solution of the Salicylate in methanol, spraying the solution on the surface, and waiting for the methanol to evaporate.

Aspirin, or **acetylsalicylic acid** (acetosal) is a drug in the family of salicylates, often used as an analgesic (to relieve minor aches and pains), antipyretic (to reduce fever), and as an anti-inflammatory. It also has an antiplatelet ("blood-thinning") effect and is used in long-term, low doses to prevent heart attacks and thrombus formation in hypercoaguable states (e.g.in cancer).

Low-dose, long-term aspirin use irreversibly blocks the formation of thromboxane A_2 in platelets, producing an inhibitory effect on platelet aggregation. This anticoagulant property makes it useful for reducing the incidence of heart attacks. Aspirin produced for this purpose often comes in 75 or 81 mg, dispersible tablets and is sometimes called "junior aspirin" or "baby aspirin." High doses of aspirin are also given immediately after an acute heart attack. These doses may also inhibit the synthesis of prothrombin and may, therefore, produce a second and different anticoagulant effect, but this is not well understood.

Its primary, undesirable side-effects, especially in higher doses, are gastrointestinal distress (including ulcers and stomach bleeding) and tinnitus. Another side-effect, due to its anticoagulant properties, is increased bleeding in menstruating women. Because there appears to be a correlation between aspirin

---

*wrt, fungal infection.

and Reye's syndrome in children under the age of about 12, aspirin is no longer used to control flu-like symptoms or the symptoms of chickenpox in minors.

Aspirin was the first-discovered member of the class of drugs known as non-steroidal, anti-inflammatory drugs (NSAIDs), not all of which are salicylates, though they all have similar effects and a similar action mechanism.

A French chemist **Charles Frederic Gerhardt** was first to prepare acetylsalicylic acid (future aspirin) in 1853. In the course of his work on the synthesis and properties of various acid anhydrides, he mixed acetyl chloride with a sodium salt of salicylic acid (sodium salicylate). A vigorous reaction ensued, and the resulting melt soon solidified. Since no structural theory existed at that time Gerhardt called the compound he obtained "salicylic-acetic anhydride" (wasserfreie Salicylsäure-Essigsäure). When Gerhardt tried to dissolve the solid in a diluted solution of sodium carbonate it immediately decomposed to sodium salts of salicylic and acetic acids. This preparation of aspirin ("salicylic-acetic anhydride") was one of the many reactions Gerhardt conducted for his paper on anhydrides, and he did not pursue it further.

Six years later in 1859 von Gilm obtained analytically pure acetylsalicylic acid (he called it "acetylirte Salicylsäure") by a reaction of salicylic acid and acetyl chloride.[5] In 1869 Schröder, Prinzhorn and Kraut (Prinzhorn is credited in the paper with conducting the experiments) repeated both Gerhardt's (from sodium salicylate) and von Gilm's (from salicylic acid) syntheses and concluded that both reactions gave the same compound—acetylsalicylic acid. They were first to assign to it the correct structure with the acetyl group connected to the phenolic oxygen.

In 1897 Felix Hoffmann a chemist at Friedrich Bayer & Co. obtained acetylsalicylic acid by a reaction of salicylic acid and acetic anhydride,[8] that is essentially repeating the Gilm/Kraut procedure but substituting acetic anhydride for acetyl chloride. This synthesis served as the basis for Bayer claims to discovery of aspirin. According to a legend publicized by Bayer, Hoffmann made some of the formula and gave it to his father, who was suffering from the pain of arthritis and could not stand the side-effects of salicylic acid.[7] Much later in 1949 another Bayer researcher Arthur Eichengrün asserted his own priority as a discoverer of aspirin (see below for the details of Hoffmann-Eichengrün controversy). This argument did much to obscure the real history of aspirin, whose origin is not the pharmaceutical industry but earlier academic research. Despite the fact that pure aspirin was synthesized by von Gilm and by Kraut's group many years before Hoffmann (see above), Bayer continues to insist that "The active ingredient in Aspirin®, acetylsalicylic acid, was synthesized for the first time in a chemically pure and thus stable form in 1897 by a young chemist working for Bayer, Dr. Felix Hoffmann.

However, the German patent office refused to grant Bayer patent for acetylsalicylic acid on the grounds that neither the product nor the process of preparation were novel.[7] In 1905 Bayer sued Chemische Fabrik von Heyden in Britain for infringing the British patent on aspirin granted to it in 1900. Hayden claimed that existing prior art, in particular Kraut's work, invalidated Bayer's patent. Bayer advanced the argument that they were first to prepare a pure form of aspirin. The judge agreed with Hayden and invalidated the Bayer patent. A similar struggle took place in the U.S. Circuit court in Chicago; however, in the US Bayer's patent was upheld in 1909. This created a situation where aspirin in the US was three times more expensive than in Canada and ten times more expensive than in Europe, resulting in rampant smuggling of aspirin, which Bayer unsuccessfully tried to contain. Bayer began marketing aspirin in July 1899. It was marketed alongside another of Hoffmann's products, an acetylated derivative of morphine called 'Heroin' that he invented 11 days after aspirin Heroin was initially the more successful of the two painkillers and it was common belief that it was healthier than aspirin. But, as heroin's shortcoming of addictiveness became more obvious, aspirin stepped to the forefront. Aspirin was originally sold as a powder (still the preferred form in many European countries) and was an instant success; in 1914, Bayer introduced aspirin tablets.

Several claims to invention of acetylsalicylic acid have arisen. Acetylsalicylic acid was already being manufactured by the Chemische Fabrik von Heyden Company in 1897, although without a brand name[citation needed]. A new pretender to the title of aspirin discoverer appeared in 1949. Arthur Eichengrün who was 81 "claimed that he had instructed Hoffmann to synthesise acetylsalicylic acid and that the latter had done so without knowing the purpose of the work".[10] In 2000, Walter Sneader of University of Strathclyde in Glasgow reexamined the case and concluded that "Arthur Eichengrün was telling the truth when he wrote that acetylsalicylic acid was synthesised under his direction and that the drug would not have been introduced in 1899 without his intervention".[10] Subsequently, in his polemic with Sneader, Axel Helmstaedter, General Secretary of the International Society for the History of Pharmacy, noted that Sneader did not credit several earlier publications which discussed the Hoffmann-Eichengrün controversy in detail, with most historians doubting Eichengrün's account. Bayer countered to Sneader in a press-release that according to the records, Hoffmann and Eichengrün held equal positions, and Eichengrün was not Hoffmann's supervisor. Hoffmann was named on the US Patent as the inventor, which Sneader did not mention. Eichengrün who left Bayer in 1908 had multiple opportunities to claim the priority and had never before 1949 done it; he neither claimed nor received any percentage of the profit from aspirin sales.[12]

It was not until the 1970s that the mechanism of action of aspirin and similar drugs called NSAIDs was elucidated (see below).

Synthesis of Aspirin

The synthesis of aspirin is classified as an esterification reaction, where the alcohol group from the salicylic acid reacts with an acid (acetyl anhydride) to form an ester. Aspirin is commercially synthesized using a two-step process. First, phenol (generally extracted from coal tar) is treated with a sodium base which generates sodium phenoxide, which is then reacted with carbon dioxide under high temperature and pressure to yield salicylate, which is acidified, yielding salicylic acid. This process is known as the Kolbe-Schmitt reaction.

OH → (CO_2, NaOH) → O, ONa, ONa → (H_2SO_4) → O, OH, OH

Salicylic acid is then acetylated using acetic anhydride, yielding aspirin and acetic acid as a byproduct. It is a common experiment performed in undergraduate organic chemistry practicals, and generally tends to produce low yields due to the relative difficulty of its extraction from an aqueous state. The trick to getting the reaction to work is to acidify with phosphoric acid and heat the reagents under reflux with a boiling water bath for between 40 to 60 minutes.

The original synthesis of aspirin from salicylic acid involved acetylation with acetyl chloride. Unfortunately, the byproduct from this is hydrochloric acid, which is corrosive and environmentally hazardous. As described above, it was then later found that acetic anhydride was a better acylating agent, with the byproduct acetic acid formed, which does not have the unwanted properties of hydrochloric acid and can also be recycled. The salicylic acid/acetic anhydride method is commonly employed in undergraduate teaching labs.

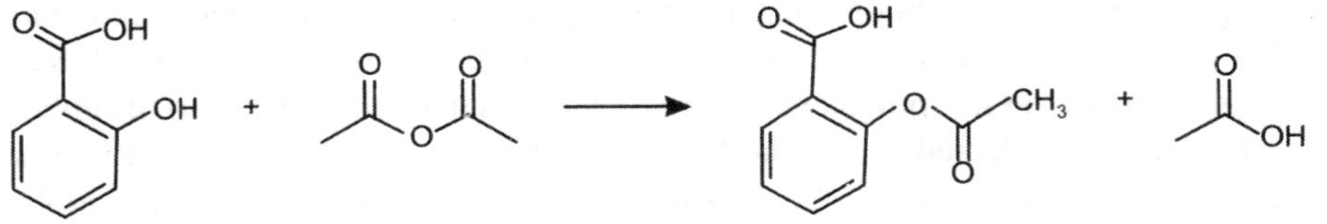

Formulations containing high concentrations of aspirin often smell like vinegar. This is because aspirin can undergo autocatalytic degradation to salicylic acid in moist conditions, yielding salicylic acid and acetic acid.

The acid dissociation constant (pK_a) for Acetylsalicylic acid is 3.5 at 25 °C. ASA, being a weak acid, dissociates as shown by the following reaction equation:

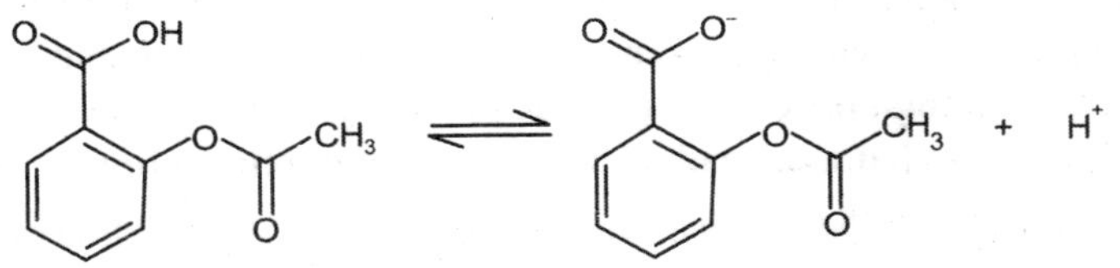

Mechanism of action

In 1971, the British pharmacologist, John Robert Vane, who was then employed by the Royal College of Surgeons in London, showed that aspirin had suppressed the production of prostaglandins and thromboxanes.[14] For this piece of research he was awarded both a Nobel Prize in Physiology or Medicine in 1982 and a knighthood.

Aspirin's ability to suppress the production of prostaglandins and thromboxanes is due to its non-competitive and irreversible inhibition of the cyclooxygenase (COX) enzyme. Cyclooxygenase is required for prostaglandin and thromboxane synthesis. Aspirin acts as an acetylating agent where an acetyl group is covalently attached to a serine residue in the active site of the COX enzyme. This makes aspirin different from other NSAIDs (such as diclofenac and ibuprofen), which are reversible inhibitors.

Prostaglandins are local hormones (paracrine) produced in the body and have diverse effects in the body, including but not limited to transmission of pain information to the brain, modulation of the hypothalamic thermostat, and inflammation. Thromboxanes are responsible for the aggregation of platelets that form blood clots. Heart attacks are primarily caused by blood clots, and their reduction with the introduction of small amounts of aspirin has been seen to be an effective medical intervention. The side-effect of this is that the ability of the blood in general to clot is reduced, and excessive bleeding may result from the use of aspirin.

There are at least two different types of cyclooxygenase: COX-1 and COX-2. Aspirin irreversibly inhibits COX-1 and modifies the enzymatic activity of COX-2. Normally COX-2 produces prostanoids, most of which are pro-inflammatory. Aspirin-modified COX-2 produces lipoxins, most of which are anti-inflammatory. Newer NSAID drugs called COX-2 selective inhibitors have been developed that inhibit only COX-2, with the hope for reduction of gastrointestinal side-effects.

However, several of the new COX-2 selective inhibitors have been recently withdrawn, after evidence emerged that COX-2 inhibitors increase the risk of heart attack. It is proposed that endothelial cells lining the microvasculature in the body express COX-2, and, by selectively inhibiting COX-2, prostaglandins (specifically PGI2; prostacyclin) are downregulated with respect to thromboxane levels, as COX-1 in platelets is unaffected. Thus, the protective anti-coagulative effect of PGI2 is decreased, increasing the risk of thrombus and associated heart attacks and other circulatory problems. Since platelets have no DNA, they are unable to synthesize new COX once aspirin has irreversibly inhibited the enzyme, an important difference with reversible inhibitors.

Furthermore, aspirin has 2 additional modes of actions, contributing to its strong analgesic, antipyretic and anti-inflammatory properties:

- It uncouples oxidative phosphorylation in cartilaginous (and hepatic) mitochondria, by diffusing from the inner membrane space as a proton carrier back into the mitochondrial matrix, where it ionizes once again to release protons. In short, aspirin buffers and transports the protons. (Note: This effect in high doses of aspirin actually causes fever due to the heat released from the electron transport chain, instead of its normal antipyretic action.)
- It induces the formation of NO-radicals in the body that enable the white blood cells (leukocytes) to fight infections more effectively. This has been found recently by Dr. Derek W. Gilroy, winning Bayer's International Aspirin Award 2005.

More recent data suggest that salicylic acid and its derivatives will modulate NFêB signaling. NFêB is a transcription factor complex that plays a central role in many biological processes, including inflammation.

Indications

Aspirin, as with many older drugs, has proven to be useful in many conditions. Despite its well-known toxicity it is widely used, since physicians are familiar with its properties. Indications for its use include:

- Fever
- Pain (especially useful for some forms of arthritis, osteoid osteoma, and chronic pain)
- Migraine
- Rheumatic fever (drug of choice)

- Kawasaki disease (along with IVIG)
- Pericarditis
- Coronary artery disease
- Acute myocardial infarction[16]

In addition, aspirin is recommended (low dose, 75-81 mg daily) for the prevention of:

- Myocardial infarction — in patients with either documented coronary artery disease or at elevated risk of cardiovascular disease. Women may possibly benefit less from aspirin than men.
- Stroke — as secondary prevention (i.e. to prevent recurrence)

Dosage

For adults doses of 300 to 1000 mg are generally taken four times a day for fever or arthritis, with a maximum dose of 8000 mg a day.[19] The correct dose of aspirin depends on the disease process that is being treated. For instance, for the treatment of rheumatic fever, doses near the maximal daily dose have been used historically. For the prevention of myocardial infarction in someone with documented or suspected coronary artery disease, doses as low as 75 mg daily (or possibly even lower) are sufficient.

For those under 12 years of age, the dose previously varied with the age, but aspirin is no longer routinely used in children due to the association with Reye's syndrome; paracetamol or other NSAIDs, such as Ibuprofen, are now being used instead. Kawasaki disease remains one of the few indications for aspirin use in children with aspirin initially started at 7.5-12.5 mg/kg body weight taken four times a day for up to two weeks and then continued at 5 mg/kg once daily for a further six to eight weeks.

Contraindications and Warnings

- Aspirin should be avoided by those known to be allergic to ibuprofen or naproxen.
- Caution should be exercised in those with asthma or NSAID-precipitated bronchospasm.
- It is generally recommended that one seek medical help if symptoms do not improve after a few days of therapy.
- Caution should be taken in patients with kidney disease, peptic ulcers, mild diabetes, gout or gastritis; manufacturers recommend talking to one's doctor before using this medicine.
- Taking aspirin with alcohol increases the chance of gastrointestinal hemorrhage (stomach bleeding).
- Children, including teenagers, are discouraged from using aspirin in cold or flu symptoms as this has been linked with Reye's syndrome.
- Patients with hemophilia or other bleeding tendencies should not take salicylates.
- Some sources recommend that patients with hyperthyroidism avoid aspirin because it elevates T4 levels.

Common Side-effects

- Gastrointestinal complaints (stomach upset, dyspepsia, heartburn, small blood loss). To help avoid these problems, it is recommended that aspirin be taken at or after meals. Undetected blood loss may lead to hypochromic anemia. The Gastrointestinal complications that may result from a dose of Aspirin are due to Aspirin inhibitory action on Cyclo-oxygenase 1 and 2.
- Severe gastrointestinal complaints (gross bleeding and/or ulceration), requiring discontinuation and immediate treatment. Patients receiving high doses and/or long-term treatment should receive gastric protection with high-dosed antacids, ranitidine or omeprazole.
- Frequently, central effects (dizziness, tinnitus, hearing loss, vertigo, centrally mediated vision disturbances, and headaches). The higher the daily dose is, the more likely it is that central nervous system side-effects will occur.
- Sweating, seen with high doses, independent from antipyretic action
- Long-term treatment with high doses (arthritis and rheumatic fever): often increased liver enzymes without symptoms, rarely reversible liver damage. The potentially fatal Reye's syndrome may occur, if given to pediatric patients with fever and other signs of infections. The syndrome is due to fatty degeneration of liver cells. Up to 30 percent of those afflicted will eventually die. Prompt hospital treatment may be life-saving.
- Chronic nephritis with long-term use, usually if used in combination with certain other painkillers. This condition may lead to chronic renal failure.
- Prolonged and more severe bleeding after operations and post-traumatic for up to 10 days after the last aspirin dose. If one wishes to counteract the bleeding tendency, fresh thrombocyte concentrate will usually work.
- Skin reactions, angioedema, and bronchospasm have all been seen infrequently.
- Patients that have the genetic disease known as Glucose-6-Phosphate Dehydrogenase deficiency (G6PD) should avoid this drug as it is known to cause hemolytic anemia in large doses depending on the severity of the disease.

Interactions

Aspirin can interact with some other drugs. Some drugs, for example those used for diabetes, are bound by plasma protein in the blood supply and are therefore not free to react with their targets. However, aspirin may displace them from plasma protein, leading to a drug overdose. Aspirin has this same effect on anticoagulants.

Overdose

Aspirin overdose can be acute or chronic. In acute poisoning, a single large dose is taken; in chronic poisoning, supratherapeutic doses are taken over a period of time. Acute overdose has a mortality rate of 2%. Chronic overdose is more commonly lethal with a mortality rate of 25%; chronic overdose may be especially severe in children.

Symptoms

Aspirin overdose has potentially serious consequences, sometimes leading to significant morbidity and mortality. Patients with mild intoxication frequently have nausea and vomiting, abdominal pain, lethargy, tinnitus, and dizziness. More significant symptoms occur in more severe poisonings and include hyperthermia, tachypnea, respiratory alkalosis, metabolic acidosis, hyperkalemia, hypoglycemia, hallucinations, confusion, seizure, cerebral edema, and coma. The most common cause of death following an aspirin overdose is cardiopulmonary arrest usually due to pulmonary edema.

Toxicity

The toxic dose of aspirin is generally considered greater than 150 mg per kg of body mass. Moderate toxicity occurs at doses up to 300 mg/kg, severe toxicity occurs between 300 to 500 mg/kg, and a potentially lethal dose is greater than 500 mg/kg. This is the equivalent of many dozens of the common 325 mg tablets, depending on body weight. Note that children cannot tolerate as much aspirin per unit body weight as adults can, even when aspirin is indicated. Label-directions should be followed carefully.

Treatment

All overdose patients must be taken to a hospital immediately. Initial treatment of an acute overdose includes gastric decontamination of the patient. This is achieved by administering activated charcoal which adsorbs the aspirin in the gastrointestinal tract. Stomach pumps are no longer routinely used in the treatment of poisonings but are sometimes considered if the patient has ingested a potentially lethal amount up to 1 hour previously. Repeated doses of charcoal have been proposed to be beneficial in aspirin overdose. A study performed found that repeat dose charcoal might not be of significant value. However, most toxicologists will administer additional charcoal if serum salicylate levels are increasing.Patients are monitored until their peak salicylate blood level has been determined. Blood levels are usually performed 4 hours after ingestion and then every 2 hours after that to determine the maximum level. Maximum levels can be used as a guide to toxic effects expected. There is no antidote to salicylate poisoning. Frequent blood work is performed to check metabolic, salicylate, and blood sugar levels; arterial blood gas assessments are performed to test for respiratory alkalosis and metabolic acidosis. Patients are monitored and often treated according to their individual symptoms, patients may be given intravenous potassium chloride to counteract hypokalemia, glucose to restore blood sugar levels, benzodiazepines for any seizure activity, fluids for dehydration, and importantly sodium bicarbonate to restore the blood's sensitive pH balance. Sodium bicarbonate also has the effect of increasing the pH of urine, which in turn increases the elimination of salicylate. Additionally, hemodialysis can be implemented to enhance the removal of salicylate from the blood. Hemodialysis is usually used in severely poisoned patients; for example, patients with significantly high salicylate blood levels, significant neurotoxicity (agitation, coma, convulsions), renal failure, pulmonary edema, or cardiovascular instability are hemodialyzed. Hemodialysis also has the advantage of restoring electrolyte and acid-base abnormalities; hemodialysis is often life-saving in severely ill patients.

If the overdose was intentional, the patient should undergo psychiatric evaluation, as with any suicide attempt.

Epidemiology

In the later part of the 20th century the number of salicylate poisonings has declined mainly due to the popularity of other over-the-counter analgesics such as paracetamol. Fifty-two deaths involving single-ingredient aspirin were reported in the United States in the year 2000. However, in all but three cases, the reason for the ingestion of lethal doses was intentional, predominantly suicides.

Research into Cancer Prevention

The role that aspirin might have in reducing the rates of certain cancers has been the subject of many studies. However, given the side-effects of NSAIDs on increased gastrointestinal bleeding and rates of heart disease, there is no current medical recommendation to use these drugs for cancer reduction.

- **Prostate :** While there is some epidemiological information suggesting a correlation between the reduction of prostate cancer and aspirin use, two recent studies (a multicentric case-control study and epidemiological study) were inconclusive.
- **Colon :** A 2005 JAMA article states that aspirin may prevent carcinoma of the colon, when taken in at least twice-daily adult (325 mg) doses. It follows an earlier study, also using the Nurses' Health Study (NHS) women, which had similar conclusions but with a smaller testing population. The conclusion of the JAMA study was that both aspirin and other non-aspirin NSAIDs used over a long time (best results seen after at least a decade of regular use) reduce risk of colorectal cancer. The risk was reduced the most at doses higher than 14 tablets per week (full-strength aspirin twice a day). However, these doses can have a higher incidence of serious gastrointestinal bleeding, and reduced doses (even the low doses recommended for cardiovascular health (81 mg/day)) also seem to have an anti-cancer effect, albeit not as much as the higher doses. The results of these studies have been correlated with experimental studies in human volunteers and in rodents. At least four clinical trials show that small doses of aspirin reduce, modestly, intestinal polyps recurrence in patients with some history of adenomas. [35] Some twenty carcinogenesis studies in rats or mice also support that aspirin can prevent cancer, also reported in the Chemoprevention of Colorectal Cancer Database.
- **Gall bladder :** There is some evidence that aspirin may increase gall bladder motility and thus be effective in treating gall bladder disease.
- **Pancreatic :** A study in 2004 showed that increases in dose and duration of aspirin may *significantly increase* the risk of pancreatic cancer in women (only women were in the study); however this did not seem to affect current, regular users.
- **Upper GI tract :** A study published in 2003 reported that people who had been taking aspirin regularly for 5 or more years seemed to have a two-thirds less risk of mouth, throat and esophageal cancer than non-users. This study was done on both cancer patients and other people in hospital for various reasons, and was done by comprehensive epidemiological questionnaires of life habits rather than empirical testing.
- **Lung :** A 20-year study published in 2002 showed that in a group of 14,000 women in New York City, regular aspirin use (defined as at least once a week, various doses) reduced their

risk of lung cancer, especially small-cell carcinoma. This may be because many lung tumors have high amounts of COX-2 enzymes expressed in them, especially in adenocarcinomas and tumors caused by asbestosis.[39] Aspirin is a known blocker of both COX-1 and COX-2 enzymes, although this study suggests that if the COX-2 link is direct, other COX-2 inhibitors may also play a similar role.

Another study in 2002 of both men and women found that risk reduction was more significant in males than females; overall, the effects of smoking were far more influential than aspirin use in determining cancer risks. Of those who smoked, those who smoked the least got the most benefit from aspirin use. Some of the heaviest smokers saw no benefit from aspirin at all.

Magnesium salicylate is a common analgesic and non-steroidal anti-inflammatory drug (NSAID) used to treat mild to moderate muscular pain. It is also used to treat headaches, general back pain, and certain joint pains like arthritis.

It is found in a variety of over-the-counter (OTC) medications as an anti-inflammatory, primarily for back-pain relief. Magnesium Salicylate can be an effective OTC alternative to prescription NSAIDs, with both anti-inflamatory and pain-relieving effects.

Though the recommended doseage is 1160 mg every six hours, per package directions of the Doan's OTC brand (580 mg magnesium salicylate tetrahydrate, equivalent to 934.4 mg anhydrous magnesium salicylate), effective pain relief is often found with a half dosage, with reduced anti-inflammatory results.

Brand Names

- Doan's®, Original (OTC)
- Extra Strength Doan's® (OTC)
- Momentum® (OTC)
- Keygesic-10® (Rx)
- Mobidin® (Rx)
- Novasal® (Rx)

Bismuth subsalicylate, with a chemical formula $C_7H_5BiO_4$, is the active ingredient in the popular medication Pepto-Bismol that is used to treat nausea, heartburn, indigestion, upset stomach, diarrhea, and other temporary discomforts of the stomach and gastrointestinal tract. It is also the main ingredient of Kaopectate (since 2003, replacing attapulgite).It displays anti-inflammatory action (due to salicylic acid) and also acts as an antacid and mild antibiotic.

It can also cause a black tongue and black stools in some users of the drug, when it combines with trace amounts of sulfur in their saliva and gastrointestinal tract. This discoloration is temporary and harmless.Children should not take medication with Bismuth subsalicylate while recovering from the flu or chicken pox as some studies have found salicylate use during these illnesses can cause Reye's syndrome.

ANISIC ACID

Introduction

It is generally obtained by the oxidation of anethole, it is isomeric with methyl salicylate.

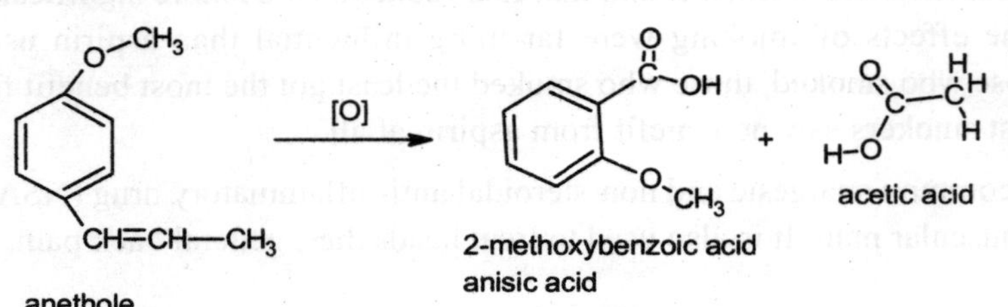

It can be prepared by the methylation of salicylic acid with methyl iodide and sodium hydroxide

CH_3I / NaOH; + NaI

salicylic acid → anisic acid

When Grignard's reagent reacts with p – bromoanisole, using magnesium and carbon dioxide anisic acid is produced

Mg / CO_2; + MgBr

1-bromo-4-methoxybenzene → anisic acid

Oxidation of p- methylanesole produces anisic acid with potassium permanganate.

$KMnO_4$ / heat

p - methyl anisole → anisic acid

Physical Properties

Molecular Formula	$= C_8H_8O_3$
Formula Weight	= 152.14732
Melting Point	= 181-186 °C
Boiling Point	= 175°C
Composition	= C(63.15%) H(5.30%) O(31.55%)
Molar Refractivity	= 39.86 cm^3
Molar Volume	= 125.9 cm^3
Parachor	= 326.0 cm^3
Index of Refraction	= 1.545
Surface Tension	= 44.9 dyne/cm
Density	= 1.207 g/cm^3
Polarizability	= 15.80 $10^{-24}cm^3$
Monoisotopic Mass	= 152.047344 Da
Nominal Mass	= 152 Da
Average Mass	= 152.1473 Da
Log P	= 1.49

Uses

4-Methoxybenzoic acid is known to have antiseptic properties. It is also used as an intermediate in the preparation of more complex organic compounds.

TYROSINE

Introduction

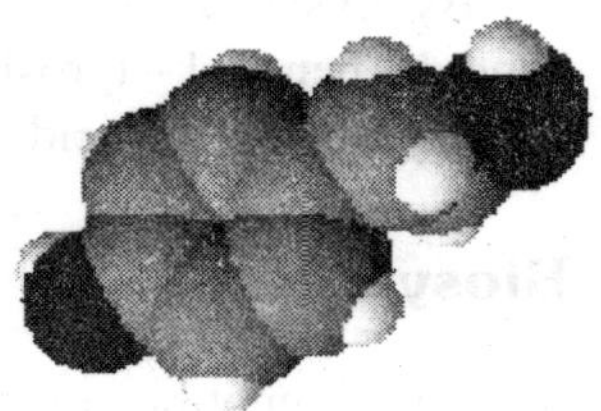

Tyrosine (from the Greek *tyros*, meaning *cheese*, as it was first discovered in 1846 by German chemist Justus von Liebig in cheese), **4-hydroxyphenylalanine**, or **2-amino-3(4-hydroxyphenyl)-propanoic acid**, is one of the 20 amino acids that are used by cells to synthesize proteins. It has a phenol side chain with a hydroxyl group. Upon the location of the hydroxyl group, there are three structural isomers of Tyr, namely para-Tyr (p-Tyr), meta-Tyr (m-Tyr) and ortho-Tyr (o-Tyr).

Enzymatically, only the first isomer (p-Tyr) is produced from L-Phe by the Phe-hydroxylase enzyme. The other two isoforms, m-Tyr and o-Tyr can be produced as a consequence of free radical attack on Phe in states with increased oxidative stress.

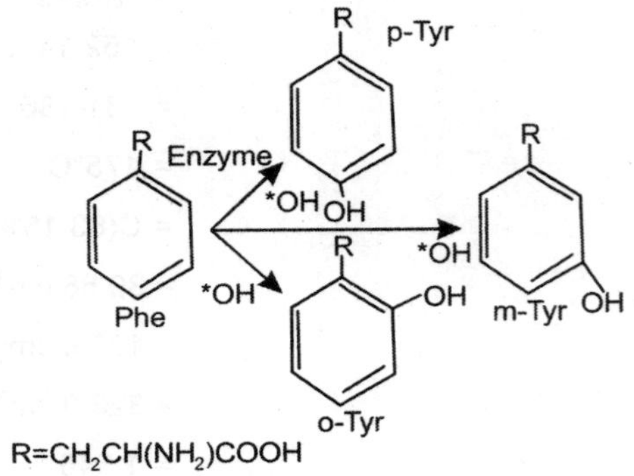

R p-Tyr
R Enzyme
*OH OH
R *OH
Phe *OH OH m-Tyr OH
o-Tyr
R=$CH_2CH(NH_2)COOH$

Preparation

It is also produced from many proteins by hydrolysis, with dilute acids, by pancreatic digestion or by purification of r – benzoyl – tyrosine can be conveniently obtained by the method Erlenmyer jun. p – Hydroxy benzaldehyde, is considered with Hippuric acid, to give p – hydroxy benzoylaminocinnamic acid, which on reduction with sodium amalgam yields benzoyl tyrosine

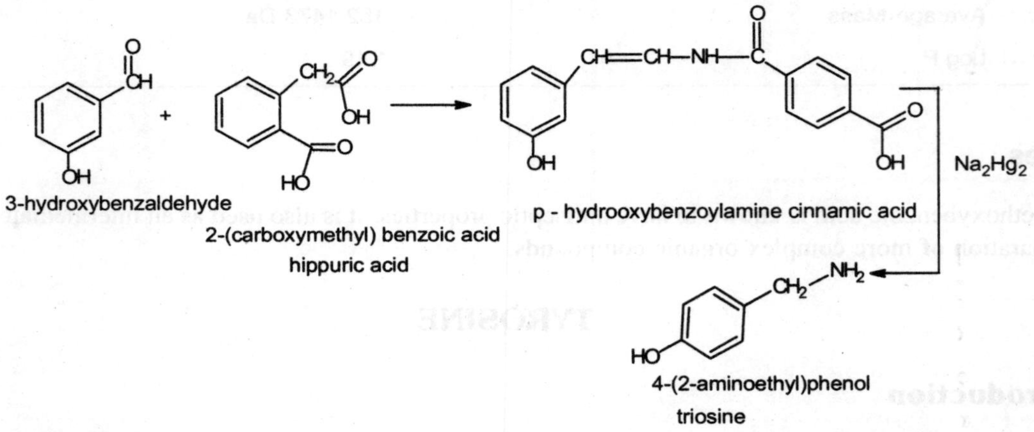

3-hydroxybenzaldehyde

2-(carboxymethyl) benzoic acid

hippuric acid

p - hydrooxybenzoylamine cinnamic acid

Na_2Hg_2

4-(2-aminoethyl)phenol

triosine

By crystallization of the brucine and cinchonine salts, Fischer succeeded in resolving this acid into l and d – benzoyl – tyrosine from the benzoyl derivatives, d and r tryosines can be prepared by heating with hydrochloride acid.

Biosynthesis

Tyrosine cannot be completely synthesized by animals, although it can be made by hydroxylation of phenylalanine if the latter is in abundant supply. It is produced by plants and most microorganisms from prephenate, an intermediate on the shikimate pathway. Prephenate is oxidatively decarboxylated with retention of the hydroxyl group to give *p*-hydroxyphenylpyruvate. This is transaminated using glutamate as the nitrogen source to give tyrosine and α-ketoglutarate.

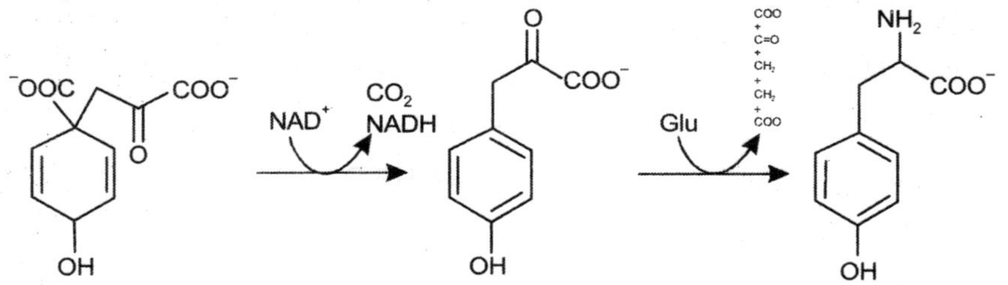

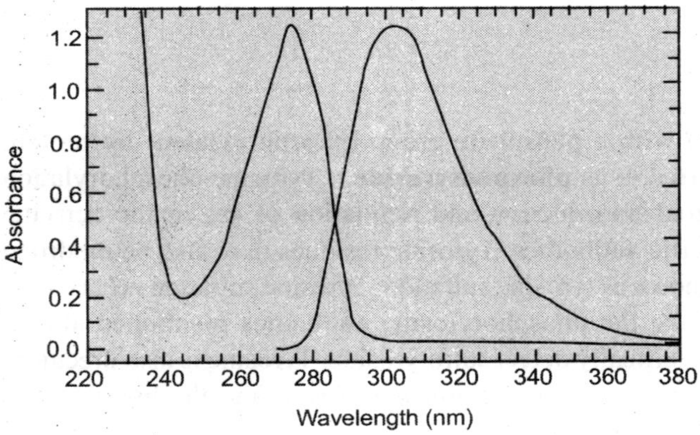

Absorbance and fluorescence of tyrosine in water/buffer

Physcial Properties

Property	Value
Molecular Formula	= $C_8H_{11}NO$
Formula Weight	= 137.17904
Composition	= C(70.04%) H(8.08%) N(10.21%) O(11.66%)
Melting Point	=343°C
Isoelectric point	= 5.66
Molar Refractivity	= 41.21 cm^3
Molar Volume	= 124.3 cm^3
Parachor	= 327.9 cm^3

Index of Refraction	= 1.577
Surface Tension	= 48.3 dyne/cm
Density	= 1.103 g/cm^3
Polarizability	= 16.34 cm^3
Monoisotopic Mass	= 137.084064 Da
Nominal Mass	= 137 Da
Average Mass	= 137.179 Da
Log P	= 0.72
pK$_a$	= 2.24

Uses

Some of the tyrosine residues can be *tagged* with a phosphate group (phosphorylated) by protein kinases. (In its phosphorylated state, it is referred to as **phosphotyrosine**.). Tyrosine phosphorylation is considered as one of the key steps in signal transduction and regulation of enzymatic activity. Phosphotyrosine can be detected through specific antibodies. Tyrosine residues may also be modified by the addition of a sulphate group, a process known as tyrosine sulfation. Tyrosine sulfation is catalyzed by tyrosylprotein sulfotransferase (TPST). Like the phosphotyrosine antibodies mentioned above, antibodies have recently been described that specifically detect sulfotyrosine. Tyrosine is also precursor to the thyroid hormones *thyroxine* and *triiodothyronine*, the pigment *melanin*, and the biologically-active catecholamines *dopamine*, *norepinephrine* and *epinephrine*.

In Papaver somniferum, the opium poppy, it is used to produce morphine.

TYROSOL

Introduction

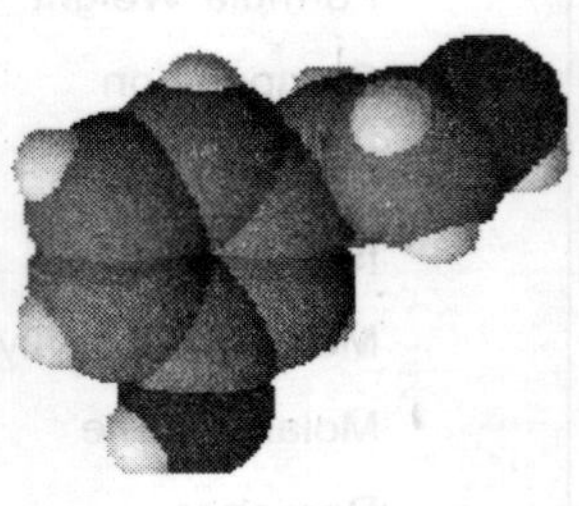

Tyrosol is a phenolic antioxidant present in a variety of natural sources. The principal source in the human diet is olive oil. Tyrosol is a derivative of phenethyl alcohol

As an antioxidant, tyrosol can protect cells against injury due to oxidation. Although it is not as potent as other antioxidants present in olive oil, its higher concentration and good bioavailability indicate that it may have an important overall effect. This effect may contribute significantly to the health benefits of olive oil and, more generally, the Mediterranean diet.

Preparation

This compound is prepared by the fermentation of tyrosine in presence of compressed yeast and sugar.

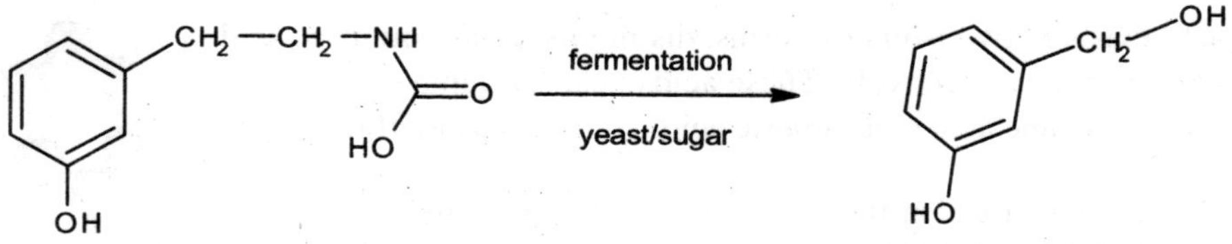

[2-(3-hydroxyphenyl) ethyl]carbamic acid

triosine

3-(2-hydroxyethyl) phenol

tyrosol

Physical Properties

Molecular Formula	**= $C_8H_{10}O_2$**
Formula Weight	= 138.1638
Composition	= C(69.54%) H(7.30%) O(23.16%)
Melting Point	= 91-92 °C
Boiling Point	= 158 °C
Molar Refractivity	= 39.21 cm^3
Molar Volume	= 118.1 cm^3
Parachor	= 315.6 cm^3
Index of Refraction	= 1.577
Surface Tension	= 50.8 dyne/cm
Density	= 1.168 g/cm^3
Polarizability	= 15.54 $10^{-24}cm^3$
Monoisotopic Mass	= 138.06808 Da
Nominal Mass	= 138 Da
Average Mass	= 138.1638 Da
Log P	= 0.62

Chemical Properties and Uses

It is the normal product of protein metabolism of yeast and hence is a by product of all kinds of yeast fermentation, being found in the majority of fermented beverages, particularly in beer and wine.

O – HYDROXYCINNAMIC ACID

Introduction

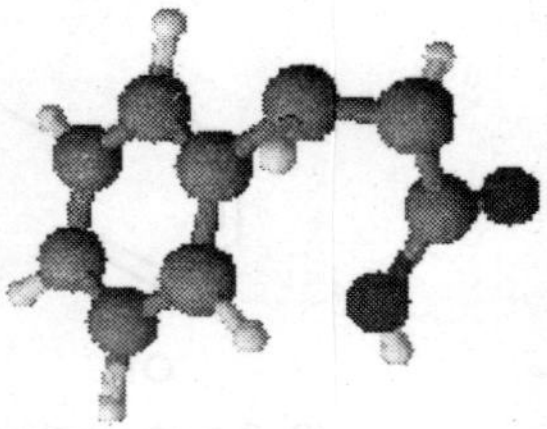

This acid exists in two isomeric forms, distinguished as coumaric acid and o – coumaric acid respectively. These acids bear the same relationship to one another as maleic acid and fumeric acids. In coumarinic II the group (cyclohexyl)–OH and COOH lie on the same side of the molecule (*cis* – form) and in the coumaric acid (i) they are on the opposite side (*trans* – form)

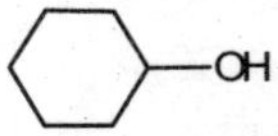

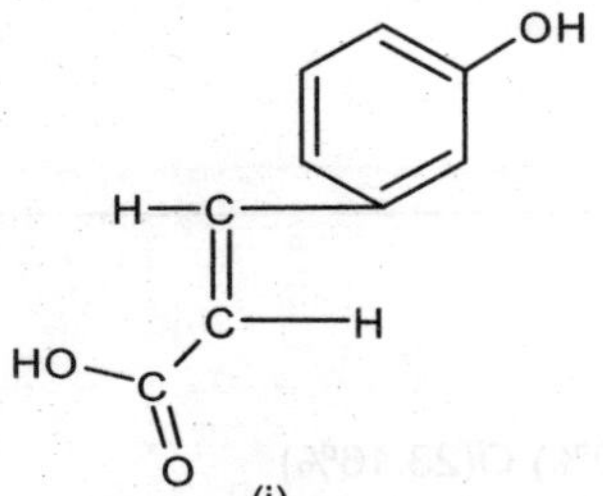

(i) 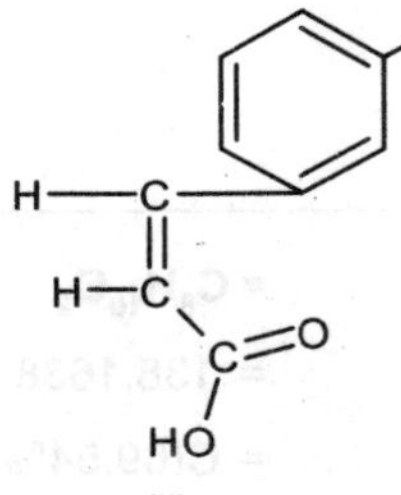 (ii) 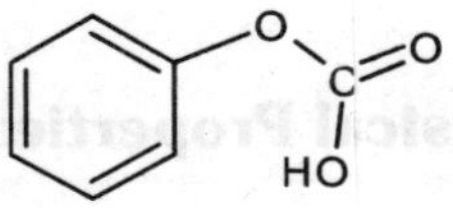 (iii)

The chief difference between these compounds is that coumarinic acid in aqueous of alcoholic solution is only stable in the form of its salts, and when liberated in the free state passes at once into the anhydride, coumerine (iii). On the otherhand o – coumaric acid is readily obtained in the free state.

O – coumaric acid occurs in melilotus, and may be prepared from o – amino cinnamic acid by way of the diazo compounds, or form coumarin by boiling with sodium ethoxide.

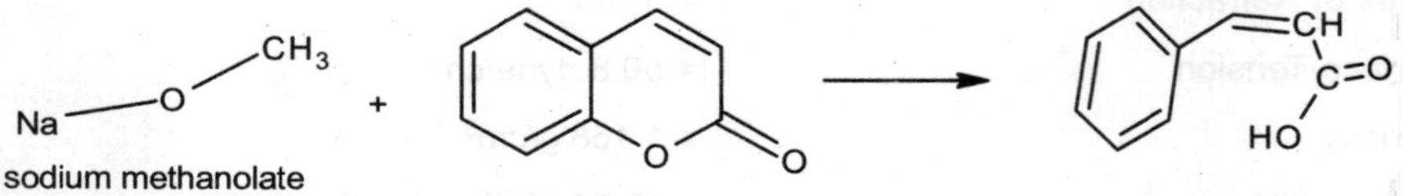

sodium methanolate
sodium ethoxide
+
2*H*-chromen-2-one
coumarin
→
(2*Z*)-3-phenylacrylic acid
p - hydroxy cinnamic acid

Physical Properties

Molecular Formula	$= C_9H_8O_2$
Formula Weight	= 148.15862
Composition	= C(72.96%) H(5.44%) O(21.60%)

Melting Point	= 208°C
Molar Refractivity	= 43.70 cm^3
Molar Volume	= 125.0 cm^3
Parachor	= 332.0 cm^3
Index of Refraction	= 1.616
Surface Tension	= 49.7 dyne/cm
Density	= 1.184 g/cm^3
Polarizability	= 17.32 $10^{-24}cm^3$
Monoisotopic Mass	= 148.052429 Da
Nominal Mass	= 148 Da
Average Mass	= 148.1586 Da

Chemcial Properties

It is readily soluble in hot water or alcohol and on reduction with sodium amalgam yields o – hydrocoumaric or meliolotic acid

ORSELLINIC ACID*

Introduction

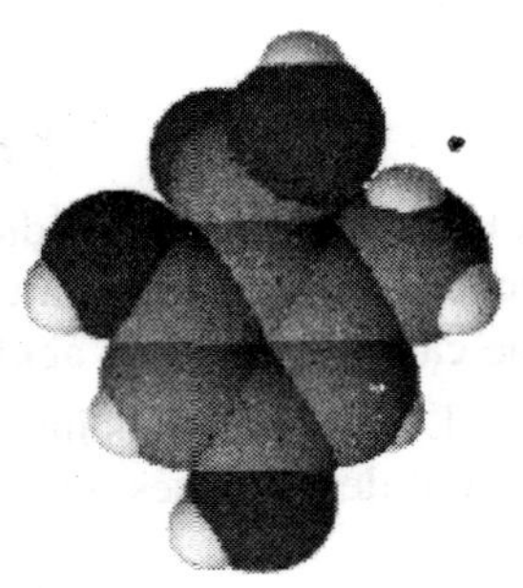

Is an important acid and seriously linked with chemistry of lichens, from which source it can be extracted.

Preparation

It is generally prepared by the oxidation of orcyl aldehydes, or from the methyl ester of dihydrorsellinic acid

On partial methylation with diazomethane it yields everninic acid. This is also produced when everninic acid and ramalic is boiled with baryta.

It forms colourless crystals in the form of needles which on rapid heating melts with decomposition in the neighborhood of 175°C.

*See Wikipedia, en.wikipedia.org for the article.

Physical Properties

Molecular Formula	$= C_8H_8O_4$
Formula Weight	= 168.14672
Composition	= C(57.14%) H(4.80%) O(38.06%)
Molar Refractivity	= 41.77 cm^3
Molar Volume	= 115.0 cm^3
Parachor	= 337.0 cm^3
Index of Refraction	= 1.645
Surface Tension	= 73.5 dyne/cm
Density	= 1.460 g/cm^3
Polarizability	= 16.55 $10^{-24}cm^3$
Monoisotopic Mass	= 168.042259 Da
Nominal Mass	= 168 Da
Average Mass	= 168.1467 Da
Log P	= 2.06

Uses

Orsellinic acid is not directly used lichens are used for the extraction of litmus

LECANORIC ACID

It is also known as p – diorsellinic acid, is as usual can be isolated from lichens and can be isolated from ramalic acid or from evernic acid (as in the case of orsellinic acid)

Lecaniric acid is insoluble in water but soluble in ether (1:30) in dry state it decomposes with melting at 166°C

EVERNIC ACID

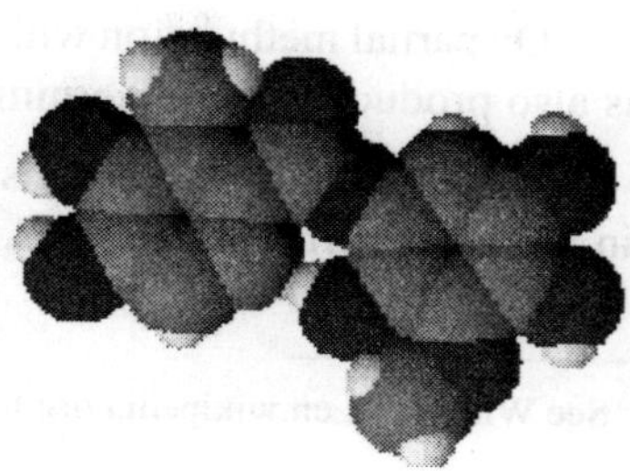

It is also known as monomethyl lecanoric acid, contains a methyoxyl group in the para position to the despide group, since the everninic acid is obtained (together with orsellinic acid) from it on hydrolysis it has shwown to be p – methyl orsellinic acid.

It can be synthesized from everninic acid as described in the lecanoric acid.

KYNURENIC ACID

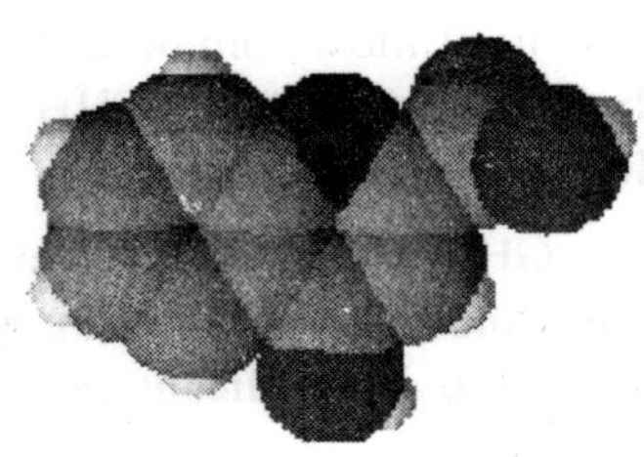

Kynurenic acid was discovered in 1853 by the German chemist Justus von Liebig in dog urine is a product of the normal metabolism of aminoacid L-tryptophan. It is formed from L-kynurenine, a reaction catalyzed by the enzyme kynurenine-oxoglutarate transaminase. It has been shown that kynurenic acid possesses neuroactive activity. It acts as an antiexcitotoxic and anticonvulsant, most likely through acting as an antagonist at excitatory amino acid receptors. Because of this activity, it may influence important neurophysiologic and neuropathologic processes. As a result, kynurenic acid has been considered for use in therapy in certain neurobiological disorders.

High levels of kynurenic acid have been identified in human urine in certain metabolic disorders, such as marked pyridoxine deficiency and deficiency/absence of kynureninase.

Physical Properties

Molecular Formula	**= $C_{10}H_7NO_3$**
Formula Weight	= 189.16748
Melting Point	= 282.5°C
Composition	= C(63.49%) H(3.73%) N(7.40%) O(25.37%)
Molar Refractivity	= 50.99 cm^3
Molar Volume	= 127.7 cm^3
Parachor	= 382.3 cm^3
Index of Refraction	= 1.730
Surface Tension	= 80.3 dyne/cm
Density	= 1.480 g/cm^3
Polarizability	= 20.21 $10^{-24}cm^3$
Monoisotopic Mass	= 189.042593 Da
Nominal Mass	= 189 Da
Average Mass	= 189.1675 Da
Log P	= 2.46

GAMMA-HYDROXYBUTYRIC ACID

Gamma-Hydroxybutyric acid (4-hydroxybutanoic acid, C_4HO_3) is both a drug and a naturally occurring substance found in the central nervous system where it might act as a neurotransmitter, as well as in

other organs such as the liver, kidneys, heart and bones. GHB is structurally related to the ketone body beta-hydroxybutyrate. As a drug it is used most commonly in the form of a salt like sodium gamma-hydroxybutyrate (Na.GHB, sodium oxybate) or potassium gamma-hydroxybutyrate (K.GHB).

GHB is also produced as a result of fermentation and so is found in small quantities in some beers and wines, particularly fruit wines such as that made from guavas. However, the amount of GHB found in wine is insignificant and not sufficient to produce any effects.

Chemical Properties

The action of GHB has yet to be fully elucidated. GHB clearly has at least two sites of action, stimulating the newly characterized and aptly named "GHB receptor" as well as the $GABA_B$. GHB, if it is indeed a neurotransmitter, will normally only reach concentrations high enough to act at the GHB receptor, as it has relatively weak affinity for the $GABA_B$. However, during recreational usage, GHB can reach very high concentrations in the brain, relative to basal levels, and can act at the $GABA_B$ receptor as well. GHB's action at the $GABA_B$ is probably responsible for its sedative effects. GHB-mediated $GABA_B$ receptor stimulation inhibits dopamine release as well as causes the release of natural sedative neurosteroids (like other $GABA_B$ agonists e.g. Baclofen). In animals GHB's sedative effects can be stopped by $GABA_B$ antagonists (blockers).

The relevance of the GHB receptor in the behavioural effects induced by GHB is more controversial. It seems hard to believe that the GHB receptor is not important when it is densely expressed in many areas of the brain, including the cortex, as well as it being the high affinity site of GHB action. There has only been limited research into the GHB receptor. However, evidence shows that it causes the release of glutamate, which is a stimulatory neurotransmitter. Drugs which selectively activate the GHB receptor but not the $GABA_B$ receptor, such as trans-4-hydroxycrotonic acid and 4-(p-chlorobenzyl)-GHB, cause convulsions in animals and do not produce GHB appropriate responding.

It does not seem that activation of the GHB receptor alone explains GHB's addictive properties; indeed research using selective $GABA_B$ agonists and analogues of GHB which are selective agonists for the GHB receptor but do not activate $GABA_B$, suggest that both the GHB receptor and the $GABA_B$ receptor are important for dopamine release and consequently abuse liability. Compounds which activate only one of the receptors but not both, do not seem to induce acute dopamine release or produce the abuse potential typical of GHB itself.

One can propose a scheme where high doses of GHB are sedative through its action at the $GABA_B$ receptor, while a lower dose is stimulatory via activation of GHB receptors. This may explain the otherwise paradoxical mix of sedative and stimulatory properties typical of GHB intoxication, as well as the so-called "rebound" effect, experienced by individuals using GHB as a sleeping agent, where they awake suddenly after several hours of GHB-induced deep sleep. That is to say, that over time, the concentration of GHB in the system decreases (because of metabolism) below a threshold for stimulating $GABA_B$ receptor function, and simply stimulates the GHB receptor, leading to wakefulness.

Recently, similar molecules to GHB, such as 4-hydroxy-4-methylpentanoic acid have been synthesised and tested on animals, in order to gain a better understanding of GHB's mode of action. Analogues of GHB such as 3-methyl-GHB, 4-methyl-GHB and 4-phenyl-GHB have been shown to produce similar effects to GHB in some animal studies, but these compounds are even less well researched than GHB itself. Of these analogues, only 4-methyl-GHB (gamma-hydroxyvaleric acid, GHV) and its prodrug form gamma-valerolactone (GVL) have been reported as drugs of abuse in humans, and on the available evidence seem to be less potent but more toxic than GHB, with a particular tendency to cause nausea and vomiting. Other prodrug ester forms of GHB have also rarely been encountered by law enforcement, including 1,4-diacetoxybutane, methyl-4-acetoxybutanoate and ethyl-4-acetoxybutanoate, but these are generally covered by analogue laws in jurisdictions where GHB is illegal and little is known about them beyond their presumably delayed onset and longer duration of action. The intermediate compound 4-hydroxybutaldehyde is also a prodrug for GHB, however as with all aldehydes this compound is caustic and is strong-smelling and foul tasting; actual use of this compound as an intoxicant is likely to be unpleasant and result in severe nausea and vomiting. GHB is also metabolised into GABA by the enzyme GABA transaminase. This is only a minor route of metabolism and its significance to the pharmacological action of GHB is not known, although there is likely to be some increase in GABA levels when GHB is consumed in quantity.

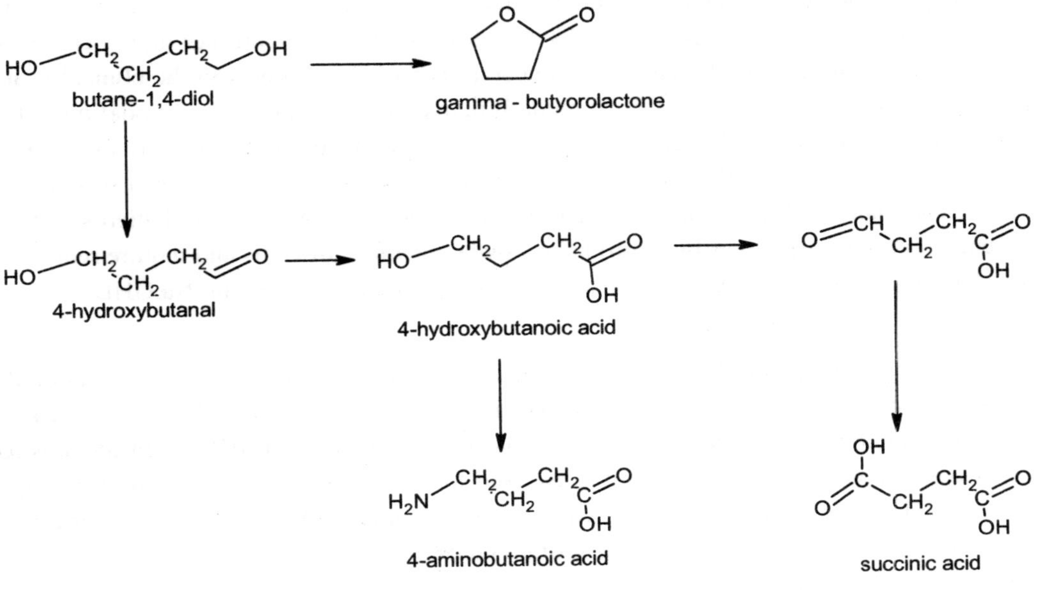

butane-1,4-diol

gamma - butyorolactone

4-hydroxybutanal

4-hydroxybutanoic acid

4-aminobutanoic acid

succinic acid

Also note that both of the metabolic breakdown pathways shown for GHB can run in either direction depending on the concentrations of the substances involved, so the body can make its own GHB either from GABA or from succinic semialdehyde. Under normal physiological conditions the concentration of GHB in the body is rather low, and the pathways would run in the reverse direction to what is shown here to produce endogenous GHB. However, when GHB is consumed for medical or recreational purposes its concentration in the body is much higher than normal, which changes the enzyme kinetics so that these pathways operate to metabolise GHB rather than producing it.

Uses

Medical

GHB has been used as a general anesthetic, and a hypnotic in the treatment of insomnia. GHB has also been used to treat clinical depression and to improve athletic performance.

In Italy, under the trade name Alcover (ATC code N07BB), GHB is used in the treatment of alcoholism (50 to 100 milligrams per kilogram per day, in 3 or more divided doses), both for acute alcohol withdrawal and medium to long-term detoxification. An author/scientist by the name of Gian Luigi Gessa has been researching alcoholism and the effects of various drugs to persons afflicted with said disease for the past ten years. His studies in 1998 note that GHB, as a pharmaceutical aid, can be much less toxic and much more effective than the leading pharmaceutical compound (disulfiram)

In the United States, the Food and Drug Administration permits the use of GHB under the trade name Xyrem to reduce the number of cataplexy attacks in patients with narcolepsy; in clinical trials Xyrem significantly reduced cataplexy attacks at a dose of 6000–9000mg per night. This is around three times the dose used recreationally, but almost all narcolepsy patients in the clinical trials were already stabilized on CNS stimulants such as modafinil; in patients not prescribed modafinil this dosage could be dangerous and should be reduced appropriately. Also the prescribing information for Xyrem states that patients should take the dose immediately before going to bed, and then a second dose 3–4 hours later. The maximum dose taken at one time should not exceed 4500 mg. Patients with hepatic insufficiency (compromised liver function) have slower clearance of GHB and require reduced doses, typically half the normal dose. Xyrem oral solution is standardised to 500 mg Na.GHB / 1 mL water, buffered to pH 7.5 with malic acid.

When GHB is used in its sodium or potassium salt form, a significant quantity of excess sodium or potassium may be consumed, which should be taken into consideration by people with heart conditions, hypertension or compromised renal function. The bioavailability of sodium GHB is considerably reduced when it is consumed with food, and so it is advised to wait at least two hours after eating before consuming the dose. Because of its strong sedative effects, patients should not drive or operate heavy machinery for at least six hours after taking sodium GHB.

Adverse effects from Xyrem in clinical trials included: headache, nausea, dizziness, nasopharyngitis, somnolence, vomiting, urinary incontinence, confusion, dyspnea, hypoesthesia, paresthesia, tremor, vertigo, and blurred vision. Out of the 717 patients and 182 healthy volunteers who took part in the trials (899 total), two of them died from drug overdoses, although only one of these involved GHB.

Acids with Side Chain

PHENYLACETIC ACID

Introduction

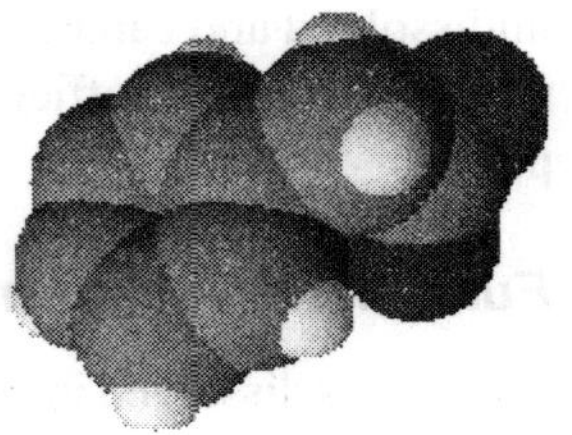

Phenylacetic acid has been found to be an active auxin (a type of plant hormone) molecule, predominantly found in fruits. However its effect is much weaker than the effect of the basic auxin molecule indole-3-acetic acid.

Phenylacetic acid is used in some perfumes, possessing a honey-like odour in low concentrations, and is also used in penicillin G production.

Preparation

Phenylacetic Acid from Benzyl Chloride

Cl —Mg→ MgCl —CO_2→ OH, O

10% of a solution of 125 g benzyl chloride in 250 ml of sodium-distilled diethyl ether is added to 24 g magnesium turnings under 100 ml ether, and a small iodine crystal is added. after start of the grignard reaction the rest of the benzyl chloride is added with stirring to maintain gentle boiling (if the reaction becomes too vigorous, useless 1,2-diphenylethane is formed) and the mixture is heated and stirred until most magnesium is dissolved. The Grignard reagent solution is poured on 1 kg water-free, crushed dry ice (solid CO_2) and stirred for 2 hrs. 200 ml warm ether is added and the mixture is heated in a water-bath until the internal temperature reaches 25°C. 200 ml 32% hydrochloric acid is added, and the heterogenous mixture stirred until any inorganic precipitate is dissolved. after filtering, the organic layer is separated, washed with cold water, and dried over 20g anhydrous sodium sulphate. The ether is distilled off, and the rest is recrystallized from water, to yield 75% - about 100g - phenylacetic acid, mp 76-77°C.

Phenylacetic acid from Benzyl Cyanide

CN —H_2SO_4→ OH, O

In a silica basin of about 35 cm. diameter, a mixture of 2.5 kilos of benzyl cyanide [3] and 7.5 kilos of 70% sulfuric acid is warmed until a few bubbles of gas appear. The heating is at once stopped, since a very vigorous action now ensues. As the vapors evolved are injurious to health, the basin is covered when the reaction begins with a sliding cover from which the vapors can be led into water. As soon as the reaction ceases, the product is poured upon crushed ice in a large, earthenware vessel. Crude phenyl acetic acid, contaminated on the surface with phenyl acetamide, is precipitated on cooling. In order to remove the amide, the acid is dissolved in lukewarm dilute soda solution. The amide remains undissolved and can be filtered off, while the phenyl acetic acid goes into solution as sodium salt and is reprecipitated in a sufficiently pure state by the addition of dilute mineral acid. After drying it is acceptably pure.

Further Purification

The acid is dissolved in five times its weight of boiling water, filtered off and allowed to cool. Phenylacetic acid separates out in an absolutely pure state in the form of flakes with a mother of pearl lustre. After filtering and drying it possesses a pleasant honey scent.

Phenylacetic acid from Mandelic Acid

HI/P

15 grams mandelic acid (0.1 mol), 2.07g KI (6.25% of theory), and 6g Red Phosphorous is dissolved in a solution of 70ml phosphoric acid and 10 ml of water, and this solution is refluxed for six hours (bp of soln 144°C). After the reaction mixture has cooled down, a little water is added to dissolve precipitated inorganic salts. The solution is extracted with ether, and the pooled extracts washed with a little dilute NaHSO3, followed by water, and is then dried over Na2SO4. The ether is removed under diminished pressure and the residue is distilled (bp 138-139 @13mmHg). The yield is 12.5 grams (90%) of phenylacetic acid melting at 76°C. 3.2 grams of phosphorous could be recovered from the aqueous solution.

Phenylacetic Acid from Acetophenone

$Pb(OAc)_4$, $BF_3.Et_2O$; NaOH

Methyl Phenylacetate

4.56g (38 mmol) acetophenone, 10ml methanol, 22g (150 mmol, 20ml) boron trifluoride etherate (BF3*Et2O) is added in one potrion to a magnetically stirred suspension of 17.7g (40 mmol) lead tetraacetate in 100ml benzene. The reaction mixture is stirred at room temp for 5h, diluted with 500ml cold water, extracted with 600ml benzene, washed with 150ml satd. NaHCO3, 200ml satd. NaCl, and then dried over sodium sulphate and filtered. Evaporation of the solvent and vacuum distillation of the residue gave methyl phenylacetate in 86% yield, bp 94-96C/5mmHg.

Phenylacetic Acid

2.4g (16 mmol) methyl phenylacetate and 10ml 2M NaOH is refluxed for 2h. The free phenylacetic acid is precipitated by the addition of conc HCl (~3ml), in 92% yield, mp 76-76.5°C.

The **Willgerodt Reaction** can also be used to transform acetophenone (or styrene) to phenylacetic acid.

Phenylacetic Acid from Ethylbenzene

$Na_2Cr_2O_7$ →

This reaction is pretty simple and cheap to do, but requires a little bit of preliminary engineering and construction. The engineering and construction consists of getting some sturdy steel pipe, welding two or three agitator vanes on the inside of the pipe, and pulling the agitation gearing out of an old washing machine and reinstalling it onto the pipe. It's no great challenge. This reaction was originally reported in 1962. It was found that when ethylbenzene was heated with sodium dichromate in water solution at a temperature of 275°C inside an autoclave, a 90% yield of phenylacetic acid resulted.

Ethylbenzene is churned out in huge quantities by the chemical industry, as it finds use as a feedstock for making styrene and also as a solvent for plastic resins. Any attempt to try to keep track of its commercial flow would be futile. Drums of it are easily available once you give yourself a corporate name. Similarly, sodium dichromate is heavily used by industry. Its largest and most common use is probably to passivate stainless steel. This process essentially consists of immersing a piece of stainless steel into a vat containing a mixture of nitric acid and sodium dichromate. The passivated stainless steel is more corrosion resistant than untreated steel. Calling yourself some kind of metal finishing or polishing operation and getting a shed or something to work in will make ordering a thousand pounds of sodium dichromate pretty easy.

An alternative source of dichromate is used chrome plating bath picked up on one of the waste exchanges. The typical used chrome plating bath will contain 32 ounces per gallon of chromic acid. The exact concentration will come with the analytic sheet. This is a 2 molar chromic acid solution. At this concentration, the great majority of the chromic acid is in the form of the dichromate. To get a sodium dichromate solution, add 20% sodium hydroxide to the chrome plating bath, slowly and with stirring, until the pH reaches 8 or 9. Then add water until the volume of the solution is double that of the starting chrome plating bath. The tramp metals in the bath, such as iron, copper or nickel, will settle out as hydroxide sludge. Decant off this sludge, and filter to get a clean .5 molar dichromate solution.

The following example is given for a .25 molar size batch, but the procedure could easily be scaled up to at least one molar and beyond by giving allowance for the longer period of time such a larger batch would require to heat up to the proper reaction temperature.

First one starts with a section of heavy pipe, of a volume of about 1500ml. It should have fine threads on each end. Obtain heavy pipe caps with matching threads for both ends. Most pipe will come with a heavy coating of zinc galvanizing on it. This must be removed by immersing the pipe and caps in 5% hydrochloric acid solution until the rapid fizzing caused by the dissolving of the zinc slows to a crawl. Then thoroughly rinse the pipe and caps in clean water, and dry to prevent rust.

Then agitator vanes should be welded onto the interior of the pipe. There should be three or more vanes, depending upon the diameter of the pipe. For example, a one gallon size pipe section for a one mole batch scale should have four vanes to get good agitation. They should run the entire length of the pipe section, and extend about halfway in from the wall of the pipe. The agitator vanes, of course, should be made of steel, and be clean and free of paint. Once the vanes are welded in, the bottom cap can be screwed on the pipe, tightened down with a pipe wrench, and preferably welded into place. Now the cap assembly is constructed. It consists of a shaft welded onto the top of the upper pipe cap. The shaft runs to the washing machine agitator gearing, preferably with a bearing about halfway up the shaft to prevent wobbling as the agitator is going.

To do the reaction, put 1000ml of water into the pipe, along with 74 grams of sodium dichromate. Alternately put 600ml of water in the pipe along with 400ml of the .5 molar dichromate solution made from the used chrome plating bath. Next add 30ml of ethylbenzene. Then screw the pipe into the upper pipe cap, and tighten it with a pipe wrench. The reaction mixture must be rapidly heated to 275°C. The best way I can think of to do this is to use a solder as a high temperature heating bath. Spools of solder can be easily picked up at the hardware store and tossed into a heavy iron pot for melting. Solder melts at about 220C, and so makes a suitable heating bath for this reaction. The solder bath should be heated to about 300C, then the pipe reaction vessel should be immersed in it to a depth about equal to the depth of the liquid inside the pipe, and agitation should then be commenced on the pipe reaction vessel. The solder bath should be of a volume at least equal to the reation pipe so that it holds a fair sized heat reservoiur. The temperature of the solder bath will quickly fall when the pipe is immersed, but it shouldn't be allowed to fall below 275°C. It will take about 15 to 20 minutes for a batch of this size to warm up to the reaction temperature of 275°C. A larger batch will take longer to warm up. The geometry of the pipe will also effect the length of time required for the reaction mixture to warm up. A longer, narrower length of pipe presents more surface area, and so warms up faster than a shorter fatter one.

Once the reaction temperature is reached, continue heating at 275°C with good effective agitation for one hour. The back and forth agitation action must be vigorous enough to force the ethylbenzene, which floats on top of the water and chromate solution, into contact with the chromate. Once it reacts and forms the sodium salt of phenylacetic acid, it freely dissolves in the water layer. If at the end of this process one should find an organic layer still floating on top of the water, then the agitation was too slow, or not enough time was allowed for the mixture to reach reaction temperature prior to counting the one hour reaction time. During the course of the reaction, there will be a steam pressure inside the pipe amounting to hundreds of pounds per square inch. For this reason, the top must be cranked down good and tight, and the pipe must be of heavy construction. As a precaution, don't hover around the batch while it cooks. Also wear a face shield, leather jacket, apron and gloves. Be watchful of globs of solder spinning off the outside of the pipe. When the one hour reaction heating period is over, the pipe agitation should be stopped, and the kettle of molten solder lowered from the reaction pipe and wheeled away. Next, cool down the reaction pipe by immersing it in a tub of water. When it has cooled down, the tub of water may be removed, and the pipe reaction vessel opened with a pipe wrench.

The green colored contents should be poured into a beaker, and then filtered to remove the chrome oxide sludge. Rinse the sludge to get any product off it. If time permits, it might be easier just to let the sludge settle to the bottom of the beaker, decant off the liquid for filtering, resuspend the sludge with a little water, and then filter this remnant. Now to the filtered reaction mixture, add 6N sulfuric acid (this is concentrated sulfuric acid diluted in half with water) slowly with stirring until the pH of the reaction mixture gets down to two or even one. This causes the sodium salt of phenylacetic acid which was produced in the reaction to convert to phenylacetic acid. Phenylacetic acid will dissolve in hot water, but crystallizes out of cold water. Using a funnel to prevent spillage, transfer the now acidified reaction mixture to a large separatory funnel. Then extract two times with 100ml portions of toluene. This solvent can be found at the hardware store in the paint section. The extracts should be transferred to a distilling flask, and the glassware rigged for fractional distillation. The first portion of toluene distilling will azeotropically dry the phenylacetic acid. It will distill with a milky appearance. Then the main body of the toluene will distill with a clear colour. When most of it has distilled, remove the residual toluene in the product by applying a vacuum. When the toluene has been distilled out, stop the vacuum and the heating, and while the phenylacetic acid is still molten in the flask, pour it out into a sturdy glass baking dish, or something similar. There it can set up into a hard, white mass that you can chip at when needed. The yield from this size run will be 25 to 30 grams. A one gallon size batch will yield over 100 grams.

Physical Properties

Molecular Formula	= $C_8H_8O_2$
Formula Weight	= 136.14792
Composition	= C(70.57%) H(5.92%) O(23.50%)
Boiling point	= 265.5°C
Melting point	= 76.5°C
Molar Refractivity	= 37.36 ± 0.3 cm^3
Molar Volume	= 116.8 ± 3.0 cm^3
Parachor	= 305.6 ± 4.0 cm^3
Index of Refraction	= 1.552 ± 0.02
Surface Tension	= 46.7 ± 3.0 dyne/cm
Density	= 1.164 ± 0.06 g/cm^3
Dielectric Constant	= Not available
Polarizability	= 14.81 ± 0.5 $10^{-24}cm^3$
Monoisotopic Mass	= 136.052429 Da
Nominal Mass	= 136 Da
Average Mass	= 136.1479 Da
Flash point	=132°C c.c.
Auto-ignition temperature	= 543°C
Octanol/water partition coefficient as log Pow	= 1.41

Chemcal Properties

When chlorinated of brominated by the direct action of the particular halogen, it replaces the hydrogen of the α- carbon of the side chain as in the fatty acids.

phenylacetic acid + Cl_2 → dichloro(phenyl)acetic acid

Oxidation of benzoic acid is prepared

phenylacetic acid →[O] benzoic acid + $H_3C{-}OH$ (methanol)

Uses

It is used in pesticide, medicine, dye and perfume, also used in manufacturing phenoxin, phenthoate ethyl, phenylacetic acid, penicillin and phenyl.

MANDELLIC ACID

Introduction

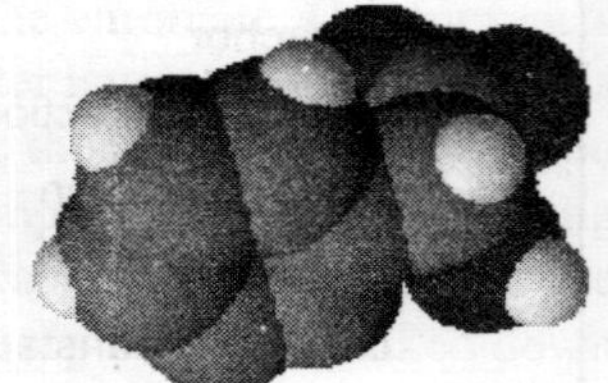

Mandelic acid was discovered while heating an extract of bitter almonds with diluted hydrochloric acid. The name is probably derived from the German “Mandel” for “almond”.

Preparation

Mandelic acid is isomeric with hydroxytoluic acid. It was originally prepared from amygdalin of bitter almonds, by boiling with mineral acids. The benzaldehyde and hydrocyanic acid of the amygdalin, which are undoubtedly present in combination, and they are hydrolyzed, the process has been imitated by forming the cyanohydrin of benzaldehyde, or mandelic nitrile and hydrolyzing the product with strong hydrochloric acid.

benzaldehyde —HCN→ cyano(phenyl)acetic acid + $2H_2O$ → hydroxy(phenyl)acetic acid

Lab preparation

Procedure

In a 125 ml Erlenmeyer flask dissolve 11 g of sodium bisulphite in 50 ml of water by brief swirling, add 10 ml. of benzaldehyde and swirl vigorously and stir until the oily aldehyde is all converted into the crystalline bisulphite addition compound. Cool to room temperature but not below, add 14 g. of potassium cyanide (caution) and 25 ml. of water (rinse down the walls), and take the mixture to a hood. Swirl and stir for about 10 min. until all but a trace of solid has dissolved (break up lumps with the rod). Mandelonitrile separates as a thick oil. Pour the mixture into a separatory funnel, rinse the flask with small amounts of ether and water (and at once wash the flask free of cyanide), and then shake the mixture vigorously for a full minute to insure complete reaction. Add 50 ml. of ether, shake, and run the aqueous solution down the drain. Wash the ether extract with 25 ml. of water and then with 25 ml. of saturated salt solution (note and account for the difference in appearance of the ether layer). Then run the solution into a 125 ml distilling flask containing 15 ml. each of coned. hydrochloric acid and water. Add a boiling stone, stopper the flask, and mount it on a steam bath for distillation through a condenser into an ice-cooled receiver. Distill off the ether and then disconnect the condenser, note the time, and continue heating on the steam bath with frequent swirling to mix the layers and promote hydrolysis. Note that the initially lighter-than-water layer of mandelonitrile gradually changes to an oil of density greater than that of the aqueous acid. In a little less than 1 hour the oil dissolves to a clear solution; however if the solution is cooled at this point it will become cloudy from separation of unhydrolyzed oil still present. Hence continue heating for one half hour more to complete the reaction and then cool to room temperature. Measure 100 ml. of benzene into a 250 ml. distilling flask and make a mark at the level of the liquid. Transfer the acid solution to a separatory funnel and rinse the flask with a little ether (5 ml). Then add to ml. of ether, shake well, let the layers separate, and draw off the aqueous layer into a second separatory funnel or a dry flask, and wait a few minutes for separation of a little more aqueous solution. Run the ethereal extract into the flask containing benzene, and reextract the aqueous solution with two further 20 ml portions of ether, and add the extracts to the flask containing benzene. Add a boiling stone, connect the flask to a condenser and an ice-cooled receiver, put a thermometer in place, and distil the solvent, cautiously at first since the flask is very full. Water is eliminated gradually by azeotropic distillation and the boiling point rises as ether and water are eliminated. Continue the distillation until the solution has cleared, water is no longer evident in the neck of the distilling flask, and the volume of the solution has been reduced to the 100 ml. mark. Disconnect the flask and inspect the bottom to see if a trace of ammonium chloride has separated, as either a solid or a gum, make sure that no flames are near, and decant the hot solution into a 250 ml. Erlenmeyer flask. Crystallization usually starts soon; yield 10.0 gram, mp 118-119 deg C. In case no crystals appear, pour the solution into a 25 ml. distilling flask, evaporate the solvent, scratch to induce crystallization, and digest the solid with benzene.*

Mandelonitrile

To a 40°C solution of 40 g $NaHSO_3$(sodium bisulphite) and 250 mL water 34 g benzaldehyde (0.324 mol) is added with vigorous stirring. After 20 minutes cool the reaction mixture down to 0°C, add 200 mL Ether and drip in a solution of 17 g NaCN(sodium cyanide in 70 mL water. Agitate for 30 minutes at 0°C, then seperate the organic phase and extract the aqueous phase twice with 50 mL portions of ether (or toluene). The combined ether extracts are washed with 2x100ml conc. NaHSO3 solution to remove traces of unreacted benzaldehyde (dont skip this step), washed 1 x 50ml brine then dried over anhydrous calcium chloride. After rotovaping the solvent under a vacuum, the remaining oil is allowed to solidify upon cooling then placed in the vacuum dessicator to remove last traces of water to give mandelonitrile in high yield.

Physcical Properties

Molecular Formula	= $C_8H_8O_3$
Formula Weight	= 152.14732
Composition	= C(63.15%) H(5.30%) O(31.55%)
Melting Point	= 119 °C
Boiling Point	= decomposes after melting on further heating
Molar Refractivity	= 38.90 cm^3
Molar Volume	= 115.1 cm^3
Parachor	= 321.0 cm^3
Index of Refraction	= 1.590
Surface Tension	= 60.4 dyne/cm
Density	= 1.321 g/cm^3
Polarizability	= 15.42 $10^{-24}cm^3$
Monoisotopic Mass	= 152.047344 Da
Nominal Mass	= 152 Da
Average Mass	= 152.1473 Da

***Note:** When mandelic acid that has been crystallized from benzene is allowed to stand in contact with the mother liquor for several days, the needles gradually change into granular, sugarlike crystals of a molecular compound containing one molecule each of mandelic acid and benzene. These crystals are stable only in contact with benzene and at temperatures below 32.6 deg C on exposure to the air the benzene of crystallization evaporates and the mandelic acid is left as a white powder.

Chemical Properties and Uses

It is an isomer of cresotinic acid (2-hydroxy-3-methylbenzoic acid) and oxymethylbenzoic acid (2-methoxybenzoic acid). Derivatives of mandelic acid are formed as a result of metabolism of adrenaline and noradrenaline by monoamine oxidase and catechol-o-methyl transferase. It is also present in certain skin care products, is an intermediate molecule in the production of other biochemicals, may be used as an analytical reagent and is a precursor in the manufacture of dyes.

Since the molecule contains a chiral carbon atom, the acid exists in either of two optically active forms. The racemic mixture of the two is known as *paramandelic acid.*

VANILLYL MANDELIC ACID

Vanillyl mandelic acid (VMA) is a chemical end product of catecholamine metabolism. It is found in urine together with homovanillic acid (HVA). In timed urine tests its quantity (concentration ìg /24 h) is assessed, along with creatinine clearance, and the concentration of cortisols, catecholamines, and metanephrines.

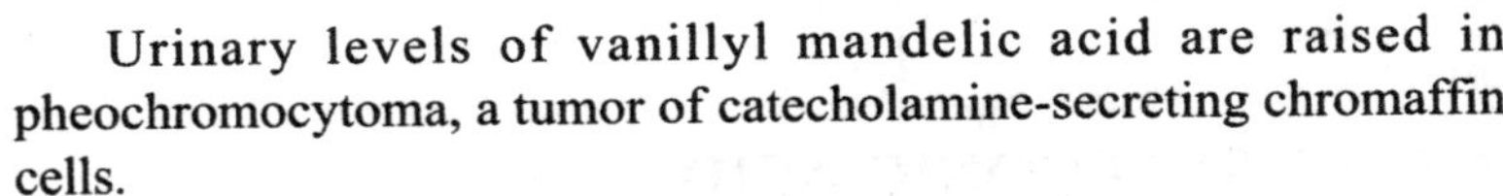

Urinary levels of vanillyl mandelic acid are raised in pheochromocytoma, a tumor of catecholamine-secreting chromaffin cells.

Aromatic Sulphonic Acids

In aromatic sulphonic acids, an aryl group has replaced a hydroxyl group of sulphuric acid $HO - SO_2$ — OH. The simple benzene sulphonic acid $C_6H_5 — SO_2 — OH$, is prepared by direct replacement of a ring hydrogen with the sulphonic acid group, - SO_2 OH, derived from sulphuric acid, H_2SO_4.

Nomenclature

Aromatic sulphonic acids are commonly named by adding sulphonic acid, as suffix to the name of the parent compound the two mentioned as a single word. Substituent in the ring is more frequently indicated by numbers than o, m and p. If sulphonic acid group and carboxylic group are both present in a compound, the former is prefix "sulpho" in the name. Some sulphonic acids, containing other functional groups have trivial names, e.g. p – amino benzene sulphonic acid $H_2N — C_6H_4 — SO_3H$, This is better known as sulphanilic acid, and its m – isomer is metanilic acid

A few examples are shown below:

Aromatic sulphonic acids are next to carboxylic acids after importance and therefore produced in large scale. The strongly polar hydrophobic, non – polar aromatic molecule to make it water soluble for e.g. in dyes. These are used as wetting, emulsifying and surface active agents, for making sulpha drugs, ion exchange resins, insect rerepellents, synthetic tanning compounds and plasticizers, saccharin etc.

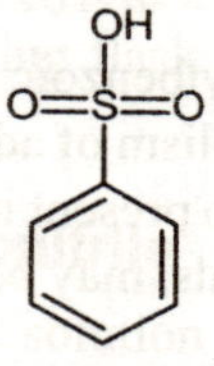

benzenesulfonic acid

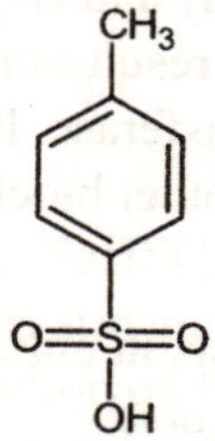

4-methylbenzenesulfonic acid

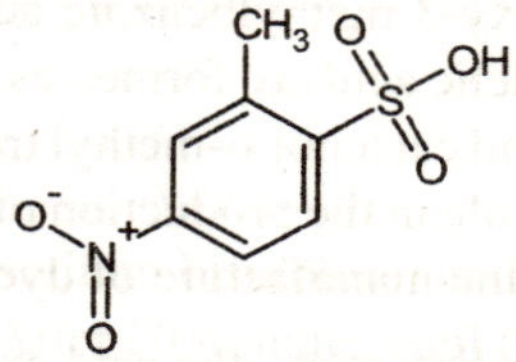

2-methyl-4-nitrobenzenesulfonic acid

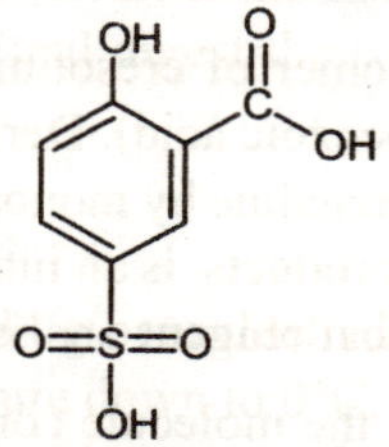

2-hydroxy-5-sulfobenzoic acid

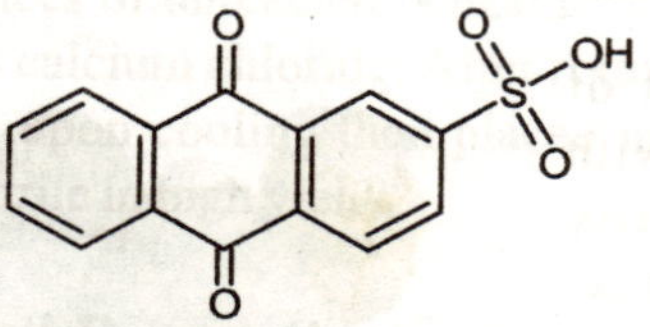

9,10-dioxo-9,10-dihydroanthracene-2-sulfonic acid

BENZENE SULPHONIC ACID

Introduction

Benzene sulphonic acid was first prepared in 1834 by Mischerlich by heating benzene with fuming sulphuric acid (solution of sulphur trioxide in sulphuric acid also known as oleum, fuming sulphuric acid, this method is still followed.

Preparation

Benzenesulphonic acid is formed on refluxing benzene and excess conc. sulphuric acid with frequent shaking for about six hours. But with fuming sulphuric acid it takes only fifteen minutes, but as much as boiling proceeds careful addition with cooling is necessary. Benzene (15cc) is slowly added with frequent shaking to fuming sulphuric acid (25cc), taken in a round bottomed flask, in about 15 mins. The flask is kept at 30 – 50 cc by cooling in water, if necessary. The reaction is complete. When all the benzene at the top layer has disappeared. The cooled acid solution, diluted with water, is partially neutralized with solid sodium bicarbonate and heated with sodium chloride (25 gm).

For salting out the product. The filtered solution deposits sodium benzosulphonate on cooling. It is washed with saturated sodium chloride solution and finally with ethanol

benzene + sulfuric acid ⇌ benzenesulfonic acid + H_2O

benzenesulfonic acid + $NaHCO_3$ → sodium benzenesulfonate + CO_2 + H_2O

Mechanism

The affinity electrophile is either SO_3 (a strong Lewis Acid) or protonated sulphur trioxide $HO - SO_2^+$ (a still stronger Lewis Acid) Even in conc. Sulphuric acid; sulphur dioxide in pyridine solution readily sulphonates benzene. The sulphur trioxide molecule is a resonance hybrid; the positive sulphur is the electrophilic site;

cannonical forms of SO_3 molecule

The steps of sulphonation benzene are:

1. benzene + $O{=}S^+(=O){-}O^-$ ⇌ σ - complex

2. σ-complex (H, SO_3) ⇌ SO_3^--benzene + H^+

3. SO_3^--benzene + H^+ → benzenesulfonic acid

Sulphonation differs from other aromatic electrophilic substitution reaction in that it is a reversible process; the second step, including elimination of proton, is the slow one. A significant primary kinetic isotropic effect K_H/K_D = about 2, this has been observed with duteriated benzene C_6D_6 (or tritium, T) being replaced more slowly than hydrogen.

Manufacture

In industry, benzene is sulphonated in vapour phase one of the continuous methods with conc. Sulphuric acid in a sulphonator (see Fig. 32.4) into which benzene vapour goes up through a reaction zone at about 150°C and meets the acid which comes down. Benzene reacts with sulphuric acid to produce benzene sulphuric acid to produce benzene sulphonic acid; benzene also continuously removes the water formed in as an azeotrope.

Sulphonation normally stops when the concentration of sulphuric acid falls below 78%, the reaction reaches an equilibrium level. When water content reaches 28% here sulphonation goes on until a few percent of free sulphuric acid remains (there is no water to dilute it) which is neutralized. Benzene, water and acid vapours are condensed; benzene is recovered and recycled. Sodium sulphate neutralizes benzenesulphonic acid, the therefore evolved sulphonic acid, is led to acidifies. Sodium sulphate precipitates from the hot filtrate. The concentrated solution of sodium benzene sulphonate is directly converted into phenol. The free acid may be obtained by extraction of the mass after sulphonation with ethylene chloride.

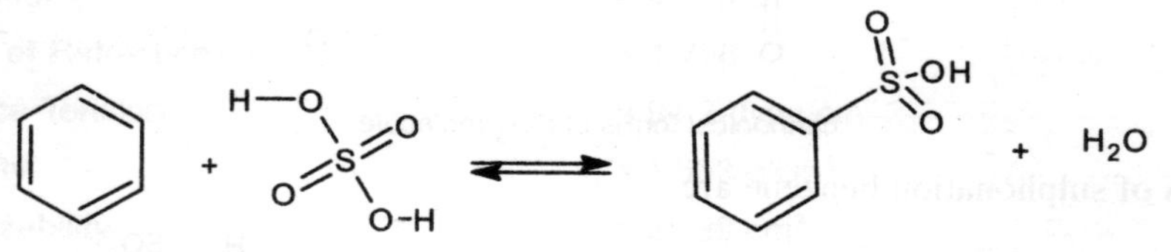

benzene + sulfuric acid ⇌ benzenesulfonic acid + H_2O

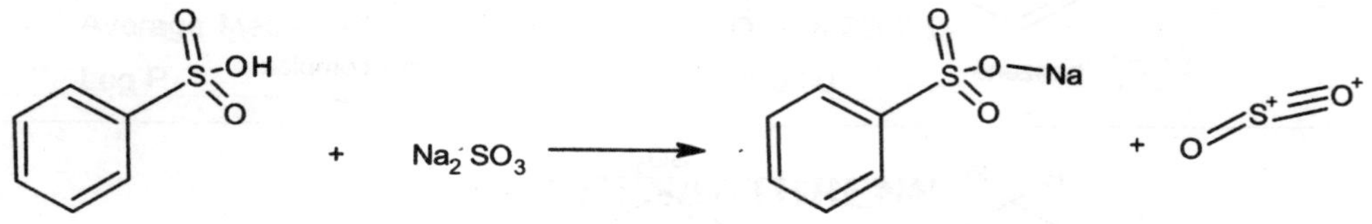

benzenesulfonic acid + Na_2SO_3 ⟶ sodium benzenesulfonate + sulphur dioxide

In the bath operation, benzene is slowly run in conc. Sulphuric acid with agitation at 65 – 75°C. The temperature is then raised to about 105°C; the reaction is complete in about 4.5 to 5 hours. The reflux condenser of the sulphonating vessel returns the unreacted benzene for sulphonation. Diphenylsulphone, $C_6H_6 - SO_2 - C_6H_5$ is insoluble in water, is a minor co – product. Besides conc. sulphuric, fuming sulphuric acid or sulphur trioxide is also used for making benzene sulphonic acid. Cast iron vessels, resistant to 75 – 100% sulphuric acid are generally used.

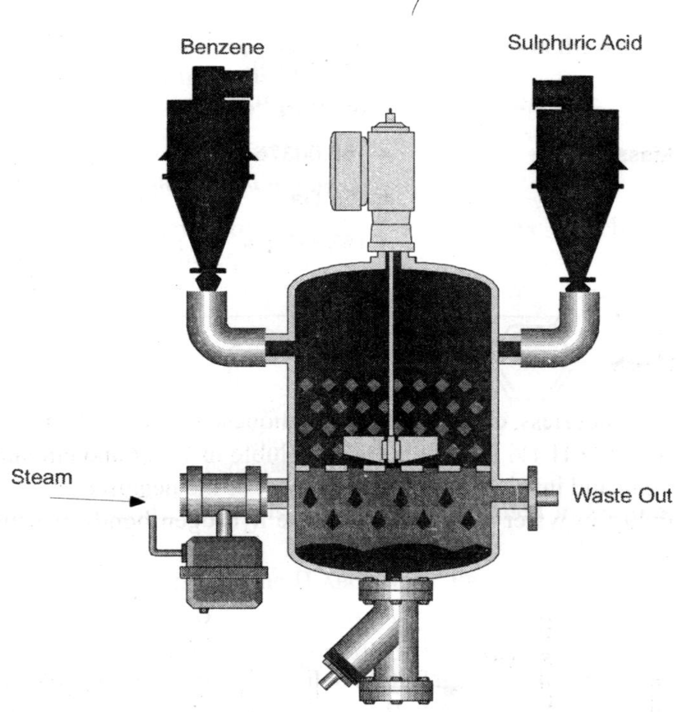

Fig. 32.4. Showing the Industrial Preparation of Benzenesulphonic acid.

Physical Properties

Molecular Formula	= $C_6H_6O_3S$
Formula Weight	= 158.17504
Melting Point	= 44°C
Melting Point	= 66°C(anhydrous)
Composition	= C(45.56%) H(3.82%) O(30.34%) S(20.27%)
Molar Refractivity	= 37.10 cm^3
Molar Volume	= 112.2 cm^3
Parachor	= 300.2 cm^3
Index of Refraction	= 1.575
Surface Tension	= 51.2 dyne/cm

Density	= 1.409 g/cm^3
Polarizability	= 14.70 10^{-24}cm^3
Monoisotopic Mass	= 158.003764 Da
Nominal Mass	= 158 Da
Average Mass	= 158.175 Da
Log P	= 0.47

Chemical Properties

Benzenesulphonic acid is colourless, corrosive, highly deliquescent, hence difficult to purify, crystalline solid. The hydrate, $C_6H_5 - SO_3H$ 1½ H_2O. it is freely soluble in water and ethanol, slightly soluble in water and ethanol, benzene and insoluble in ether. It's most of the negative charge is spread over three oxygen atom, the solubility in water is due to extensive hydrogen bond formation by these oxygen atom.

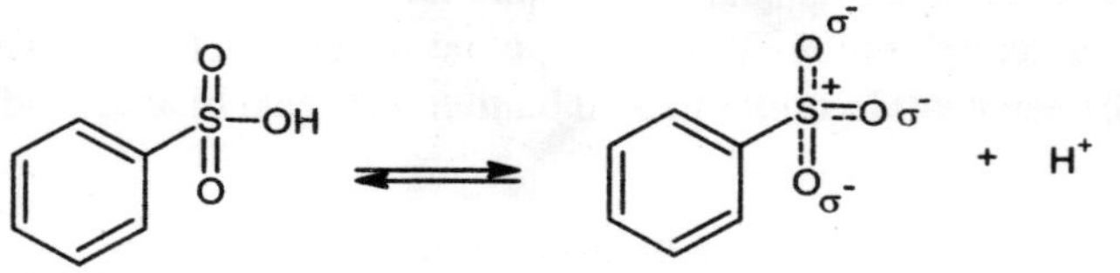

The acid can be distilled, which remains unchanged under high vacuum (136°C under 0.01 mm pressure. Metal sulphonates including those lead, calcium and barium are soluble in water.

Reactions

I. Formation of Phenol: The stable C – S bond is cleaved only under vigorous conditions; e.g. benzenesulphonic acid is hydrolyzed to phenol on passing steam into its solution in sulphuric acid about 175°C, fused with solid sodium hydroxide at 300°C, the sodium salt forms sodium penoxide; thus phenol is manufactured.

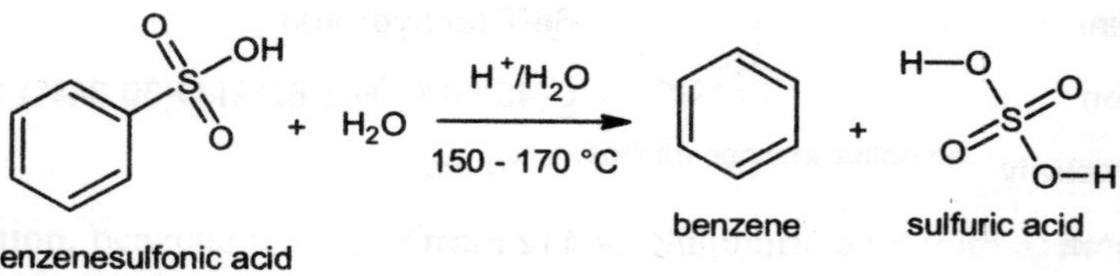

benzenesulfonic acid

II. Formation of Sulphonyl Chloride: Benzene sulphonyl chloride, is produced when benzenesulphonic acid reacts with phosphorous pentachloride or thionyl chloride, this compound benzene sulphonyl chloride is used in Hinsberg methods for separation of primary secondary and tertiary amines.

benzene + chlorosulfuric acid —(-HCl)→ benzenesulfonic acid —($Cl-SO_3H$, $-H_2SO_4$)→ Bezene sulphonyl chloride

Benzene sulphonyl chloride can be converted into an ester by reacting with phenol and into an amide with ammonia, but these compounds are not directly available from benzene sulphonic acid.

benzenesulfonic acid + PCl_5 —(Δ)→ benzenesulfonyl chloride + $POCl_3$ + HCl

benzenesulfonyl chloride —(C_2H_5OH, pyridine)→ ethyl benzenesulfonate + HCl

III. Desulphonation: As the name suggests this is the reverse process of the sulphonation, using excess water (as dilute acid, and removal of benzene; with steam, yields benzene; the equilibrium position thus shifts to the left. It is an electrophilic displacement by a proton, an excess of H^+ (from mineral acid) accelerates the hydrolysis. m – Xylene may be separated from coal – tar, xylene by sulphonation (o – and p – isomers do not react under the condition with concentrated sulphuric acid at 60°C followed by acidic hydrolysis.

benzenesulfonic acid + H_2O —(H^+/H_2O, 150 - 170 °C)→ benzene + sulfuric acid

IV. Formation of Benzonitrile: Benzenesulphonic acid (as sodium salt) and Sodium cyanide form phenyl cyanide or Benzonitrile on fusion.

sodium benzenesulfonate + sodium cyanide (Na—C≡N) —(fuse)→ benzonitrile + Na_2SO_3

Poison

V. Formation of Aniline: Sodium benzosulphonate yields aniline on fusion with sodamide, it is of academic interest

sodium benzenesulfonate + sodium azanide → aniline + Na_2SO_3

VI. Further Sulphonation: The – SO_3H group deactivates the ring and is m- directing* fuming sulphuric acid in excess converts benzene sulphuric acid to benzene m- disulphonic acid to benzene – m – disulphonic acid, on prolonged heating at 85°C. Resorcinol is made from it by caustic fusion also (see page 1481).

benzenesulfonic acid + sulfuric acid —(85 °C, I hr)→ benzene-1,3-disulfonic acid + H_2O

VII. Nitration: Benzene sulphonic yields mostly m – nitrobenzenesulphonic acid with conc. Nitric acid and sulphuric acid.

benzenesulfonic acid + nitric acid —(H_2SO_4)→ 3-nitrobenzenesulfonic acid 72% + H_2O

Uses

Benzenesulphonic acid is mostly converted into phenol. It is sulphonated to benzene – m – disulphonic acid for making resorcinol. This acid acts as intermediate for may dye stuff. This acid also acts in the formation of some surface active agents or synthetic detergents like sodium – alkyl – phenyl sulphonates. It also finds is use in the preparation of sulpha drugs e.g. sulphaguanidine, sulphadiazine, sulphathiazole, etc. those containing the $H_2N - C_6H_4 - SO_2 - NH$

*We will read about this property in the chapter of organic reaction and mechanism.

p-TOLUENESULFONIC ACID

p-Toluenesulfonic acid is an organic compound with the formula $CH_3C_6H_4SO_3H$. TsOH, as it is abbreviated, is a white solid that is soluble in water, alcohols, and other polar organic solvents. Most often, TsOH refers to the monohydrate, $TsOH{\cdot}H_2O$.TsOH is a strong organic acid, about a million times stronger than benzoic acid. It is one of the few strong acids that is solid and, hence, conveniently weighed. Also, unlike some other strong acids, TsOH is non-oxidizing.

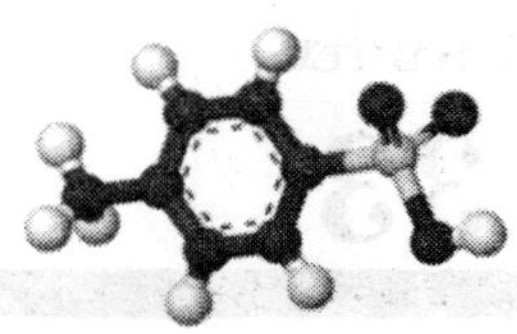

Preparation

TsOH is prepared on an industrial scale by the sulfonation of toluene. It hydrates readily. Common impurities include benzenesulfonic acid and sulfuric acid. Impurities can be removed by recrystallization from concentrated hydrochloric acid followed by drying.

Physical Properties

Molecular Formula	= $C_7H_8O_3S$
Formula Weight	= 172.20162
Melting Point	= 106-107 °C 103-106 °C (monohydrate)
Boiling Point	= 140 °C at 20 mmHg
Composition	= C(48.82%) H(4.68%) O(27.87%) S(18.62%)
Molar Refractivity	= 41.72 ± 0.4 cm^3
Molar Volume	= 128.4 ± 3.0 cm^3
Parachor	= 338.5 ± 6.0 cm^3
Index of Refraction	= 1.562 ± 0.02
Surface Tension	= 48.2 ± 3.0 dyne/cm
Density	= 1.340 ± 0.06 g/cm^3
Polarizability	= 16.54 ± 0.5 $10^{-24}cm^3$
Monoisotopic Mass	= 172.019414 Da
Nominal Mass	= 172 Da
Average Mass	= 172.2016 Da

Uses

Toluenesulfonic acid finds use in organic synthesis as an "organic-soluble" acid catalyst. It is also used as a catalyst with 3,4 dihydro-2H pyran in alcohol protection.

CHAPTER

33

Aromatic Compounds with Condensed Nuclei

Introduction

In this chapter we are going to read about those aromatic compounds having condensed nuclei.

NAPHTHALENE

History

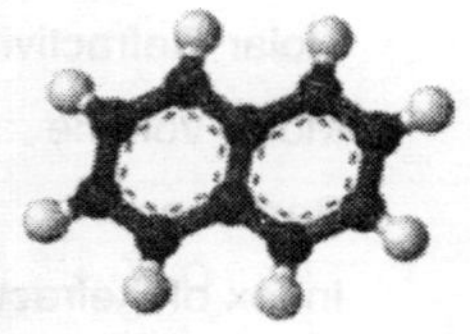

In 1819-1820, at least two chemists reported a white solid with a pungent odor derived from the distillation of coal tar. In 1821, John Kidd described many of this substance's properties and the means of its production, and proposed the name *naphthalene*, as it had been derived from a kind of naphtha (a broad term encompassing any volatile, flammable liquid hydrocarbon mixture, including coal tar). Naphthalene's melting point is 79 to 83 degrees Celsius and has a density of 1.14 g/cm^3. Along with being highly flammable, naphthalene is harmful to self (Xn) and to nature (N).

Naphthalene's chemical formula, $C_{10}H_8$, was determined by Michael Faraday in 1826. The structure of two fused benzene rings was proposed by Emil Erlenmeyer in 1866, and confirmed by Carl Graebe three years later.

Preparation

Naphthalene consists of middle oil distillate of coal tar, from which on standing, a portion frequently crystallizes, a further quantity is obtained by fractionating the same oil after the separation of phenol with sodium hydroxide. When uncrystallizable, (which first distillates) is removed, and the subsequent distillated solidifies. This is naphthalene but is impure and therefore needs purification.

Industrial Preparation of Naphthalene

Coke-Oven Gas At present, the only commercial source of naphthalene is the gas formed in the high-temperature carbonization of bituminous coal in by-product coke ovens. The coal tar which deposits on cooling the hot coke-oven gas is the principal source of naphthalene. Depending on the nature of the coal and on the carbonization conditions, coal tar contains 5-11% of naphthalene, which is the most abundant of all compounds found in coal tar. In the majority of the by-product plants in the U.S., a tar containing about 10% naphthalene is obtained. The tar is distilled and a fraction containing about 40% naphthalene is used for processing.

The amounts of naphthalene recoverable from the water of the final cooler and from the residue from the light-oil distillation are of minor importance. The method of naphthalene recovery from the light-oil residue is, however, the same as from the middle oil of the tar distillation and the two raw materials are therefore combined and used for the production of naphthalene.

Petroleum, although naphthalene occurs in some mineral oils, especially in Borneo and California crudes, the concentration is too low to permit economic recovery. The modern methods of gasoline production (catalytic cracking, re-forming) may, however, make it economically feasible to produce naphthalene as a by-product, and some petroleum companies in the U.S. are said to plan the recovery of naphthalene from cracked petroleum stocks. Fractions boiling in the range 200-230°C contain about 5-15% naphthalene. The higher-boiling fractions contain naphthalene horologes which are more abundant than naphthalene itself. An approximate ratio of naphthalene: methylnaphthalenes: dimethylnaphthalenes is 1: 7 : 10.

Recovery

Crude naphthalene (m.p. 74-78°C.) can be recovered from coke-oven gas at three points of the by-product-recovery system; see Diagram 1. The largest quantities of naphthalene are obtained from the tar which separates by cooling of the gas. The tar is fractionated and practically all naphthalene is present in the cut distilling be-tween 200 and 250°C. ("middle oil," also called "creosote oil" or "chemical oil"). The middle oil is freed from tar bases (by washing with sulfuric acid) and from tar acids (by treatment with sodium hydroxide). The naphthalene is then obtained from the treated oil by crystallization or by distillation. The *crystallization* is carried out in shallow steel pans with discharge holes which are closed by wooden plugs. Depending on the season, it takes from 3 to 10 days until the naphthalene has crystallized from the oil. The plugs are then removed to let the oil drain into storage tanks. The crystals are centrifuged and washed with water in the centrifuge to remove as much of the adhering oil as possible. The naphthalene usually has a melting point of about 68°C and is used for producing refined naphthalene (see next page). A salable crude naphthalene of m.p. 74-76 °C is obtained if the adhering oil is removed by hot pressing instead of by centrifuging. The crystallization plant uses much floor space and labor costs are high. The process is therefore used where the quantity of recoverable naphthalene is not large or where only the older installations are available.

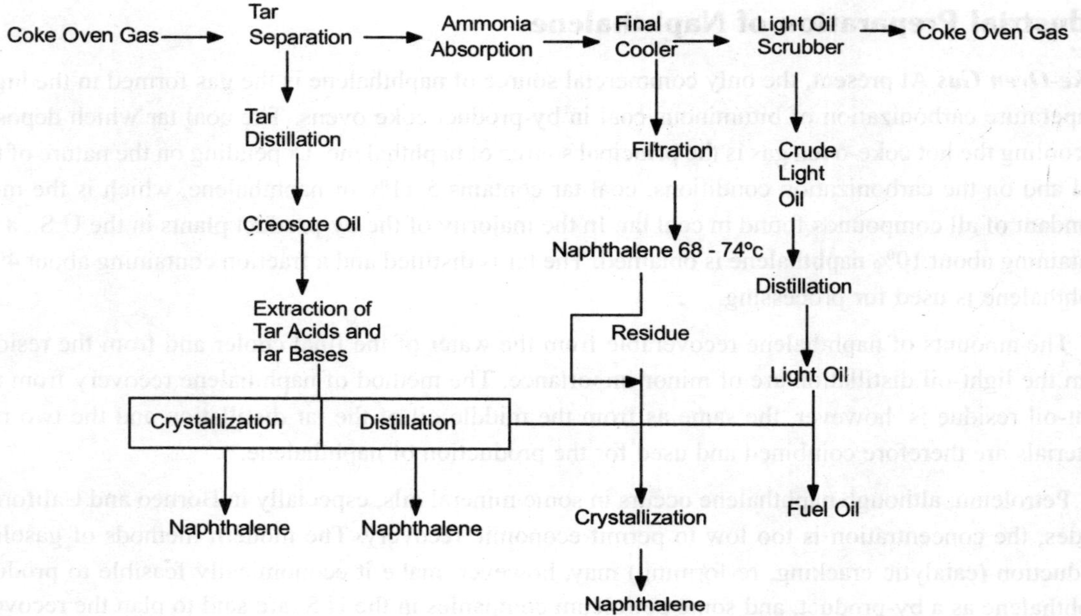

Fig. 33.1. The Recovery of Naphthalene from Coke Oven Gas

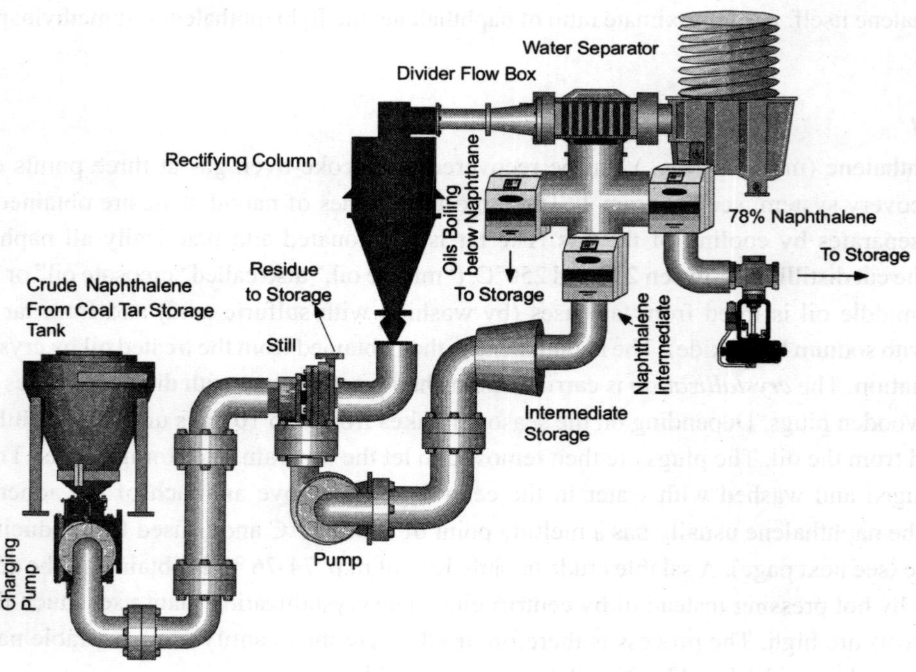

Fig. 33.2. The Refining of Crude Naphthalene from coal Tar

The recovery of naphthalene by *distillation* is shown in Figure 4. The extracted (i.e., washed) middle oil from the tar fractionation is charged to the still, which is operated batch wise. The forerun is added to the crude heavy solvent fraction obtained in the distillation of the tar. The intermediate fraction is stored, and then fractionated again in a separate operation. The final fraction has a melting point of 78°C and therefore represents the highest grade of crude naphthalene. Treatment of this crude naphthalene with concentrated sulfuric acid, washing, and finally re-distillation will yield refined naphthalene of a minimum melting point of 79.6°C. The lines leading from the crude condenser to the intermediate tank and to the naphthalene tank must be well insulated to prevent crystallization of naphthalene. The two tanks are insulated and heated to keep the naphthalene molten for separation of water. The naphthalene distillation could be carried out at reduced pressure and therefore at lower temperatures. It is usually; however, so difficult to keep the vacuum lines and the valves unobstructed by naphthalene deposits that vacuum distillation is rarely carried out.

The water from the final coolers contains some naphthalene which can be obtained by filtration. The melting point is usually 68-74°C and the naphthalene is therefore added to the residue from the light-oil distillation. *Light oil* is obtained from coke-oven gas by scrubbing the gas with wash oil and distilling the light oil from the enriched wash oil. The light oil contains some naphthalene and other higher-molecular weight hydrocarbons which form the residue in the light-oil fractionation. The naphthalene is recovered from the residue either by crystallization or by distillation as described above. Smaller coke-oven plants and some of the larger plants are not equipped for processing of tar. These plants sell the tar and the light-oil residue to tar-distilling plants which recover from the tar all compounds of commercial interest.

Naphthalene is extremely volatile, even far below its boiling point, so that in the coal gas manufacture a little of it passes through the scrubber and purifiers and finds its way into the gas pipes where it occasionally accumulates in sufficient quantity to obstruct the flow of gas, it burns with a luminous flame with a smoky appearance and is utilized in increasing the illuminating power (or carbureting power) of the coal gas in its passage to the burner can be directed into a small metal chamber containing solid naphthalene which is wormed by the heat of the gas flame. A little naphthalene volatilizes and mixes with the coal gas adding considerably to its luminosity.

Naphthalene acts as vermin* killer and a mild antiseptic; but its chief industrial uses are in the manufacture of phthalic anhydride, from which many dye stuffs are made.

The formula of naphthalene is $C_{10}H_8$ its chemical properties resembles benzene(see chemistry of naphthalene) it can be sulphonated halogenated and nitrated in the same way and produces similar products.

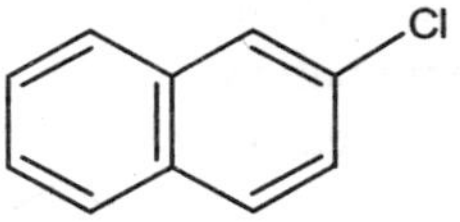

2-chloronaphthalene
$C_{10}H_7Cl$

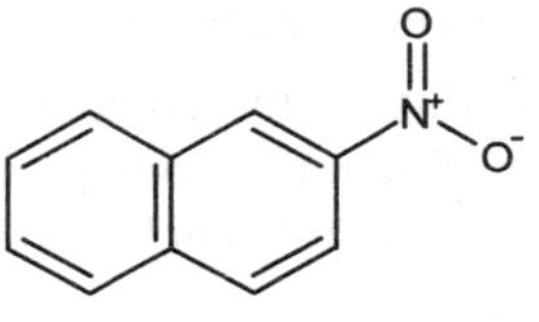

2-nitronaphthalene
$C_{10}H_7NO_2$

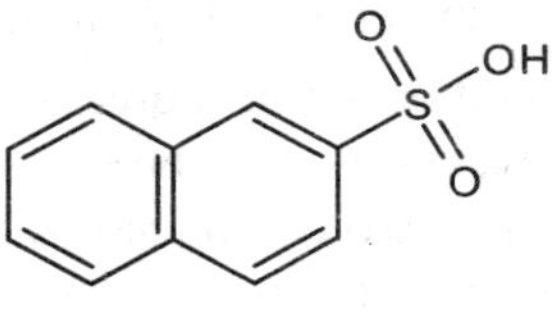

naphthalene-2-sulfonic acid
$C_{10}H_7SO_3H$

*pests.

2,3-dichloronaphthalene $C_{10}H_7Cl_2$

2,3-dinitronaphthalene $C_{10}H_7(NO_2)_2$

naphthalene-2,3-disulfonic acid $C_{10}H_7(SO_3H)_2$

Naphthalene produces amino compounds (See derivatives of benzene) these naphtha amines like amino derivatives of benzene can be diazotized.

Naphthalene also forms additive products with hydrogen and chlorine Tetrahydronaphthalene ($C_{10}H_{12}$) and Decahydronaphthalene ($C_{10}H_{18}$) are obtained by the catalytic reduction of naphthalene by hydrogen under pressure in the presence of nickel. They are called terilin and decalin, respectively and are used as substitutes for turpentine Naphthalene dichloride is prepared by adding hydrochloric acid to a mixture of naphthalene and potassium chlorate (prepared moist and then dried) naphthalene tetrachloride is prepared by passing chlorine into a solution of chloroform and naphthalene. The first is a yellow liquid and second a solid.

Structure of Naphthalene

According to present views, Naphthalene contains two benzene nuclei which is not distinct, but have two carbon atoms in distinct, but have two carbon atoms in common, these phenomenon is known as **condensed nuclei.**

From the close analogy existing between the naphthalene and benzene nucleus and this view is apparently confirmed by the behaviour of naphthalene on oxidation, for it readily yields phthalic acid, when oxidized by air in presence of vanadium pentoxide.

Phthalic acid contains a carbon skeleton of atoms, 6 in the benzene ring and 2 in the ortho position outside it

Now if the carbon atoms of naphthalene outside the benzene ring are joined as to complete a second ring of 6 carbon atoms and if to each of the 8 outlaying carbon atoms of the two rings a hydrogen atom is attached, such a structure would give the necessary number of carbon and hydrogen atoms required by the formula for naphthalene.

It would account, moreover for the stability of the compound and its structural formulae represents two benzene rings, yet different from any previous combination, in as much as two carbon atoms are common to the two nuclei, and this is the reason, for naming the structure as condensed nuclei.

Let us now examine the experimental evidence upon which the structure rests —

We may first refer to the reaction of synthesis of naphthol, by Fittig's reactions, by heating phenylisocrotonic acid.

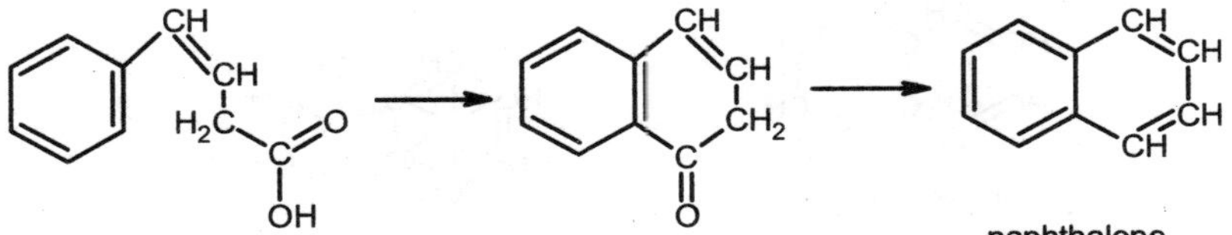

Phenylisocrotonic acid → naphthalen-1(2*H*)-one → naphthalene

Another synthesis has been affected by passing phenylbutylene or, phenylbutylene bromide, over red – hot soda – lime.

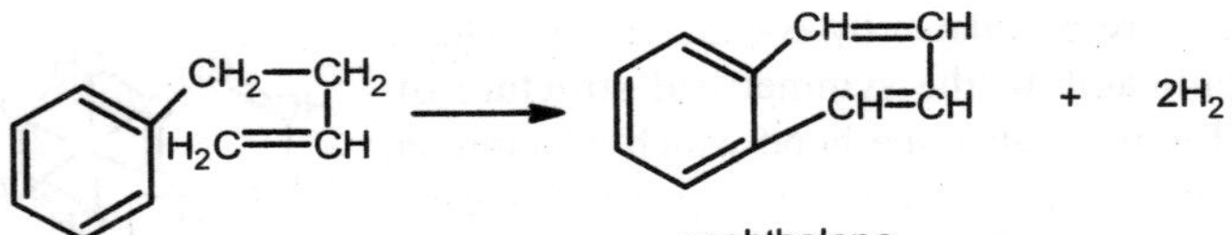

but-3-en-1-ylbenzene
Phenylbutylene → naphthalene + $2H_2$

Both synthesis, however, only point with certainty to the presence of only one benzene ring in the compound.

More conclusive and complete is the experimental evidence of Graebe.

It has been stated that when naphthalene is oxidized, it forms Phthalic acid, which is a benzene derivative, in the same way of nitro – naphthalene is oxidized, it yields nitro – phthalic acid. The benzene ring (A), stamped, as it were with the nitro group remains intact. If, however, the nitro – compound is reduced, the product naphthylamine gives phthalic acid on oxidation. In this case it is the stamped benzene ring (A), which has been destroyed, the second benzene ring (B) has been preserved. The changes are represented in the reaction equation below:

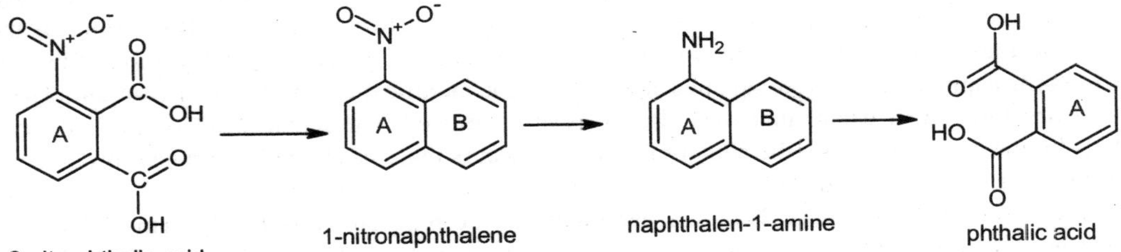

3-nitrophthalic acid → 1-nitronaphthalene → naphthalen-1-amine → phthalic acid

It follows that naphthalene is a combination of two benzene nuclei, that nation of two 6 – membered rings, consequently, the two carboxyl group of phthalic acid must be ortho to one another, otherwise one of the rings of naphthalene would consist of more than six carbon atoms, hence the xylene, which gives rise to Phthalic acid.

In naphthalene, as in benzene we are met by the difficulty of disposing of the fourth carbon bond. The formula with alternate double bond, proposed by Kekule for benzene, was applied by Erlenmeyer to naphthalene, and seems a natural and logical consequences of the relation of the two hydrocarbons—

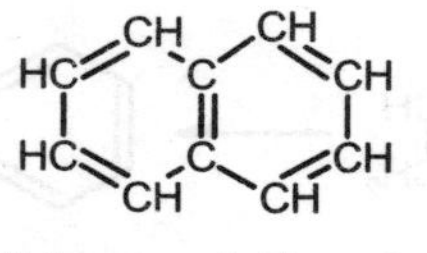

Erlenmeyer's Formula for Naphthalene

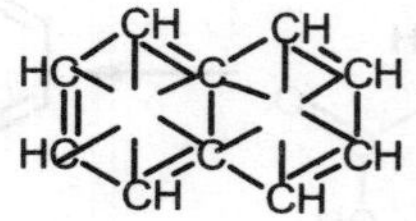

Bamberger's centric formula

But a similar set of objections to those advanced against the olefinic (Kekule) formula for benzene may be brought against Erlenmeyer formula for naphthalene.

A modification of the centric formula was therefore postulated by Bamberger but where as the centric formula might possibly be applicable to the symmetrical structure of benzene, the problem of naphthalene is very different, since the bond which is common to both nuclei is now disrupted. An early attempt to rectify this mistake was made by condensing the one centric nuclei with one of the Kekule thus:

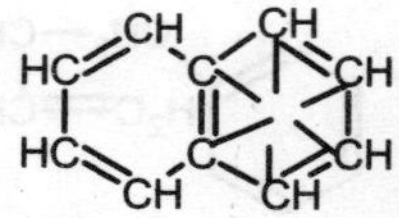

It will be seen that the latter is not a benzenoid ring, since it contains only two double links and should therefore be definitely olefin in character like benzene dihydride more commonly cyclohexadiene C_6H_8.

Experiments shows that naphthalene is in fact more unsaturated in character than benzene and the presence of two double bonds is confirmed by the addition of hydrogen to form first the 1 – 4 – dihydride and then the 1: 2: 3: 4 – tetrahydride, one ring being completely reduced before the other is attached; also the fact that molecules of ozone affords additional proof of the presence of the two double bonds.

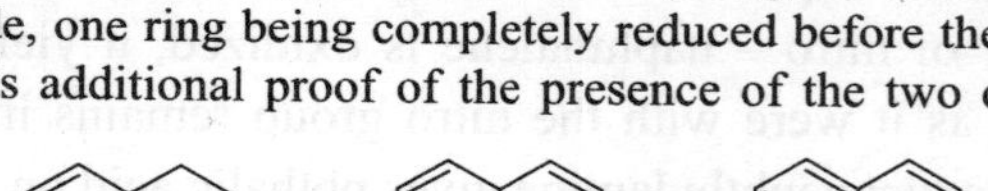

Moreover, a study of the products of naphthalene shows that there is no evidence of any difference of any difference in the constitution of the two nuclei such as results from the modified centric formula. They seem to have the same properties. No difference has been noted between the α and β positions of the other ring until substitution has accurately fallen into place. As might be expected the condensation of two aromatic nuclei to form naphthalene is accompanied by some slight modification in chemical reaction. The greater degree of unsaturation of naphthalene has already been mentioned. The replacement of the hydroxyl of the naphthols by the amino group by the action of ammonia occurs more easily and at a much lower temperature than in the case of phenol, and the naphthols until the phenols, react easily with alcohol in the presence of hydrogen chloride to form ethers. Now the naphthals can be reduced by sodium in amyl alcohol to give tetrahydrides which more closely resemble phenol but at the same time other products are formed which have completely lost their phenolic character and are in aromatic alcohols (similar to benzyl alcohol)

Thus α - naphthol gives two compounds represented by the following formulæ.

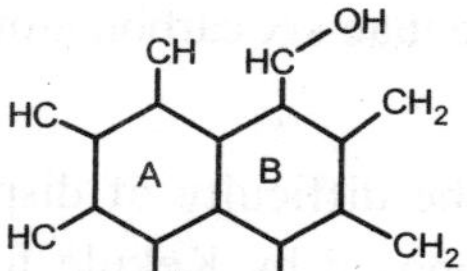

ac - Tetrahydronaphthal

1

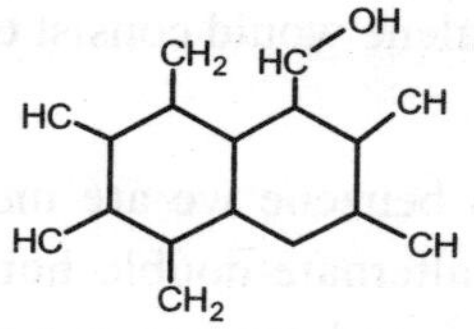

ar - tetrahydronaphthol

2

In structure 1 the nucleus marked (A) retains its aromatic character, while that marked B has become aliphatic though still cyclic and is therefore known as alicyclic (ac) Ac – tetrahydronaphthol is a secondary alcohol. In the second structure reduction has taken place in ring A, which becomes aliphatic, while B remains aromatic and in the fact that the compound as tetrahydro naphthal has reverted more closely to the character of phenol. The two naphthols as the two naphthylamines behave similarly in this respect each yielding 2 tetrahydrides, the α - compounds giving ar – derivatives and the β compounds ac – derivatives as the chief products. Thus it appears that each of the two nuclei can revert to the benzenoid by the reduction of the other. We may infer therefore that a hybrid structure like that of the modified centric formula is not justified. The centric formula for naphthalene has therefore been abandoned; nor it is favoured for benzene itself, since it conflicts with the planer configuration of the 12 atoms of benzene that has been deduced from X – Ray analysis. Hence formulæ of the Kekule type have not been superseded, since they are superior to all the others and hence devised, and can be used to illustrate the chemical of the aromatic compounds. They are not however not altogether satisfactory, since they do not account for all the physical properties. For example, the distance between the adjacent atoms in a molecule can now be measured accurately, and it has been found that its value is neither that of single bond nor that of a double bond, but lies between them, and is moreover the same at all parts of the molecule.

The modern view is that an entire new type of structure, called as resonance structure is needed to interrupt the behaviour not only of aromatic nuclei, but also of a great many other groupings, such for example, as the carboxyl group. The resonance structure of benzene appears to involve a modification of both double and single bonds of Kekule structure to something intermediate in character but completely different from the centric bonds of the Armstrong Bayer formula. No satisfactory method has yet been devised of formulating a resonance structure and its discussion is beyond the scope of this book, for example practical purposes we can still use the Kekule' type formula or just the plain hexagons.

It should be noted that practical reduction of naphthalene to a dihydride leaves in one ring only one double bond which is now definitely olefinic in nature, whilst further reduction to tetralin gives a product resembling it its properties an aromatic nucleus with two side – chains in the ortho substituent in aromatic nucleus are easily prepare corresponding derivatives of naphthalene.

Physcial Properties

Molecular Formula	$= C_{10}H_8$
Formula Weight	= 128.17052
Melting Point	= 80.2 °C
Boiling Point	= 218 °C
Composition	= C(93.71%) H(6.29%)
Molar Refractivity	= 44.09 cm^3
Molar Volume	= 123.5 cm^3
Parachor	= 311.1 cm^3
Index of Refraction	= 1.632
Surface Tension	= 40.2 dyne/cm

Density	= 1.037 g/cm^3
Dielectric Constant	= 2.82 ± 0.1
Polarizability	= 17.48 cm^3
Monoisotopic Mass	= 128.0626 Da
Nominal Mass	= 128 Da
Average Mass	= 128.1705 Da
Log P	= 3.45

Chemical Properties

Reactions

The reactions of naphthalene are of three distinct types: substitution, addition, and rupture of the rings. All three types of reactions are industrially important.

Oxidation. The largest part of the naphthalene production is used for making phthalic anhydride. This important chemical became available at low cost when during World War I the air oxidation of naphthalene in the presence of a vanadium catalyst was developed. Naphthalene vapor mixed with air is passed over the catalyst at 400—460°C. and phthalic anhydride is separated from the exit gas in good yield. Phthalic anhydride is a starting material for alkyd resins, for plasticizers such as dibutyl phthalate, and the anthraquinone dyes

Hydrogenation. Hydrogenation of naphthalene proceeds more readily than hydrogenation of benzene, with formation of di-, tetra-, hexa-, octa-, and decahydronaphthalenes. Only tetrahydro- and decahydronaphthalene are important commercially (see under "Derivatives").

Sodium metal in ethyl alcohol reduces naphthalene to the 1,4-dihydride and in amyl alcohol to the 1,2,3,4-tetrahydride. Hydrogen at high pressures and temperatures gives mainly tetrahydronaphthalene. In the presence of catalysts.hydrogenation proceeds at lower-pressures and temperatures. Platinum and palladium are active at room temperature and low pressure. Nickel catalysts are very active and are used commercially. The kinetics were investigated by Lozovo. Besides nickel, mixed catalysts (Ni-Co-Cu-Mn) or molybdenum catalysts are used in certain cases. The temperature is an important factor in catalytic hydrogenation. Above 400°C. destructive hydrogenation takes place, and the hydrogenated naphthalene decomposes to aromatic hydrocarbons, cyclohexanes, and cyclopentanes. The destructive hydrogenation of naphthalene was used in Europe to produce high-grade motor fuel.

Sulphur and sulphur compounds deactivate nickel catalyst~ rapidly. Technical naphthalene contains approximately 0.25% sulphur and must therefore be purified. A commercially used method consists in treating molten naphthalene with about 1% sodium metal followed by distillation of the naphthalene under reduced pressure. Removal of sulphur compounds was reported recently by partial chlorination of crude naphthalene at about 90°C. Thianaphthene reacts preferentially and the high-boiling chlorothianaphthenes form the residue in the subsequent distillation. J. Varga found that crude naphthalene can be hydrogenated at 120 atm. with a mixture of sulphur and molybdic acid as catalyst.

Sulphonation. Concentrated sulphuric acid yields predominantly the 1 – naphthalenesulphonic acid at low temperature and the 2-naphthalenesulphonic acid at high temperature. Kaphthalenepolysulphonic acids (di-, tri-, and tet.ra-) are obtained by using sufficient fuming sulphuric acid containing up to 70% sulphur trioxide. The naphthalenesulphonic acids *(q.v.)* are used primarily as intermediates for synthetic organic dyes. Alkylnaphthalenesulphonic acids are used for making synthetic detergents, wetting agents and emulsifiers. N aphthalenesulphonic acids can be converted to the important naphthols by fusion with an alkali (sodium hydroxide). The alkali salt of the naphthol which is formed is dissolved in water, and the naphthol is precipitated by passing an acidic gas (carbon dioxide, sulphur dioxide) into the solution. Regarding recent developments in sulphonation.

Nitration Naphthalene reacts easily with nitric acid, and mono- and polynitronaphthalenes are formed readily. Reaction of naphthalene with a mixture of concentrated nitric acid and 80% sulphuric acid at 50°C. yields I-nitronaphthalene and only very small amounts of 2-nitronaphthalene The ll-nitronaphthalene cannot be made by nitration but has to be prepared by indirect methods. Reduction of I-nitronaphthalene gives 1,1-naphthylamine, an important intermediate for organic dyes. The 2-naphthylamine has to be made by indirect methods such as from 2-naphthol (Bucherer reaction).

Experiment: 1. Add to a few grams of 1-naphthalamine a quantity of insufficient to dissolve hydrochloric acid and pour off the mixture, add a solution of ferric chloride, a blue solution is obtained slowly. To the separated mixture, add alcohol and a few drops of acetic acid and add drop by drop sodium nitrite a yellow solution is obtained, which on addition to dilute hydrochloric acid changes to violet

Halogenation. In the absence of catalysts, chlorine reacts mainly by addition, to give naphthalene di- and tetrachloride. On heating, these compounds lose hydrogen chloride and form I-chi 01'0- and 1,4-dichloronaphthalene. In the presence of a catalyst, such as ferric chloride, substitution takes place. All the hydrogen atoms of the naphthalene molecule can be replaced. Chlorination of naphthalene dissolved in chlorobenzene yields l-chloronaphthalene, and in carbon tetrachloride solution yields 1,4- and 1,5-dichloronaphthalenes. Chlorination of molten naphthalene gires polychloronaphthalenes up to octachloronaphthalene. Commercially important are the products consisting essentially of hexa-, hepta- or octachloronaphthalenes, which are nonflammable waxy materials of high melting points (100-190°C.) and high dielectric constants

Alkylation. The alkylation of naphthalene proceeds readily with many kinds of alkylating agents such as alkyl halides, olefins or alcohols, in the presence of various catalysts, for example sulphuric acid, aluminum chloride and other metal halides (ferric chloride, zinc chloride, etc.), hydrofluoric acid and phosphoric acid. Aluminum chloride is the most active catalyst but it forms a complex with alcohols and must then be used in large quantities (more than 100 mole % of the alcohol). A reactive alkylating agent should be used because naphthalene alone reacts in the presence of aluminum chloride, forming dinaphthyls preferentially if the alkylating agent (for example methyl chloride) has a relatirely low reactivity. Reactions with aluminum chloride are usually accompanied by formation of resinous by-products, while other catalysts, such as sulphuric or hydrofluoric acid, give alkylates of good color which are practically free of resinous matter. Sulphuric acid is the preferred catalyst for alkylations with alcohols. If acid of sufficient strength is used, alkylation and sulphonation (in the ring) take place simultaneously and alkylnaphthalenesulphonic acids are produced. This method is used commercially on a large scale in the manufacture of detergents.

Alkylnaphthalenes are more readily alkylated than naphthalene and therefore mixtures of mono- and polyalkylated naphthalenes are formed in practically all alkylations. It is, however, important to produce alkylnaphthalenes with a limited total number of carbon atoms in the side chains to obtain in subsequent sulfonations the monosulphonic acid of high detergent activity. Alkylnaphthalenemonosulphonic acids and their alkali salts having up to ten carbon atoms in the alkyl groups are water-soluble but the solubility decreases sharply when this number is exceeded. Examples are the sulphonic acids derived from the amylnaphthalenes, which are produced from naphthalene and pentenes. If the number of carbon atoms in the alkyl groups is appreciably higher than 10, the monosulphonic acids are soluble in organic solvents and in petroleum oils. Examples are the kerylnaphthalenes made from keryl chloride and naphthalene in the presence of a catalyst. "Keryl chloride" is an alkyl chloride of about 14 carbon atoms produced by monochlorination of kerosene.

Special alkylating agents of technical interest are unsaturated acids and alcohols. The former react at the double bond; thus oleic acid gives naphthylstearic acid. Unsaturated alcohols (for example oleyl alcohol) can react at the double bond or at the hydroxyl group. When the reaction takes place at the double bond, an alkylnaphthalene is formed which contains an alcoholic hydroxyl in the side chain. Sulphuric acid can react with these compounds at the naphthalene ring to form a sulphonic acid and at the hydroxyl group to form a sulphuric ester.

The scientific and patent literature on alkylation of aromatic hydrocarbons is very extensive.

Alkylated naphthalenes are almost exclusively used for manufacturing sulphonic acids and their salts. The sulphonates have a wide field of application as surface active agents.

Chloromethylation. Naphthalene in glacial acetic acid solution reacts with paraformaldehyde and concentrated hydrochloric acid in the presence of sirupy phosphoric acid with formation of 1-(chloromethyl) naphthalene. The (chloromethyl)-naphthalene is a lacrimator and vesicant* and must be handled with caution. Using paraldehyde in place of paraformaldehyde, chloroethylation can be effected, especially with methoxynaphthalenes. The (chloroethyl)naphthalene splits off hydrogen chloride readily and vinylnaphthalenes are formed.

Diels-Alder Reaction. Naphthalene does not undergo the Diels-Alder reaction but vinylnaphthalenes react easily as shown in the following example with maleic anhydride:

HC=CH$_2$ + HC—C(=O)—O—C(=O)—CH → (VIII) → (IX)

1-vinylnaphthalene (VII)   furan-2,5-dione   (VIII)   (IX)

*forming blisters.

The compound (VIII) rearranges in boiling acetic acid to the isomer (IX).

Friedel-Crafts Reaction. Naphthalene reacts with alkyl halides (except methyl halides), with acid chlorides and with acid anhydrides in the presence of anhydrous aluminum chloride. The products are alkylnaphthalenes or alkyl naphthyl ketones, respectively. Naphthalene is very reactive and, if the alkylating agent is relatively unreactive, the naphthalene reacts readily in the presence of aluminum chloride with formation of dinaphthyls.

Naphthols are the naphthalene representatives of the phenols, and share their general characters, although they exhibit some minor differences in chemical behaviour. The hydroxyl group is more readily replaced by ammonia, as we have seen in the formation of the naphthylamines and the ethers are prepared by the combined action of alcohol and sulphuric acid after the fashion of ethyl ether. A number of the α and β-naphthol monosulphonic acids, as well as disulphonic acids, are used in the manufacture of azo-dyes.

α-NaPhthol, $C_{10}H_7(OH)$, is obtained by fusing naphthalene monosulphonic acid with caustic soda or from oc-naphthyl. amine by means of the diazoreaction Its synthesis from phenyl isocrotonic acid. a-NaPhthol, is sparingly soluble in water and has a phenolic smell, it is volatile. Ferric chloride gives a violet of dinaphthol, which is an oxidation product and has the following structure in the right.

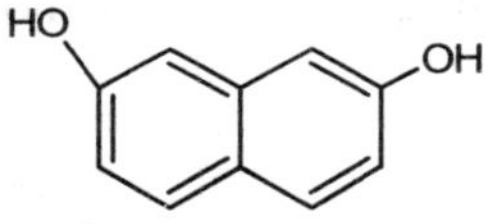

HO OH

Experiment: Dissolve a little a-naphthol in very dilute alcohol an to the hot solution add ferric chloride solution, the violet coloured di naphthol is thrown down, to the another portion of the same sample when cold add a solution of sodium hypochlorite a green colouration is produced when further sodium hypochlorite is added the colour changes to blue.

Naphthol forms with nitrous acid an *a-* and β-nitroso-α,-naphthol,or naphthaquinonoxime

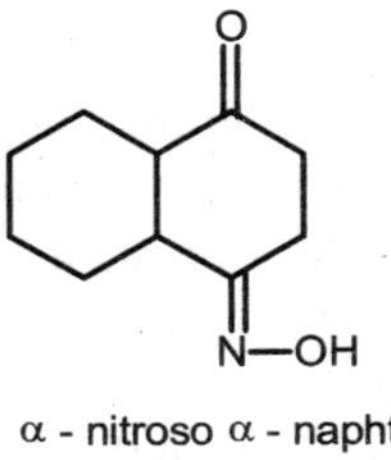

O
N—OH
α - nitroso α - naphthol

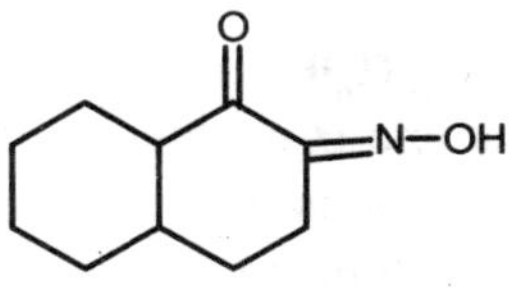

O
N—OH
β - nitroso α - naphthol

α-Naphthol melts at 94° and boils at 280°. By the action of nitric acid it is converted into nitro-derivatives, which correspond in chemical properties and colour to picric acid, and like picric acid are used as dyes for silk and wool.

Martius' yellow, $\mathbf{C_{10}H_5(NO_2)_2ONa + H_2O}$, is the sodium salt of dinitro-ex-naphthol, and is obtained by the action of strong nitric acid upon ex-naphthol.

Naphthol yellow is the potassium salt of dinitro-ex-naphthol-sulphonic acid, and is prepared by the action of nitric acid on α-naphthol-trisulphonic acid.

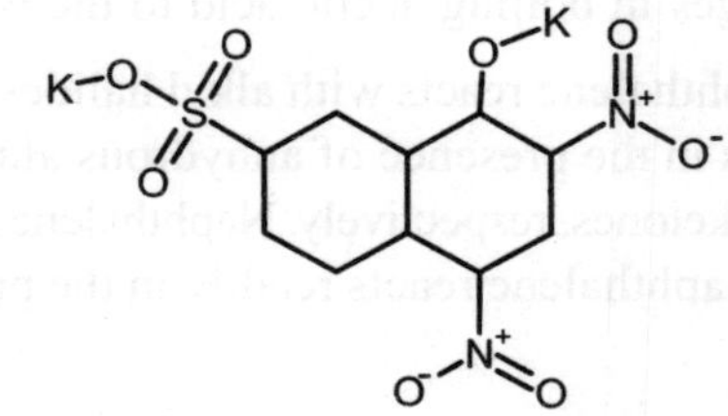

β-Naphthol, $C_{10}H_7(OH)$, is prepared from β-naphthalene sulphonic acid by fusion with caustic soda, which is the commercial process, or by the action of nitrous acid on β-naphthylamine. It crystallizes in leaflets; it melts at 122° and boils at 286°.β-Naphthol is coloured green by ferric chloride, and also yields a precipitate of β-dinaphthol of a similar constitution to the α-compound. β-Naphthol and especially the sulphonic acid derivatives have an extensive application in the production of azo-dyes.

Naphthaquinones - α Naphthaquinone $C_{10}H_6O_2$ corresponds exactly to benzoquinone, both in its properties and method of preparation.

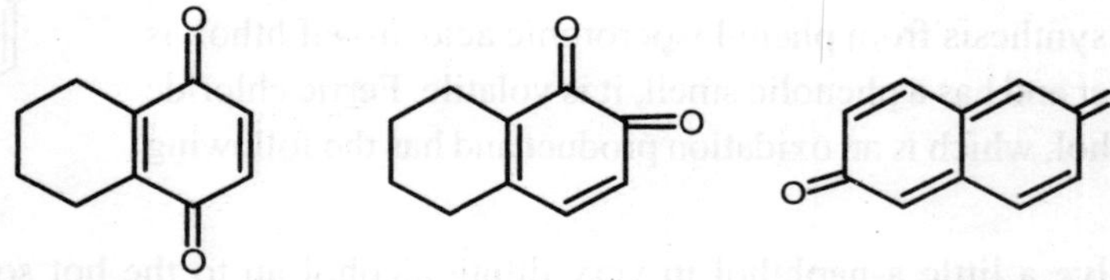

a - Naphthoquinone β- Naphthoquinone amphi- Naphthoquinone

It is obtained by oxidising α-naphthylamine, 1-4-diamino- or dihydroxy-~aphthalene, or 1-4-aminonaphthol, and also, though less readily, by the oxidation of naphthalene itself with chromic acid in acetic acid solution. It cfystallises in yellow plates which sublime at 1000 and are volatile in steam. It possesses, moreover, an unmistakable quinone smell. Sulphurous acid reduces it to 1-4 -dihydroxynaphthalene.

β-Naphthaquinone, $C_{10}H_60_2$ corresponds to ortho-quinone among the benzene derivatives. It is obtained by oxidising β-amino-α-naphthol with ferric chloride. Like ortho-quinone it crystallizes in red needles, is without smell, and is non-volatile. It also resembles phenanthraquinone It is termed an *ortho-quinone,* or ortho-diketone, to distinguish it from the ordinary, or *para-quinone.* By the action of hydroxylamine each of the qui nones yields a quinonoxime, or nitro so naphthol, which are identical with the compounds obtained by the action of nitrous acid on α-naphthol.

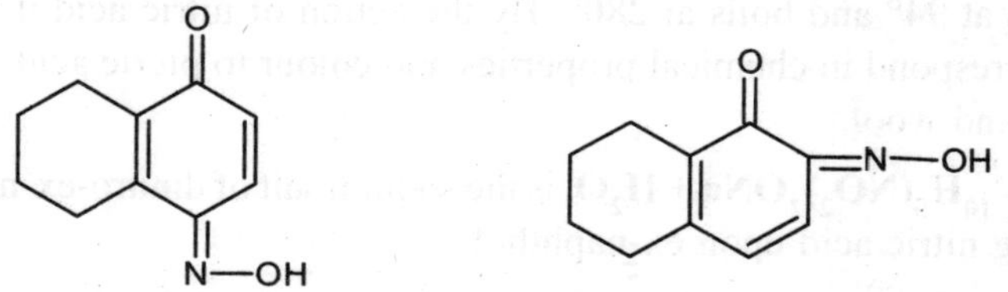

α - Naphthoquinonoxime or α - nitroso α - naphthol

β - Naphthoquinonoxime or β - nitroso α -naphthol

Uses

In the second half of the 19th century the main use for naphthalene was as an insecticide and as a repellent for clothes moths. The U.S. consumption of refined naphthalene for protecting fabrics against clothes moths and carpet beetles was 23 million pounds in 1940. In agriculture naphthalene is used as a fumigant against the carrot fly and the ,vire worm, further as louse powder for live stock and poultry. The agricultural consumption of naphthalene in the U.S. was estimated as 3.2 million pounds in 1943. Shepard gives a detailed description of the uses of naphthalene and naphthalene compounds as insecticides.

At present the largest use of naphthalene is as a chemical raw material for a wide range of organic chemicals as shown schematically in Diagram 2.

Oxidation yields phthalic anhydride, which is one of the most important raw materials for alkyd resins and for plasticizers. The present (1951) scarcity of naphthalene is due to the rapidly expanding plastics industry and the consumption of naphthalene for oxidation to phthalic anhydride accounts for about two thirds of the supply.

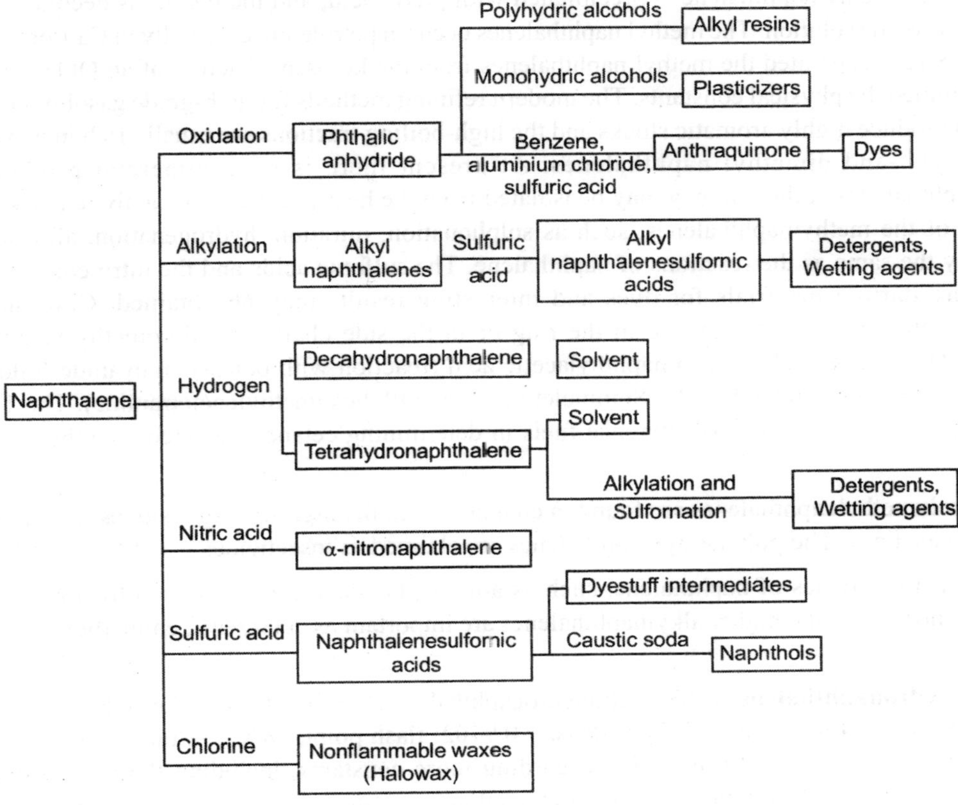

Fig. 33.3. The Different uses of Naphthalene

Derivatives

The naphthalene horologes are naphthalenes with alkyl groups such as methyl- or isobutylnaphthalene. The number of isomers is very large. Monosubstituted naphthalenes,exist only in two isomeric forms, 1- and 2-, but the disubstituted naphthalenes can exist in ten isomeric forms if the alkyl groups are alike and in fourteen when they are different. The alkylnaphthalenes can be synthesized by treating a bromonaphthalene with an alkyl iodide and sodium metal, or from naphthalene and an alkyl chloride in the presence of aluminum chloride or other catalysts. An exception is 2-methyl-naphthalene, which could not be made by either of these two methods.

Methylnaphthalenes, $C_{11}H_{10}$ or $CH_3C_{10}H_7$. The physical constants of I-methyl-naphthalene are: m.p., -30.77°C.; boiling point 244.78;°C.; density, 1.0163;. The physical constants of 2-methylnaphthalene are: m.p., 34.44°C.; boiling point, 241.14°C.; density, 0.99045; refractive index 1.60192. The 1- and 2-methyl-naphthalenes occur in coal tar in concentrations of about 1 and 1.5% respectively and can be separated from a tar distillation cut boiling between 230 and 250°C. The 2-methylnaphthalene can be obtained from the cut by crystallization due to its high melting point. The low-melting I-methylnaphthalene is precipitated with picric acid, and the picrate is decomposed with sodium hydroxide solution. The methyl naphthalenes occur in petroleum, especially in California crudes. Mair and Streiff separated the methyl naphthalenes from the kerosene fraction of an Oklahoma crude and determined the physical constants. The modern refining methods for high-grade gasolines (catalytic cracking) produce highly aromatic stocks and the high-boiling fractions are usually rich in naphthalene and methyl- -and dimethyl naphthalenes. At present there is no commercial production of methylnaphthalenes, although they may be isolated from the heavy residues of catalytic cracking. The reactions of the methy\naphthalenes, such as sulphonation, nitration, hydrogenation, alkylation, are practically the same as the reactions of naphthalene. The sulfonic acids and the nitro compounds can be used as starting materials for dyes and interesting results may be obtained. Chlorination of methylnaphthalenes may take place in the ring or in the side chain. l-(Chloromethyl)naphthalene, $ClCH_2C_{10}H_7$, can be converted to naphthylacetic acid (reaction with potassium cyanide followed by hydrolysis) or to 1-naphthaldehyde (Sommelet reaction with hexamethylenetetramine). The 1- and 2-methylnaphthalenes are used as reference fuels in determining cetane and cetene numbers of Diesel oils.

The polymethylnaphthalenes are found in coal tar and in the high-boiling residues from cracked or reformed gasolines. The polymethylnaphthalenes are claimed as insecticides.

Higher homologues of naphthalene, such as amylnaphthalenes, are produced by alkylation. The alkali Sulphonates of the higher alkylnaphthalenes are important as detergents, emulsifiers and wetting agents.

Tetrahydronaphthalene (1,2,3,4-tetrahydronaphthalene, Tetralin); $C_{10}H_{12}$; formula weight, 132.20; m.p., -35.80°C.; boiling point 207.57°C density 0.9702; flash point, 78°C.; refractive index, 1.54135; vapor pressure at 25°C., 0.40 mm. Hg. Regarding more constants, including the compositions and densities of naphthalene-tetrahydronaphthalene solutions, vapor pressures of the systems benzene-tetrahydronaphthalene, hexane-tetrahydronaphthalene, and benzene-naphtha-lene-tetrahydronaphthalene,. Tetrahydronaphthalene is miscible with liquid sulphur dioxide, but sulphur dioxide vapor does not dissolve in tetrahydronaphthalene. It is a good solvent for sulphur, iodine, aromatic hydrocarbons (naphthalene, anthracene, etc.), and for oils, fats, and resins. Due to the presence of only one ammatic ring, tetrahydronaphthalene does not form a picrate.

A convenient method for indicating subtitutions at undetermined positions in the alicyciic or aromatic rings employs *ac-* (alicyclic) to denote that the substitution is in the hydrogenated ring. Similarly, *ar-* (aromatic) shows that the substituent is in the aromatic ring. For example, ac-chlorotetrahydronaphthalene and ar-chlorotetra hydronaphthalene.

The reactions of tetrahydronaphthalene resemble those of alkyl benzenes. Chlorine or bromine substitute in the alicylic ring at high temperatures but at low temperatures in the presence of iodine, chlorine forms 5- and 6-chlorotetrahydronaphthalene. A mixture of concentrated nitric and sulphuric acids gives 5- and 6-nitrotetra-hydronaphthalene.

Oxidation with oxygen or air at less than 90°C. gives tetrahydronaphthalene hydro peroxide which can be used as additive for ignition acceleration in Diesel oils (5). Tetrahydronaphthalene is oxidized by potassium permanganate to phthalonic acid (o-carboxyphenylglyoxylic acid), C_6H_4((COOH)COCOOH, and by nitric acid to phthalic acid. Vapor-phase oxidation over a vanadium catalyst yields phthalic anhydride. Hydrogenation with hydrogen under pressure in the presence of a nickel catalyst at 150-200°C. gives a mixture of *cis* and *trans* decahydronaphthalene. Tetrahydronaphthalene dehydrogenates to naphthalene at 200-300°C. in the presence of a catalyst. Thermal dehydrogenation takes place at about 450°C. and is accompanied by decomposition (cracking) of the tetrahydronaphthalene to aromatic hydrocarbons such as toluene and xylene. At temperatures of 500-550°C. in the presence of a chromium catalyst the formation of benzene, toluene, and xylene is the main reaction.

Table 33.1: The Properties of Tetrahydronaphthalene

	Cis isomer	Trans isomer	Commmrcial product
M.p., °C	–43.26	–30.68	–124
B.p., °C	194.5	185.8	189-191
d_4^{20}	0.8967	0.8700	0.8857
nd	1.480820	1.469620	1.475315
Speific heat (15-18°C), cal./gram.	—	—	0.395
Heat of combustion of liquid, kg.-cal./mole	—	—	1.503
Flash point, °C	—	—	57—58.5

Tetrahydronaphthalene is alkylated at the benzene ring by alkylating agents (alkyl chlorides, olefins or alcohols) in the presence of catalysts .

Tetrahydronaphthalene is used mainly as a solvent especially in the paint and lacquer industry. It is the best solvent for naphthalene (Fig. 33.3) and is used (mostly in European practice) for removing and preventing naphthalene deposits in gas-distributing systems. The high boiling point and the high solvent power make tetrahydronaphthalene a suitable absorbent for vapor recovery of compounds such as benzene, acetone, ethyl alcohol and ethyl acetate.

Decahydronaphthalene (bicyclo[4.4.0]decane, Decalin) $C_{10}H_{18}$, formula weight 138.24, is a colourless liquid, which burns with a luminous smokeless flame. The structure of the compound indicates that it should exist in two isomers, a cis and a trans isomer. In the cis isomer both the hydrogens on the carbon atoms common to both rings are above the plane of the molecule while in the trans isomer one hydrogen is above and one hydrogen below the plane of the molecule. The isomers have been obtained in a pure state and the physical constants determined; see Table above. The commercial decahydronaphthalene is a mixture of both isomers.

Decahydronaphthalene is insoluble in water and in liquid ammonia, only slightly soluble in methyl and ethyl alcohols, and miscible with butyl alcohol. It is only slightly soluble in liquid sulphur dioxide, but gaseous sulphur dioxide is absorbed in decahydronaphthalene in appreciable amounts. Vapors of organic compounds such as benzene, ethyl alcohol, ethyl acetate, and acetone are soluble in decahydronaphthalene, Which can therefore be used for recovery of these vapors.

At elevated temperatures decahydronaphthalene is subject to autoxidation; small amounts of 1,8- or 2,7-diaminonaphthalene, however, inhibit the oxidation. Decahydronaphthalene is chemically inert; it cannot be sulfonated or nitrated under ordinary conditions. Halogens substitute: chlorine in the presence of small amounts of iodine forms mono- and polychlorodecahydronaphthalene. Decahydronaphthalene can be dehydrogenated to tetrahydronaphthalene and to naphthalene. Destructive hydrogenation (heating with hydrogen under pressure in the presence of an aluminum iron oxide catalyst at 440°C.) converts decahydronaphthalene into aliphatic and aromatic hydrocarbons.

Decahydronaphthalene is a very good solvent for many organic compounds and is used extensively in the manufacture of lacquers and paints as solvent and thinner. It is the best known solvent for silicone materials.

Naphthoic Acids.-Naphthalene forms a series of carboxylic or naphthoic acids, which, however, afford fewer points of interest than the corresponding benzene derivatives, for they rarely occur in the products of plant life, nor have they as yet found any practical application. .

α-Napthonic acid, $C_{10}H_7$,COOH, is obtained by hydrolysis of the cyanide, C1oHiCN, and by the action of chloroformamide on naphthalene in presence of aluminium chloride, precisely as in the synthesis of benzoic acid from benzene (p. 487)' It melts at 160°.

β,-Napthonic acid is prepared from the (?,-cyanide and by the oxidation of the b-alkylnaphthalenes. It melts at 182°, Both acids resemble benzoic acid and give naphthalene when distilled with soda-lime.

Acenaphthene, $C_{12}H_{10}$, is a curious derivative of naphthalene which is found in coal-tar. It yields on oxidation a naphthalene dicarboxylic acid known as *naphthalic acid,* which contains the two carboxyl groups in the peri-positions. From the fact that this acid gives an anhydride on heating, and that acenaphthene is obtained by passing a - ethylnaphthalene through a red-hot tube, its structure has been interpreted as follows:

H_2C—CH_2

dodecahydroacenaphthylene
acenaphthene

OH O OH

decahydronaphthalene-1,8-dicarboxylic acid
Naphthalic acid

Isomerism of Naphthalene Derivatives

If we number the positions of the hydrogen atoms arranged round the double hexagon 1, 2, 3, 4, 6, 7 and 8, we notice that they may be divided into two sets of 4 each, which are symmetrically situated, viz 1, 4, 5, 8, (these adjoin one of the central carbon atoms) and 2, 3, 6, 7 (these are separated by one carbon atom from one of the central carbon atom

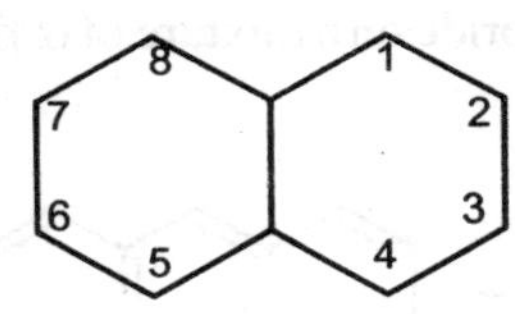

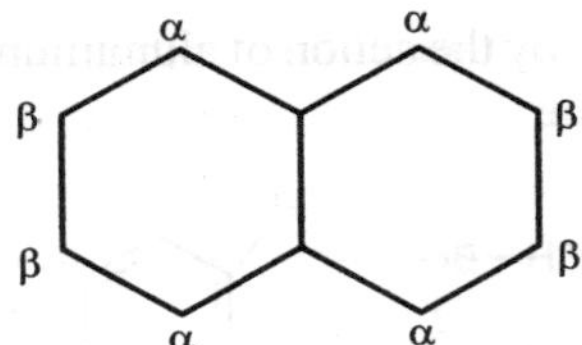

It follows that two mono- derivatives of naphthalene should exist, in which one of the first set, or one of the second set of position is occupied by the new element or group. The case is this:

The isomers of most of the non- derivatives of naphthalene are known

There are 2 monochloro monochloro – monobromo 2 naphthaamines and 2 naphthols etc, the first species is known as alpha, the second series as beta compounds. The number of di derivatives us easily estimated. Theoretically there are ten, which number also agrees with the experimental results, although the complete set has only been obtained in a few cases. They are indicated by the above numbers or by the Greek letters α and β. The 1, 2; 1, 3; and 1, 4 positions in the same ring are sometimes called ortho – meta and para positions, the positions 1, 8 is termed as the peri position and resembles as ortho position for it carboxyl group occupy these two position, the substance behaves like phthalic acid, and forms and anhydride.

ANTHRACENE

Introduction and Occurrence

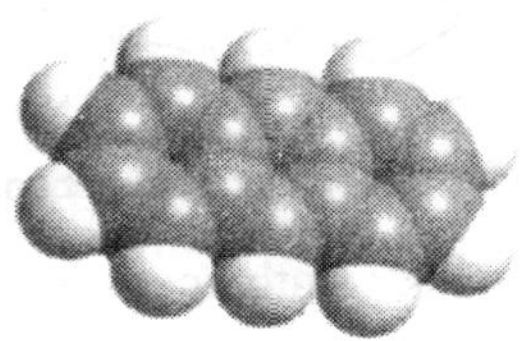

Anthracene occurs with its isomer phenathrene and a variety of other compounds in the last portion of distillate form coal tar known as anthracene oil (ανθρακενω, anthrakeno – Greek Coal). The dark coloured liquid deposits

on standing a light brown crystalline mass consisting of anthracene mixed with phenanthrene and carbazole is filtered in a filter press and washed with solvent naptha free from adhering oil. The crystalline pepper coloured mass contains about 50% of anthracene and is called commercial anthracene. This forms the raw material which is used on an extensive scale in the manufacture of alizarin and similar dyes.

The crude anthracene is purified by distillation with the addition of a little solid potassium hydroxide, which combines with carbazole producing potassium carbonate. The phenanthrene is removed with carbon disulphide in which it is much more soluble than anthracene.

In a concise view anthracene is prepared from the following reactions.

1. By heating benzene with tetrabromoethane in the presence of aluminium chloride

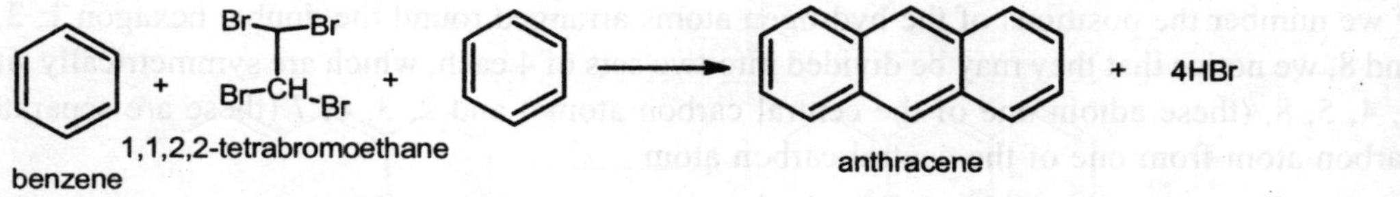

benzene + 1,1,2,2-tetrabromoethane + benzene → anthracene + 4HBr

2. In similar manner, by the action of aluminium chloride on a mixture of α β - dibromoethylbenzene and bromobenzene

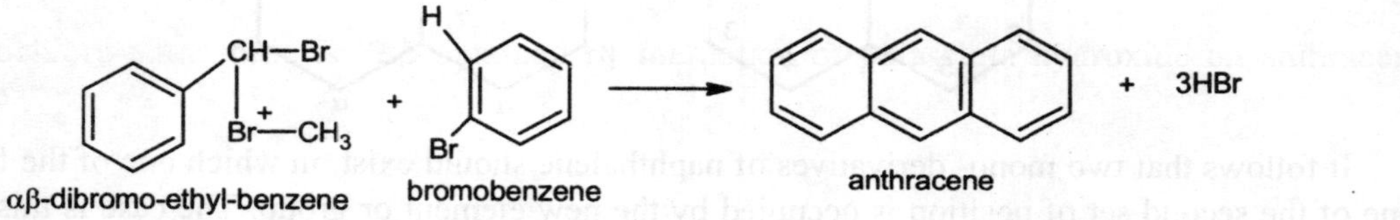

αβ-dibromo-ethyl-benzene + bromobenzene → anthracene + 3HBr

These two syntheses prove that the middle group C_2H_2 of anthracene is linked to two benzene nuclei, but give no information as to the actual points of union.

3. Anthracene, however, is also produced when o – bromo – benzyl bromide is treated with sodium. Hence it follows that the middle group is attached to o – carbon atoms of each benzene ring. This reaction the first product is Dihydro – anthracene, which is readily converted into anthracene by oxidation.

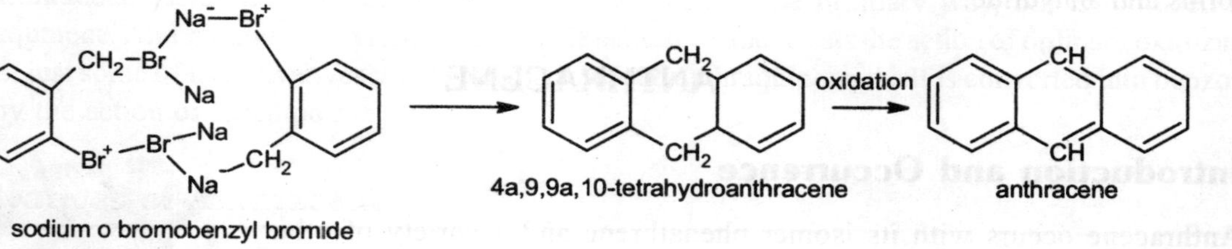

sodium o bromobenzyl bromide → 4a,9,9a,10-tetrahydroanthracene —oxidation→ anthracene

4. Anthracene is also obtained by heating o –tolyl phenyl ketone with zinc dust.

-H_2O

o - tolyl ketone → anthracene

Estimation: Dissolve 1 gm of anthracene in 45 cc glacial acetic acid, and allow to boil with reflux condenser, for an hour, then pour and rinse the contents into water and dry and then weigh. Sublime the anthraquinone, by heating it in a basin, and estimate the loss of weight, which is that of pure anthraquinone. The anthracene is calculated from the amount anthraquinone.

Physical Properties

Molecular Formula	$= C_{14}H_{10}$
Formula Weight	= 178.2292
Composition	= C(94.34%) H(5.66%)
Melting Point	= 218°C
Boiling Point	= 340°C
Molar Refractivity	= 61.93 cm^3
Molar Volume	= 157.6 cm^3
Parachor	= 414.9 cm^3
Index of Refraction	= 1.714
Surface Tension	= 47.9 dyne/cm
Density	= 1.130 g/cm^3
Dielectric Constant	= 3.08
Polarizability	= 24.55 $10^{-24}cm^3$
Monoisotopic Mass	= 178.07825 Da
Nominal Mass	= 178 Da
Average Mass	= 178.2292 Da
Log P	= 4.68

Chemical Properties

Anthracene exhibits certain chemical properties which is similar to benzene, and more particularly to naphthalene. It is converted by sulphuric acid into mono – and di sulphonic acid, by the action of sodium amalgam in alcoholic solution, it forms anthracene hydride,

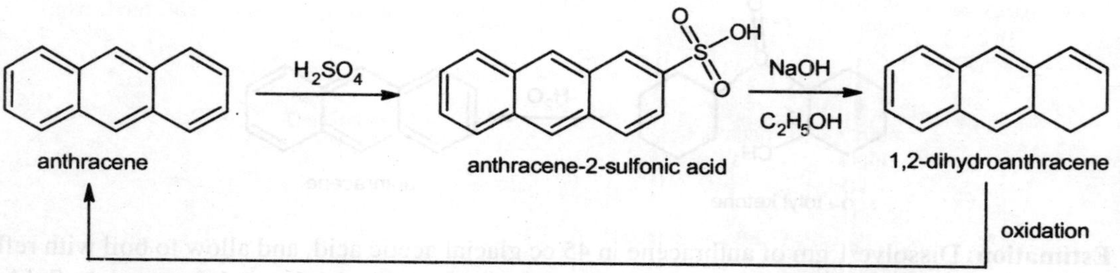

H_2SO_4 NaOH / C_2H_5OH

anthracene → anthracene-2-sulfonic acid → 1,2-dihydroanthracene → (oxidation) → anthracene

With chlorine both addition and substitution products are formed. The addition compound anthracene dichloride is obtained by passing chlorine into a cold solution of anthracene and carbon disulphide.

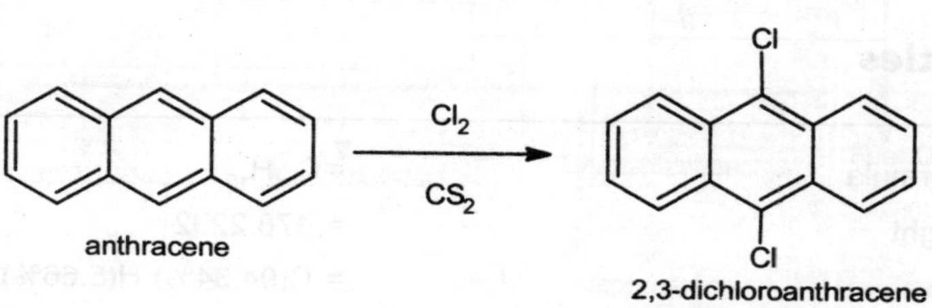

Cl_2 / CS_2

anthracene → 2,3-dichloroanthracene

Monochloro anthracene is also obtained by the action of potassium hydroxide on anthracene dichloride.

2,3-dichloroanthracene + KOH → 2-chloroanthracene + KCl + H_2O

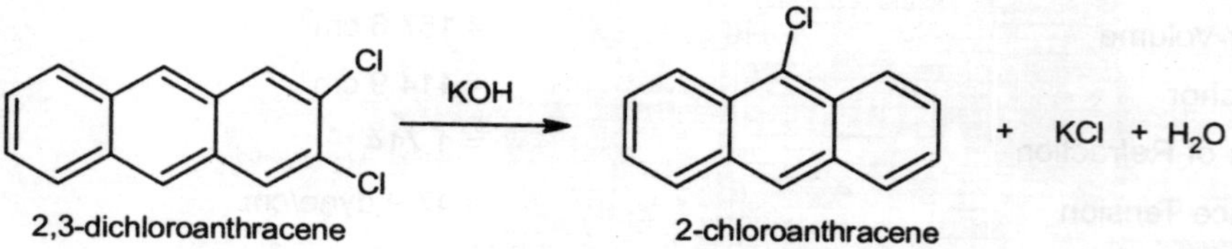

Anthracene gets readily oxidized to anthraquinone. The action is similar to the conversion of naphthalene to naptha quinone (also an oxidation process) but in the case of anthracene the process is easier. The resemblance between the substances is only apparent for we shall see presently that anthraquinone has few of the properties of a true quinone.

Anthracene yields no nitro derivatives, when nitrated in the ordinary way, but is oxidized to anthraquinone. Anthraquinone is a remarkable stable substance and resists the action of ordinary oxidizing agents; just some of its derivatives yield phthalic acid, and anthraquinone itself is converted into benzoic acid by the action of fused potash.

Derivatives of Anthracene

Now we are going to discuss about the derivatives, among these the most important is Anthraquinone.

ANTHRAQUINONE

Introduction

Anthraquinone (9,10-dioxoanthracene) is an aromatic organic compound. It is a derivative of anthracene. It has the appearance of yellow or light gray to gray-green solid crystalline powder.

Its other names are 9,10-anthracenedione, anthradione, 9,10-anthrachinon, anthracene-9,10-quinone, 9,10-dihydro-9,10-dioxoanthracene, and trade names Hoelite, Morkit, Corbit, and others.

Natural Occurrences

Anthraquinone naturally occurs in some plants (e.g. aloe, senna, rhubarb, and Cascara buckthorn), fungi, lichens, and insects, where it serves as a basic skeleton for their pigments. Natural anthraquinone derivatives tend to have laxative effects.

Preparation

Anthraquinone is prepared by heating o – benzolybenzoic acid, with concentrated sulphuric acid at 150°C.

2-benzoylbenzoic acid $\xrightarrow{\Delta}$ Anthraqinone + H_2O

The various syntheses just described point unmistakably to the existence in anthracene and its derivatives of a framework of two benzene nuclei, joined together by two central carbon atoms, which are attached to adjacent or ortho – carbon.

Seeing that anthracene is converted by reduction into anthracene hydride and that anthracene is changed by oxidation into anthraquinone, the relation of the three is very simply expressed by the following formula:

Anthracene Hydride (CH_2, CH_2) — anthracene — Anthraqinone

Moreover, as mono and di – chloro – anthracene both yield anthracene both yield anthraquinone on oxidation, the chlorine atoms, which disappear from the compound, must be attached to the central carbon atoms.

From the synthesis described it follows that the central pair of carbon atoms is linked to at least one nucleus in the ortho – position. By a similar series of reaction bromophthalic anhydride can be converted into bromoanthraquinone

In this case the central pair of carbon atoms is attached to the substituted nucleus in the ortho – position.

Now, bromoanthraquinone, which on oxidation yields phthalic acid, the substituted nucleus being destroyed.

Physical Properties

Molecular Formula	= $C_{14}H_{10}$
Melting Point	= 286 °C
Boiling Point	= 379.8°C
Formula Weight	= 178.2292
Composition	= C(94.34%) H(5.66%)
Molar Refractivity	= 61.93 cm^3
Molar Volume	= 157.6 cm^3
Parachor	= 414.9 cm^3
Index of Refraction	= 1.714
Surface Tension	= 47.9 dyne/cm
Density	= 1.130 g/cm^3
Dielectric Constant	= 3.08
Polarizability	= 24.55 X $10^{-24}cm^3$
Monoisotopic Mass	= 178.07825 Da
Nominal Mass	= 178 Da
Average Mass	= 178.2292 Da
Log P	= 4.68

Chemical Properties

From the synthesis described it followed that the central pair of carbon atom is linked to at least one nucleus in the ortho – position. By a similar series of reactions bromophthalic anhydride can be converted into bromoanthraquinone.

In this case the central pair of carbon atoms is attached to the subsequent nucleus in the ortho position.

Now, bromoanthraquinone, when fused with potash gives hydroxide of anthraquinone.

bromoanthraquinone + KOH → anthraquinone hydroxide + KBr

Anthraquinone hydroxide on oxidation yields phthalic acid, there by destroying the substation nucleus being destroyed.

Consequently the central carbon atoms are also linked to the unsubstituted nucleus in the ortho – position.

anthraquinone hydroxide → phthalic acid

Anthraquinone exhibits the properties of a ketone in its behaviour with hydroxylamine, with which it forms an oxime.

anthraquinone hydroxide + H_2N—OH (hydroxylamine) → 1*H*-isoindole-1,3(2 *H*)-dione

Moreover on reduction with zinc dust and sodium hydroxide, it yields a secondary alcohol, oxanthrol, which forms a red compound, thus affording a delicate test for anthraquinone.

anthraquinone hydroxide + NaOH —(Zinc dust)→ (2-{2-[(hydroxy-kO)methyl]benzoyl}phenyl-kC[1])sodate(1-) + ZnO

red compound

EXPERIMENT: Add a little sodium hydroxide solution to a small quantity of anthraquinone and then a little zinc dust, on heating to boiling, an intense red colouration is produced which disappears on shaking. This arises from the sodium oxoanthranolate, becoming oxidized to anthraquinone on exposure to open air.

With tin and hydrochloric acid, anthral is produced; it is a substance possessing phenolic properties.

anthraquinone hydroxide + $2H_2$ —(tin, hydrochloric acid)→ anthranol + H_2O

Reduction of anthraquinone

With more strong reducing agents like boiling with zinc dust transfers it back to anthracene.

Anthraquinone possesses few of the characteristics of benzoquinone and naphthaquinone

PURPURIN

Introduction

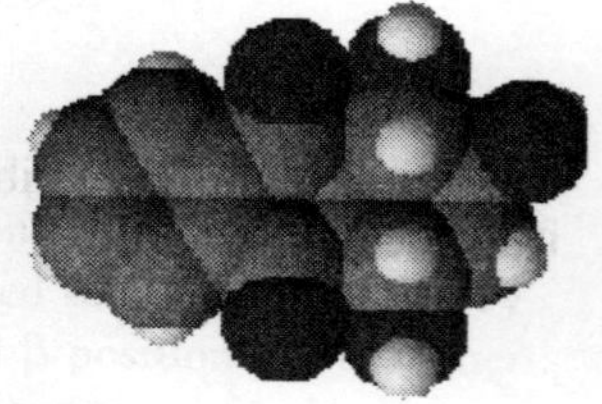

Purpurin, or **1,2,4-trihydroxyanthraquinone**, is a naturally occurring red/yellow dye in the roots of the plant madder (or known as Rubia tinctorum L). Purpurin also occurs in madder with alizarin. Purpurin is a crystalline compound with the formula of $C_{14}H_5O_2(OH)_3$ that is colourless until dissolved in alkali solutions. It is soluble in ethanol (becomes red) and is soluble in water at boiling and alkalis water (becomes yellow). It has the appearance of dark red needles. It is also called **Verantin, Smoke Brown G, Hydroxylizaric acid**, and **C.I. 58205**.

Purpurin is a fast dye for cotton printing and forms complexes with various metal ions.

As relating to biochemistry, purpurin is a glycosaminoglycan binding protein as well as a retinol binding protein. Purpurin's pattern of sequences has shown it to belong to the lipocalin family of proteins.

Physical Properties

Molecular Formula	$= C_{14}H_{14}O_5$
Formula Weight	= 262.25796
Composition	= C(64.12%) H(5.38%) O(30.50%)

Molar Refractivity	= 64.57 cm^3
Molar Volume	= 172.1 cm^3
Parachor	= 514.5 cm^3
Index of Refraction	= 1.673
Surface Tension	= 79.7 /cm^3
Density	= 1.523g/cm^3
Polarizability	= 25.59 cm^3
Monoisotopic Mass	= 262.084124 Da
Nominal Mass	= 262 Da
Average Mass	= 262.258 Da
Log P	= 0.82

Anthrapurpurin

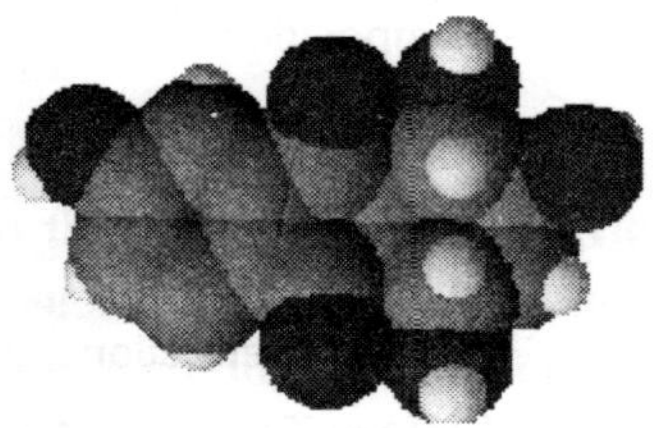

This is obtained from the anthraquinone, 1, 2, 6 disulphonic acid by fusion with sodium hydroxide and potassium chlorate in the same manner that alizarin is obtained from the monosulphonic acid.

Physical Properties

Molecular Formula	= $C_{14}H_{14}O_6$
Formula Weight	= 278.25736
Composition	= C(60.43%) H(5.07%) O(34.50%)
Molar Refractivity	= 66.45 cm^3
Molar Volume	= 170.6 cm^3
Parachor	= 529.8 cm^3
Index of Refraction	= 1.706
Surface Tension	= 92.9 dyne/cm
Density	= 1.630 g/cm^3
Polarizability	= 26.34 cm^3
Monoisotopic Mass	= 278.079038 Da
Nominal Mass	= 278 Da
Average Mass	= 278.2574 Da
Log P	= 1.04

FLAVOPURPURIN

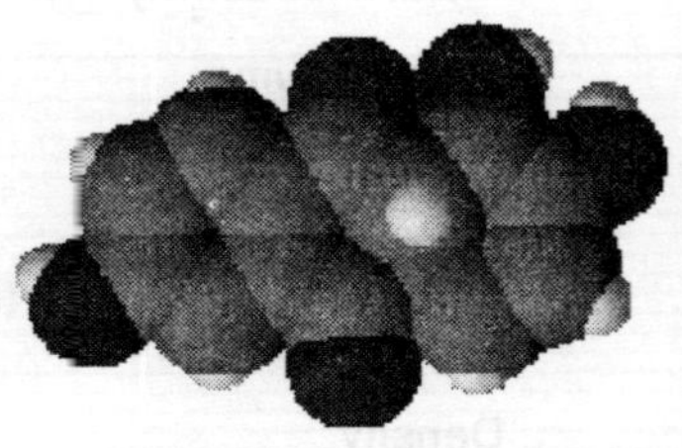

This is also called 1, 2, 6 Trihydroxyanthraquinone, and is formed like anthrapurpurin from 1, 6 anthraquinone – disulphonic acid.

All these compounds dye a brilliant red with alumina mordent, but of a slightly yellowish shade than alizarin. They are the chief constituents of commercial alizarin sold under the name alizarin yellow.

It is the interesting fact that among the ten dihydroanthra-quinones and the numerous trihydroxy – compounds which have been prepared, only those can be used as dyes, which contain the two hydroxyl groups in the α - β - position in the same nucleus.

Physcial Properties

Molecular Formula	$= C_{14}H_{10}O_5$
Formula Weight	= 258.2262
Composition	= C(65.12%) H(3.90%) O(30.98%)
Molar Refractivity	= 64.06 cm^3
Molar Volume	= 156.2 cm^3
Parachor	= 486.6 cm^3
Index of Refraction	= 1.756
Surface Tension	= 94.2 dyne/cm
Density	= 1.653 g/cm^3
Polarizability	= 25.39 cm^3
Monoisotopic Mass	= 258.052823 Da
Nominal Mass	= 258 Da
Average Mass	= 258.2262 Da
Log P	= 3.00

PHENANTHRENE

Introduction

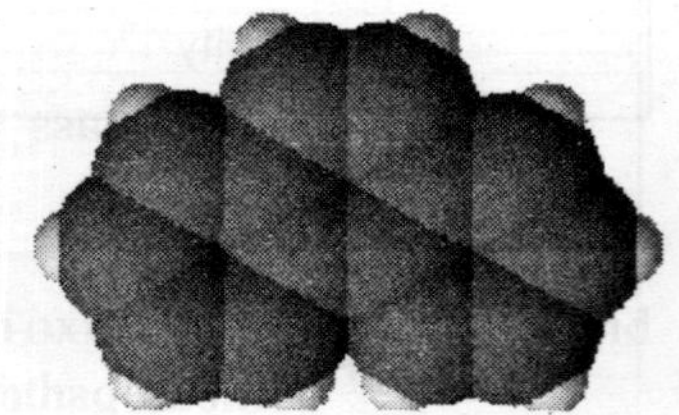

Phenanthrene is a polycyclic aromatic hydrocarbon composed of three fused benzene rings—as the above formula shows. The name *Phenanthrene* is a composite of phenyl and anthracene. It provides the framework for the Steroids. In its pure form, it is found in cigarette smoke and is a known irritant, photosensitising skin to light.

The compound with a phenanthrene skeleton and nitrogens at the 4 and 5 position, so IUPAC name 4,5-diazaphenanthrene, is known as phenanthroline.

Preparation

Phenanthrene is prepared by the following processes:

1. *By boiling o – bromobenzyl bromide with metallic sodium*: Which also yields anthracene hydride in the manner shown below in the equation.

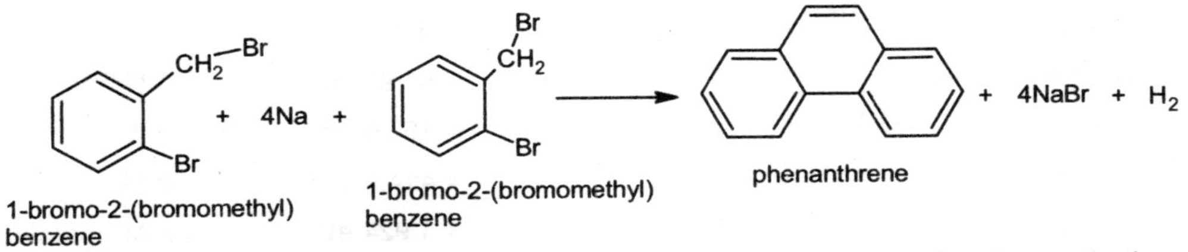

1-bromo-2-(bromomethyl) benzene + 4Na + 1-bromo-2-(bromomethyl) benzene → phenanthrene + 4NaBr + H_2

The phenanthrene hydride, which is probably first formed in the reaction looses hydrogen and gives phenanthrene.

2. From 9,10 Diphenylphenanthrene [LABORATORY PROCESS]

In a 250-mL, three-necked, round-bottomed flask equipped with a magnetic stirrer, a thermometer, and a septum is suspended 10.0 g (27.3 mmol) of benzopinacol in 40 mL of dry toluene under a nitrogen atmosphere. The suspension is cooled to 0°C using an ice-acetone bath, and 40 mL of freshly distilled trifluoromethanesulfonic acid (triflic acid) and is added in one portion via syringe. The cold bath is removed, and the mixture is stirred for 24 hr at room temperature. The resulting heterogeneous mixture is poured over 100 g of ice, the organic layer is separated, and the aqueous layer is extracted with two 100-mL portions of toluene. The organic layers are combined, and washed with water (50 mL) and with brine (50 mL). The original solution is dried with anhydrous magnesium sulfate , filtered, and the filtrate is concentrated under reduced pressure to afford 9.2 g of crude product (mp 227-231°C) as a pale yellow solid. The crude product is recrystallized from 450 mL of toluene-ethanol (1:1) to give 6.7 g (74%) of pure 9, 10-diphenylphenanthrene as a white solid. This product is allowed to distill in the ethereal solution of sodium. The phenanthrene is obtained.

Reaction Equation

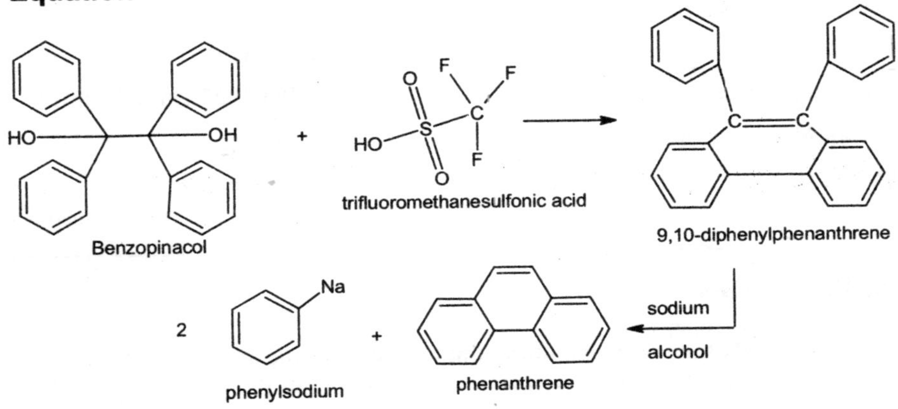

Benzopinacol + trifluoromethanesulfonic acid → 9,10-diphenylphenanthrene

9,10-diphenylphenanthrene —(sodium, alcohol)→ 2 phenylsodium + phenanthrene

Physcial Properties

Molecular Formula	= $C_{14}H_{10}$
Formula Weight	= 178.2292
Composition	= C(94.34%) H(5.66%)
Melting point	= 100.5 °C
Boiling point	= 338 °C
Molar Refractivity	= 61.93 cm^3
Molar Volume	= 157.6 cm^3
Parachor	= 414.9 cm^3
Resonance energy	= 1.924 eV
Heat of vaporization	= 52.7 KJ / mol
Molecular connectivity X	= 4.815
Water solubility	= 7.2 x 10^{-3} mmol / l
Molecular volume	= 169.5 $cm^3 mol^{-1}$
Molar volume	= 199 $cm^3 mol^{-1}$
Vapour pressure at 25°C	= 1.8+0.1 x 10^{-2} Pa
Enthalpy of sublimation at	= 25°C : 92 5+2 KJ / mol
Enthalpy of fusion	= 16.7+0.3 KJ / mol
Henry Constant	= 2.7 x 10^{-4} Atm.m^3 / mol
Index of Refraction	= 1.714
Surface Tension	= 47.9 dyne/cm
Density	= 1.130 g/cm^3
Dielectric Constant	= 3.08
Polarizability	= 24.55 cm^3
Monoisotopic Mass	= 178.07825 Da
Nominal Mass	= 178 Da
Average Mass	= 178.2292 Da
LogP	= 4.68

Chemcial Properties

Phenanthrene forms a diketone or phenanthraquinone and diphenic acid on oxidation with chromic acid (aqueous chromic oxide), which bears a close resemblance with β - naphthaquinone.

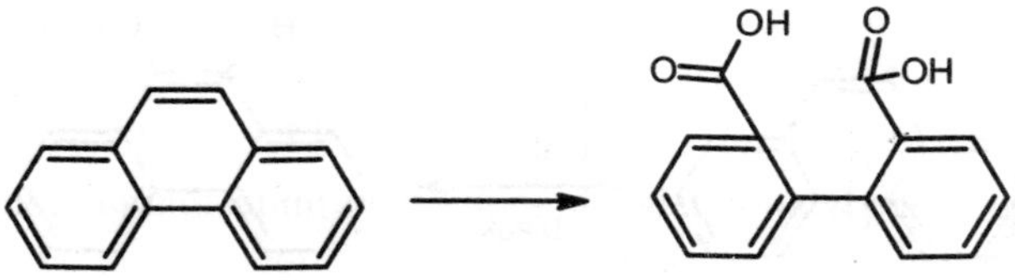

phenanthrene → diphenic acid

It crystallizes in orange needles. This gets reduced by sulphurous acid, into bisulphate.

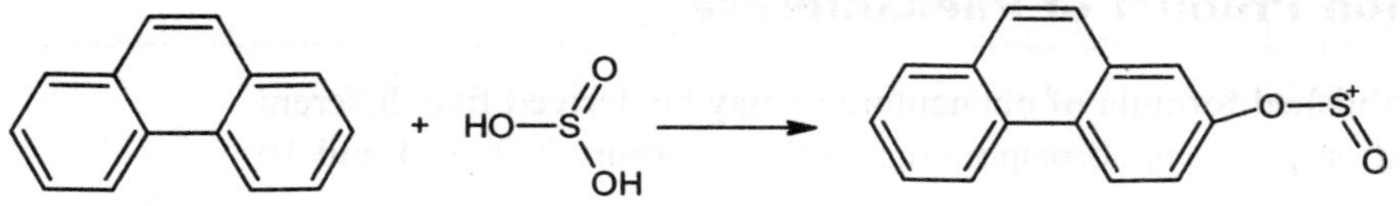

phenanthrene + sulfurous acid → phenanthrene bisulphate

When reacts with hydroxylamine it produces dioxime

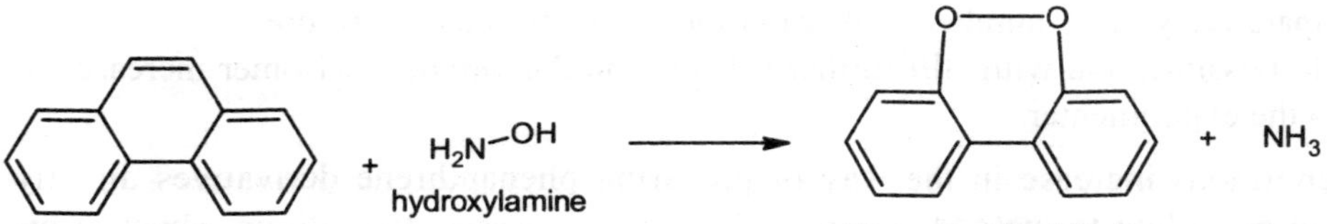

phenanthrene + hydroxylamine → dibenzo[*c,e*][1,2]dioxine + NH_3

The relation between phenanthrene and naphthalene is illustrated by following synthesis from a derivative of naphthalene,

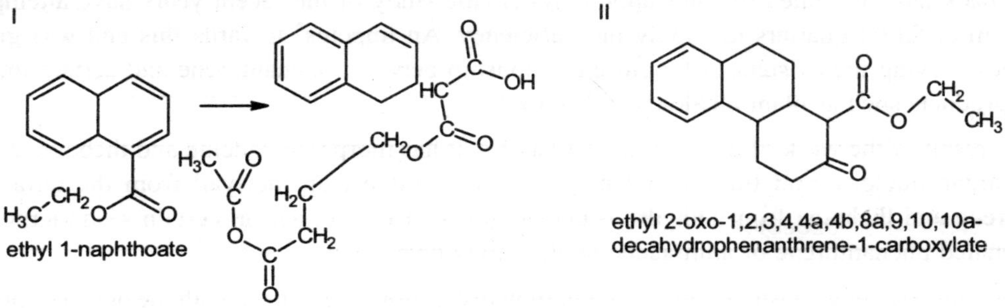

I

ethyl 1-naphthoate → 3-[(5-acetoxy-5-oxopentyl)oxy]-2-(1,2-dihydronaphthalen-2-yl)-3-oxopropanoic acid

II

ethyl 2-oxo-1,2,3,4,4a,4b,8a,9,10,10a-decahydrophenanthrene-1-carboxylate

I Ethyl 1 naphthoate condenses with ethyl acetoacetate to give 3-[(5-acetoxy-5-oxopentyl)oxy]-2-(1,2-dihydronaphthalen-2-yl)-3-oxopropanoic acid which on distillation with zinc dust is converted into phenanthrene.

In presence of platinum black, hydrogen unites with phenanthrene to form 9, 10 dihydrophenanthrene.

HO OH

phenanthrene —platinum black→ phenanthrene-9,10-diol

Substitution Product of Phenanthrene

From the established formula of phenanthrene may be derived five different monosubstitution products corresponding to the positions 1, 2, 3, 4 and 10 respectively.

It will readily be seen that the hydrogen atoms in the bridge are similarly situated; hence the 9 and 10 substitution products are saved.

A comparatively large number of distribution derivatives are therefore theoretically possible, and with still further substitution the number of isomer increases in a manner alarming to the experimenter.

This enormous increase in the way of preparing phenanthrene derivatives directly from the hydrocarbon according to methods usual in the aromatic series. In addition, physical properties of some phenanthrene compounds are no favourable for investigations.

For these reasons and more particularly owing to the fact that, contrary to expectation it has not been possible to use these compounds on any considerable scale in the dyeing industry; chemists for decades back have hesitated to enter upon a systematic study of the recent years have attempt been made from different quarters to supply the deficiency. An impetus* towards this end was given by researches proving the existence of a close relationship between phenanthrene and certain important natural products such as plant alkaloids and steroids.

As a result of the work of different investigator's, it has morphine codeine and thebaine contain a phenanthrene nucleus, and further that morphine and codeine are derived from the tetrahydro – phenathrene and thebaine from Dihydro – phenanthrene structure, and in certain sesquiterpenoids a hydrogenated phenanthrene or anthracene nucleus may be present.

Following the above postulations a number of investigations were made with the object of preparing new derivatives of phenanthrene either synthetically or directly from the hydrocarbon. Thus, for example, 3 – nitrophenanthrene and 9 nitrophenanthrene have been prepared by direct nitration and from these the corresponding amino – phenanthrene has been obtained by reduction.

Nevertheless the preparation of nitrophenanthrenes by direct nitration is tedious, owing to the resinification** which so easily occurs. Halogenation causers some substitution products which are

*motivated

**formtion of residue

less troublesome to nitrate, and when the 9 – bromo – phenanthrene described earlier is heated with nitric acid it yields 9 – bromo – 10 nitro – phenanthrene.

An observation of generation interest has been made in connection with these compounds. When heated in a closed tube to 320°C with ammonium chloride.

It is converted comparatively smoothly into 9:10 dichloro-phenanthrene.

NH_4Cl

9-bromo-10-nitrophenanthrene

9,10-dichlorophenanthrene

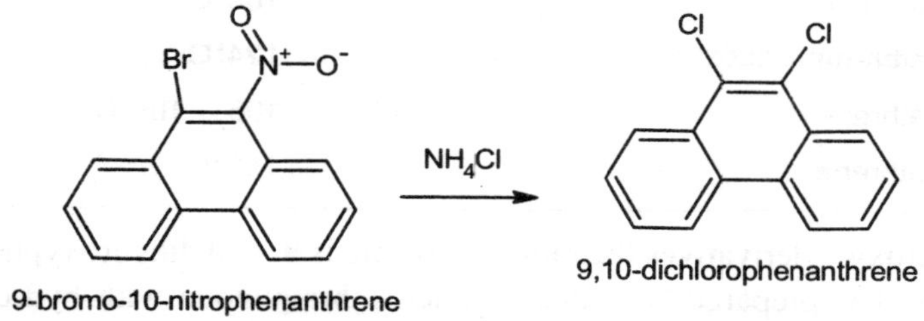

The reaction may also be applied to other aromatic compounds for example, o – bromo – nitrobenzene under the same conditions yields o – dichloro – benzene

A condition for the success of this is the o – position of the substituent, since m and p bromo nitrobenzene remain unchanged.

Sulphonification of phenanthrene lead to the production of 3 – 2 and 9 phenanthrene sulphonic acids.

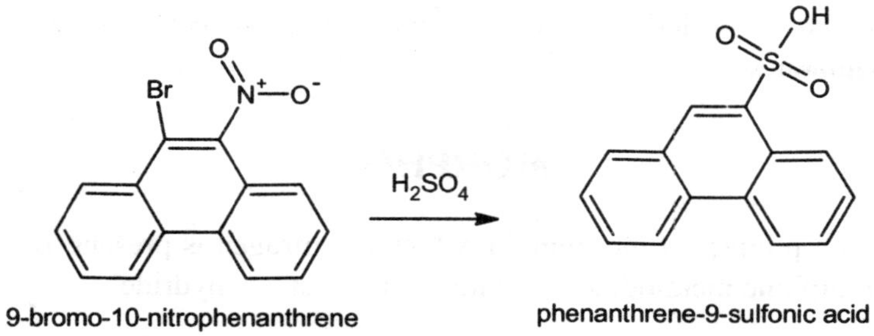

H_2SO_4

9-bromo-10-nitrophenanthrene

phenanthrene-9-sulfonic acid

Hydroxy Phenanthrene

Of more importance than the compounds just described are the hydroxy – phenanthrene certain of which, as will be seen later, have been obtained as degradation products of the opium alkaloids, morphine codeine and thebaine. They therefore are of interest in connection with the constitution of these substances. Hydroxy – phenanthrenes may be the said constitution of these substances.

Hydroxy – phenanthrenes are prepared from the sulphonic acids by fusion with potash, from the amino derivatives by the way of the diazo – reaction, or by the synthetic method of Pschorr, in this method phenanthrenes are prepared from the sulphonic acid by fusion with potash, from the amino derivatives by the way of the diazo – reaction, or by the synthetic method of Pschorr, in this method phenanthrene and its derivatives are prepared without application of high temperature.

All of the five possible mono – hydroxy – phenanthrene are known either in the free state or as the methyl ethers.

Name	Melting point
1 – Hydroxyphenathrene	156°C
2 – Hydroxyphenathrene	168°C
3 – Hydroxyphenathrene	124°C
4 – Hydroxyphenathrene	106 – 109°C
9 – Hydroxyphenathrene	153°C

Among the dihydroxy – derivatives the most accessible is 9 – 10 dihydroxyphenanthrene or hydro – phenanthraquinone, it is prepared by reducing phenanthraquinone with hydrogen Sulphide, or a calculated amount of alcoholic phenyl hydrazine

In similar manner the nitro – derivatives of phenanthraquinone may be converted into corresponding hydro – phenanthroquinones.

MORPHOL

Morphol is a disruption product of morphine in which no nitrogen is present. It was prepared from morphine methiodide by heating with acetic anhydride

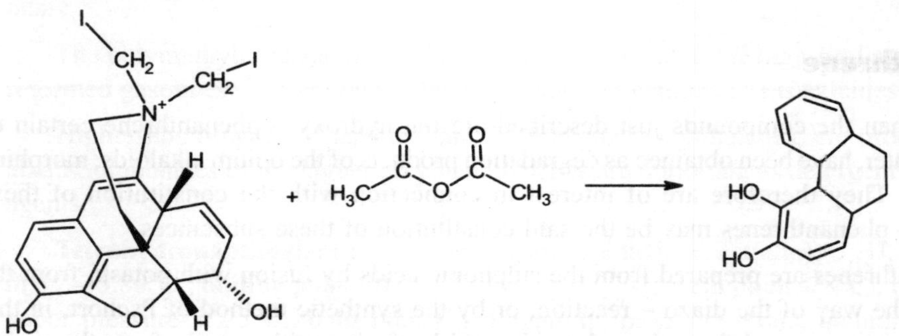

and for long could only be identified generally as a dihydroxyphenanthrene or phenanthrene 3,4 diol . Subsequently, some information as to the position of the hydroxyl group was obtained from the observation; the dihydroxy phenanthraquinone prepared from Morphol possessed dyeing properties, similar to those alizarin, and hence should contain the hydroxy groups in the ortho position

phenanthrene-3,4-diol

After the relationship between Morphol and porphenol has been made clear by reducing the morphenol to Morphol, the structure of dihydroxy – phenanthrene was proposed for Morphol. Confirmation of this formula was supplied by the synthesis of dimethyl – Morphol described below and latter the Barger synthesis of morphol itself.

Synthesis of Morphol

This was effected from 3 – phenanthrol 4 aldehyde (obtained by the interaction of 3 – phenanthrol, hydrogen cyanide and hydrogen chloride in the presence of aluminium chloride by treating it with hydrogen peroxide and potassium hydroxide aqueous pyridine solution. The morphol or 3: 4 – dihydropxy – phenathrene obtained melts at 142°C.

1,4-dihydrophenanthren-3-ol —(HCl, HCN)→ 3-hydroxy-1,4-dihydrophenanthrene-4-carbaldehyde —(H_2O_2 and KOH pyridene)→ 1,4-dihydrophenanthrene-3,4-diol

DIMETHYL MORPHOL

This compound is the methylated form of morphol. This is synthesized by the following process. The starting point of this synthesis was the methyl ether of o – nitro Vaniline (I) which was condensed with sodium salt of Phenylacetic acid (II) by Perkin's reaction to give a - phenyl – 2 nitro 3: 4 dimethoxy cinnamic acid (III). The diazo compound (IV) of the corresponding amino acid in nitrogen and water with the formation of 3: 4 dimethyloxyphenanthrene 9 – carboxylic acid, which on distillation gave carbon dioxide and dimethaoxypehnanthrene (vi). This synthesis establishes the 3: 4 positions of the two methoxyl groups as it is doubtful that any intramolecular rearrangement could take place under the above conditions of experiment

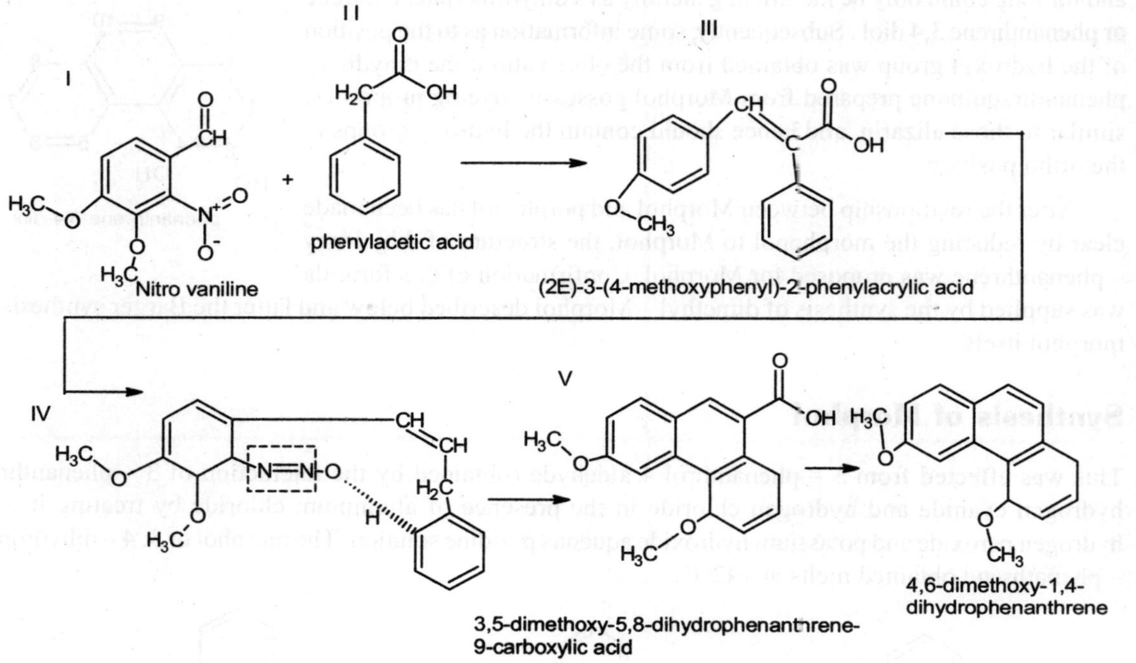

Nitro vaniline

phenylacetic acid

(2E)-3-(4-methoxyphenyl)-2-phenylacrylic acid

3,5-dimethoxy-5,8-dihydrophenanthrene-9-carboxylic acid

4,6-dimethoxy-1,4-dihydrophenanthrene

This compound crystallizes from alcohol in colourless leaflets, with picric acid it yields a double compound crystallizing in red crystals.

The synthetic 3:4 di methoxyphenathrene is identical with the dimethyl morphol prepared from methyl morphol, a degradation product of codeine.

In connection with the constitution of morphine, the molecule of which contains as alcoholic hydroxyl and phenolic group it was of importance to decide which of the hydroxyls of morphol correspond to the phenolic hydroxyls. Now codeine, methyl ether of morphine in which the methoxyl group takes place of the phenolic hydroxyl of the latter compound, has been degraded to acetyl – methyl morphol. All that was necessary, therefore, was to determine the position of the methoxyl group in the acetyl derivative. The point was settled by Pschorr's synthesis of acetyl a derivative of 3 – hydroxy – 4 methoxy – phenanthrene by the above method, using vicinal o – nitrovanilin in place of the corresponding ether.

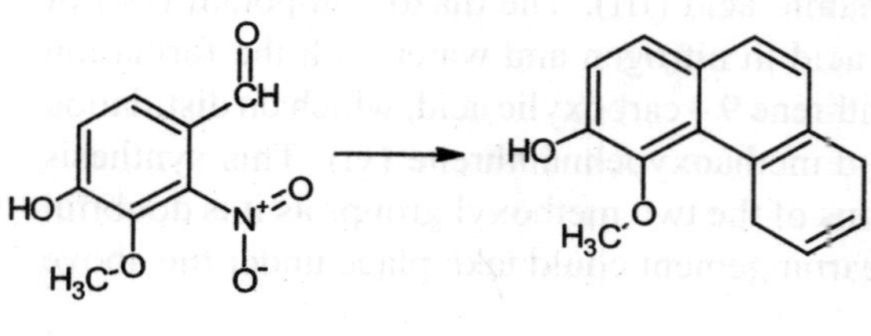

The synthetic 3 – hydro – 4 – methoxyphenanthrene prepared in this manner differs considerably from the methyl morphol obtained as disruption product of a α - methyl-morphomethane from codeine; hence it was concluded that the latter is represented by the structures 4 hydroxy – 3 methoxy phenanthrene. This inference was shortly afterwards confirmed by the synthesis of 3 – methoxy – 4 – acetoxy – phenanthraquinone, which was found to be identical with acetyl – methylphenanthroquinone from morphine.

MORPHENOL

Introduction

Which represent the molecular skeleton of morphine and thebaine, yields on fusion with alkali 3:4: 5 trihydroxy phenanthrene.

The following two hydroxy derivatives of phenanthrene have been obtained from the opium alkaloid thebaine under different experimental conditions.

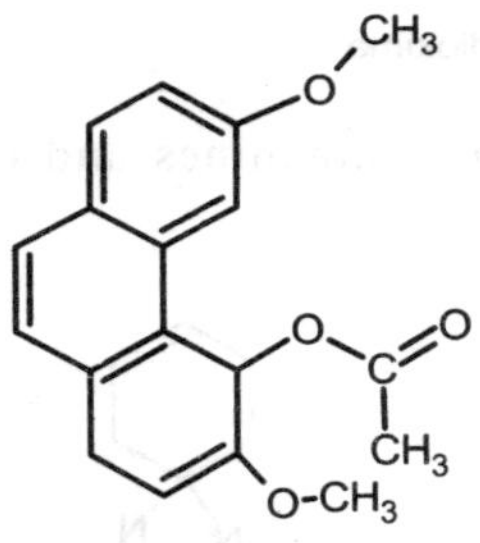

acetyl thebaol

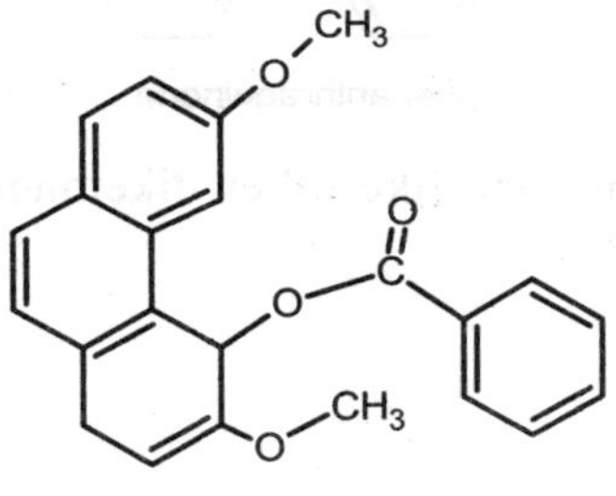

Benzoyl thebaol

Acetyl – thebaol: - This compound is prepared by heating thebaine with acetic anhydride. When treated with sodium ethoxide, it is converted into thebaol. 3:6 dimethoxy – 4 – hydroxyphenathrene.

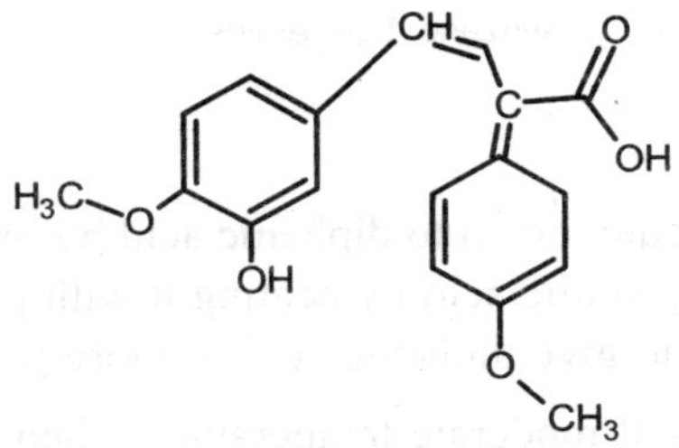

The constitution of thebaol was established by Pschorr, who synthesized it form vicinal o – nitro – isovanilin and p – methoxyphenyl acetic acid by the method of hydroxy phenanthrene. The first step in the process is the formation of the compound indicated.

This synthesis also established the position of the two methoxy group of thebaine, Benzoyl – thebaol has been prepared in the form of colourless needles by the action of bezoyl chloride on thebaine.

Phenanthraquinone and its Derivatives

This compound is generally prepared by oxidation of phenanthrene with chromic acid in glacial acetic acid.

When a solution of phenanthraquinone in glacial acetic acid is treated with sulphuric acid and toluene containing thiotoluene, a blue green colouration is developed; after dilution with water and extraction with ether the colour changes to violet (Laubenheimer's reaction)

Phenanthraquinone is closely related to the α- diketone in properties, with hydroxylamine (as stated earlier produces monoxime or a dioxime).

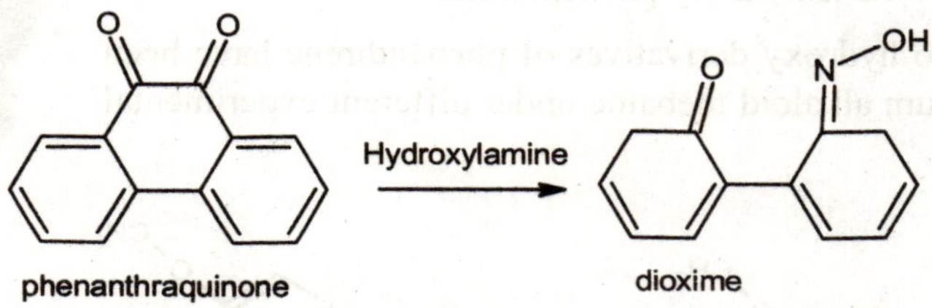

phenanthraquinone — Hydroxylamine → dioxime

Phenanthraquinone like other diketones, react with o – dimanines and condenses to form phenanthraphenazine.

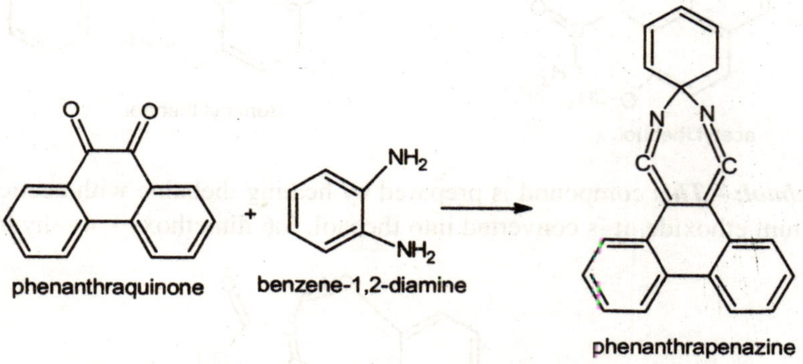

phenanthraquinone + benzene-1,2-diamine → phenanthrapenazine

The conversion of phenanthraquinone into diphenic acid by oxidation with chromic acid mixture, and the formation of dipenylene glycolic acid by heating it with potassium hydroxide, with alcoholic potash phenanthraquinone reacts to give diphenic acid or more complex products.

On treatment with nitric acid at moderate temperatures phenanthraquinone yields a mononitrate and with excess bromine at low temperatures a di bromo addition product considering the ease with which these compounds loose nitric acid and bromine respectively, it is probable that they are oxonium compounds for the following structures.

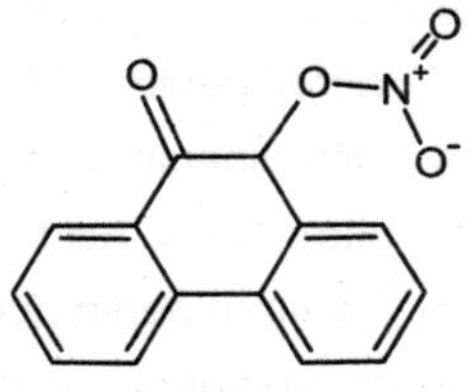

phenanthraquinone nitrate

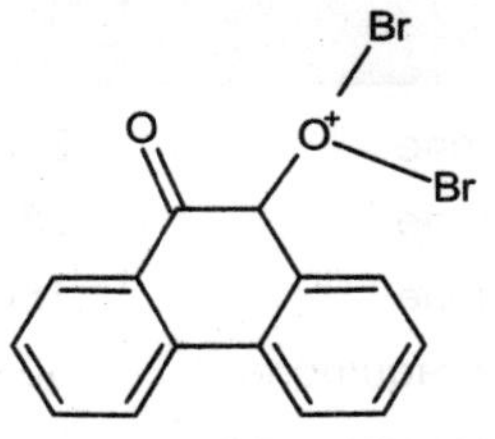

phenanthraquinone di bromide

Nitroderivatives of Phenanthraquinone

All the possible mono – nitro derivative are known, with exception of 1 – nitro – phenanthraquinone. Of the di nitro compounds the symmetrical 2: 7 and 4: 5 di nitro phenanthraquinone have been prepared.

3 – Nitrophenanthrene has been obtained by Schmidt and Kampf and its structure confirmed by the oxidation of 3 nitro – phenanthrene. The remaining nitro – compounds have been prepared by direct nitration of phenanthraquinone. When boiled for two or three minutes with concentrated nitric acid, the quinone is converted mainly into a mixture of 2 – and 4 nitro - phenanthraquinone s. Energetic nitration yields exclusive, the above symmetrical dinitro derivatives.

There is a tendency for the symmetrical dinitro derivatives to be formed by the nitration of any of the mono nitro compounds so far known. The pronounces tendency for substitution to assume the 2: 7 positions is also apparent in other substitution reactions.

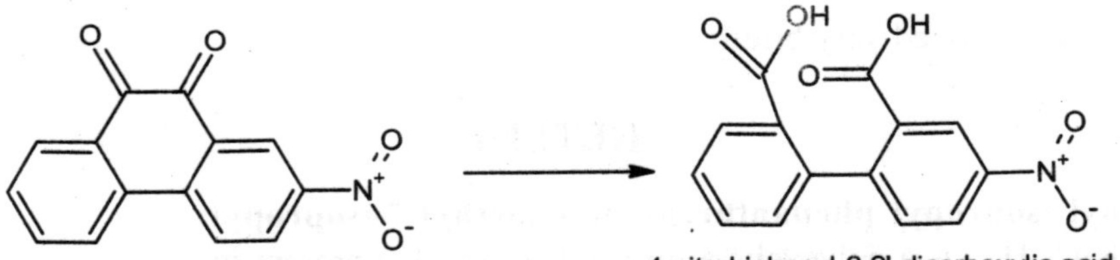

2-nitrophenanthrene-9,10-dione phenanthroquinone

4-nitrobiphenyl-2,2'-dicarboxylic acid p - p Dinitro - diphenic acid

On oxidation with potassium dichromate and sulphuric acid all nitro – derivatives of phenanthrene yield the corresponding nitro diphenic acids.

In the following table we will find the melting points of all known nitro derivatives of phenanthraquinone together with those of the corresponding diphenic acids.

On reduction with tin and hydrochloric acid the nitro compounds yield the corresponding amino – hydro phenanthraquinone, which are generally very unstable in the free state and readily undergo oxidation to aminoquinones. The latter can be diazotized and converted in the usual manner into various products of substitution of phenanthraquinone.

Name	m.p	Name	m.p
2 – nitro phenanthraquinone	275°C	p – nitrodipenic acid	220°C
3 – nitro phenanthraquinone	280°C	m - nitrodipenic acid	267°C
4 – nitro phenanthraquinone	180°C	o - nitrodipenic acid	249°C
2 – 7dinitronitrophenanthraquinone	302°C	pp - dinitrodipenic acid	253°C
4 – 5 dinitronitrophenanthraquinone	228°C	oô dinitrodipenic acid	303°C

The have, for example, been for service in determining the constitution of bromo – derivative. These can be obtained partly by direct bromination and by the oxidation of bromo – phenanthraquinones, on oxidation they yield the corresponding bromodiphenic acids as shown in the table below:-

Name	m.p	Name	m.p
2 – bromophenanthraquinone	233°C	p – bromo-diphenic acid	239°C
3- bromophenanthraquinone	268°C	m - bromo-diphenic acid	257°C
4 – bromophenanthraquinone	126°C	*pp'* - dibromo-diphenic acid	277°C
2:7 Dibromo-phenanthraquinone	323°C		

Some more Compounds with Condensed Nuclei

In addition to naphthalene, anthracene and phenanthrene, a number of compounds are there with condensed nuclei, these generally occur in the fraction of coal tar boiling above 360°C

These may be described very briefly

RETENE

Retene, **methyl isopropyl phenanthrene** or **1-methyl-7-isopropyl phenanthrene**, $C_{18}H_{18}$, is a polycyclic aromatic hydrocarbon present in the coal tar fraction, boiling above 360°CCelsius. It occurs naturally. It also occurs in the tars obtained by the distillation of resinous woods. It crystallizes in large plates, which melt at 98.5 °C and boil at 390 °C. It is readily soluble in warm ether and in hot glacial acetic acid. Sodium and boiling amyl alcohol reduce it to a tetrahydroretene, whilst if it be heated with phosphorus and hydriodic acid to 260°C, a dodecahydride is formed. Chromic acid oxidizes it to retene quinone, phthalic acid and acetic acid. It forms a picrate which melts at 123-124 °C. Its CAS number is [483-65-8] and its SMILES structure is Retene is derived by degradation of specific diterpenoids biologically produced by conifer trees.

The presence of traces of retene in the air is an indicator of forest fires; it is a major product of pyrolysis of conifer trees. It is also present in effluents from wood pulp and paper mills.

Retene, together with cadalene, simonellite and ip-iHMN, is a biomarker of higher plants, which makes it useful for paleobotanic analysis of rock sediments. Ratio of retene/cadalene in sediments can reveal the ratio of the genus *Pinaceae* in the biosphere.

Physcial Properties

Molecular Formula	$= C_{18}H_{18}$
Formula Weight	= 234.33552
Composition	= C(92.26%) H(7.74%)
Molar Refractivity	= 80.94 cm^3
Molar Volume	= 224.0 cm^3
Parachor	= 567.6 cm^3
Index of Refraction	= 1.642
Surface Tension	= 41.2 dyne/cm
Density	= 1.046 g/cm^3
Dielectric Constant	= 2.85
Polarizability	= 32.09 cm^3
Monoisotopic Mass	= 234.140851 Da
Nominal Mass	= 234 Da
Average Mass	= 234.3355 Da
Log P	= 6.48

CHRYSENE

Chrysene is a polycyclic aromatic hydrocarbon (PAH) with the molecular formula $C_{18}H_{12}$ that consists of four fused benzene rings. It is a natural constituent of coal tar, from which it was first isolated and characterized. It is also found in creosote, a chemical used to preserve wood.

Chrysene is formed in small amounts during the burning or distillation of coal, crude oil, and plant material.

Physcial Properties

Molecular Formula	$= C_{18}H_{12}$
Formula Weight	= 228.28788
Composition	= C(94.70%) H(5.30%)

Melting Point	= 254°C
Boiling Point	= 448°C
Molar Refractivity	= 79.78 cm^3
Molar Volume	= 191.7 cm^3
Parachor	= 518.7 cm^3
Index of Refraction	= 1.771
Surface Tension	= 53.5 dyne/cm
Density	= 1.190 g/cm^3
Dielectric Constant	= 3.24 ± 0.1
Polarizability	= 31.62 cm^3
Monoisotopic Mass	= 228.0939 Da
Nominal Mass	= 228 Da
Average Mass	= 228.2879 Da
Log P	= 5.91

Use

Chrysene is used in the manufacture of some dyes.

FLUORANTHENE

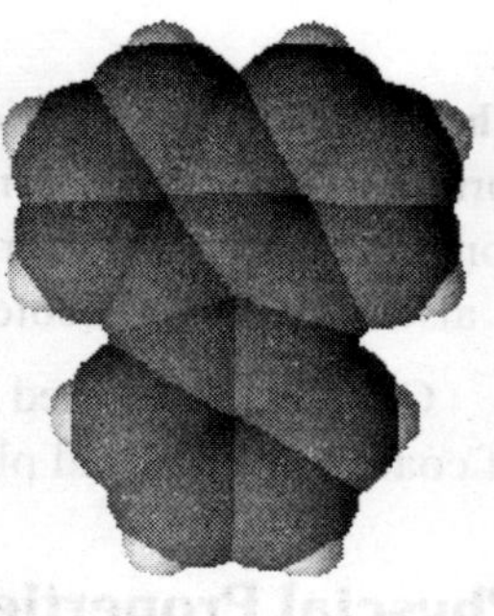

Fluoranthene is a polycyclic aromatic hydrocarbon (PAH) consisting of a naphthalene and a benzene unit connected by a five-membered ring. It is a member of the class of PAHs known as non-alternant PAHs because it has rings other than those with six carbon atoms. It is an isomer of the alternant PAH pyrene. It is not as thermodynamically stable as pyrene because its electrons cannot resonate throughout the complete structure as the corresponding ones in pyrene.

Fluoranthene is found in many combustion products, along with other PAHs. Its presence is an indicator of less efficient or lower-temperature combustion, since non-alternant PAHs are less preferred in formation than alternant PAHs. Fluoranthene is one of the U.S. Environmental Protection Agency's 16 priority pollutant PAHs and is a carcinogen.

Fluoranthene has been isolated from coal tar pitch. Its name derives fluorescence under UV light.

Physical Properties

Molecular Formula	$= C_{16}H_{10}$
Formula Weight	= 202.2506
Composition	= C(95.02%) H(4.98%)
Melting Point	$= 110^0C$
Boiling Point	$=375^0C$
Molar Refractivity	$= 72.46\ cm^3$
Molar Volume	$= 161.9\ cm^3$
Parachor	$= 449.6\ cm^3$
Index of Refraction	= 1.851
Surface Tension	= 59.4 dyne/cm
Density	$= 1.248\ g/cm^3$
Dielectric Constant	= 3.41
Polarizability	$= 28.72\ cm^3$
Monoisotopic Mass	= 202.07825 Da
Nominal Mass	= 202 Da
Average Mass	= 202.2506 Da
Log P	= 5.17

TETRACENE

Tetracene, also called **naphthacene**, **benz[b]anthracene** and **2,3-benzanthracene**, is a polycyclic aromatic hydrocarbon. It has the appearance of a pale orange powder. Tetracene is used as a sensitiser in chemoluminescence.

It is a molecular organic semiconductor, used in OFETs and OLEDs. In May 2007, researchers from two Japanese universities, Tohoku University in Sendai, and Osaka University, have reported an ambipolar light-emitting transistor made of a tetracene single crystal. Ambipolar means that the electric charge is transported by both, positively charged molecular cations and negatively charged molecular anions. Tetracene can be also used as a gain medium in dye lasers. Jan Hendrik Schön during his time at Bell Labs (1997-2002) claimed to have developed an electrically pumped laser based on tetracene. However his results could not be reproduced and it is considered scientific fraud.

Tetracene is the four ringed member of the series of acenes, the previous one being anthracene (tricene) and the next one being pentacene.

Physical Properties

Property	Value
Molecular Formula	= $C_{18}H_{12}$
Formula Weight	= 228.28788
Composition	= C(94.70%) H(5.30%)
Melting point	= 357 °C
Molar Refractivity	= 79.78 cm^3
Molar Volume	= 191.7 cm^3
Parachor	= 518.7 cm^3
Index of Refraction	= 1.771
Surface Tension	= 53.5 dyne/cm
Density	= 1.190 g/cm^3
Dielectric Constant	= 3.24
Polarizability	= 31.62 $10^{-24}cm^3$
Monoisotopic Mass	= 228.0939 Da
Nominal Mass	= 228 Da
Average Mass	= 228.2879 Da
Log P	= 5.91

HEXACENE

Hexacene is an aromatic molecule consisting of six linearly-fused benzene rings. Hexacene and its derivatives are investigated for potential applications as organic semiconductor.

Hexacene is one of a series of linear acenes, the previous one being pentacene (five fused benzene rings) and the next one being heptacene (seven fused benzene rings).

Hexacene is very unstable, being highly reactive in the 6, 15 positions due to pi electron localization. Its spectra must be collected on freshly purified samples handled in an inert atmosphere such as nitrogen or argon. Only with very large substituents it is possible to isolate an hexacene for instance the compound *6,15-bis(tri-t-butylsilylethynyl)hexacene* which melts with decomposition at 96°C.

The first reported synthesis of hexacene itself dates back to 1942. In 1955 the compound is synthesized by dehydrogenation of *hexacosadehydrohexacene* by palladium on carbon [4]. In 1982 hexacene is reported as a bluish green compound with decomposition at 380 degrees [5]. In 2007, a research group claimed the first reproducible unsubstituted hexacene synthesis (invalidating the previous claims) based on photochemical decarbonylation of a dione precursor.

hv, 395 nm

- 2 CO

O_2

The compound cannot be isolated it dimerizes even at concentration as low as 10^{-4} M and reacts in solution with oxygen to the organic peroxide. In a PMMA polymer matrix on the other hand these side-reactions are limited and the compound can survive up to 12 hours. The next homologue heptacene also studied by this group, is even more unstable (decomposes within 4 hours).

Physical Properties

Molecular Formula	= $C_{26}H_{16}$
Formula Weight	= 328.40524
Composition	= C(95.09%) H(4.91%)
Molar Refractivity	= 115.46 cm^3
Molar Volume	= 260.0 cm^3
Parachor	= 726.4 cm^3
Index of Refraction	= 1.842
Surface Tension	= 60.9 dyne/cm
Density	= 1.263 g/cm^3
Dielectric Constant	= 3.45
Polarizability	= 45.77 cm^3
Monoisotopic Mass	= 328.125201 Da
Nominal Mass	= 328 Da
Average Mass	= 328.4052 Da
Log P	= 8.37

BENZPYRINE

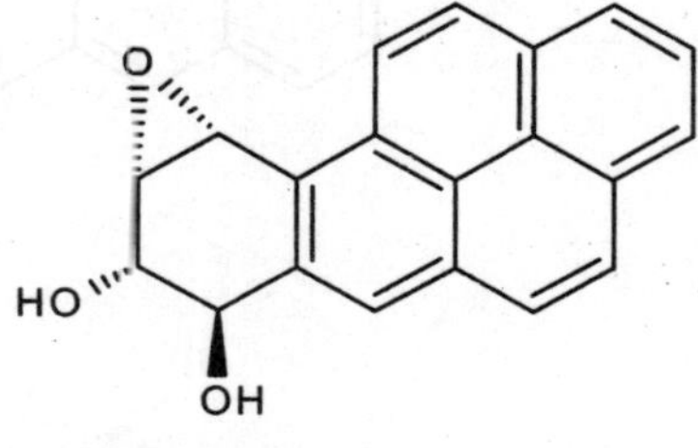

Benzo[a]pyrene, $C_{20}H_{12}$, is a five-ring polycyclic aromatic hydrocarbon that is mutagenic and highly carcinogenic. It is a crystalline yellow solid. Benzo[a]pyrene is a product of incomplete combustion at temperatures between 300 and 600 °C. Benzo[a]pyrene was determined in 1933 to be the component of coal tar responsible for the first recognized occupation-associated cancers, the sooty warts (cancers of the scrotum) suffered by chimney sweeps in 18th century England. In the 19th century, high incidences of skin cancers were noted among fuel industry workers. By the early 20th century, malignant skin tumors were produced in laboratory animals by repeatedly painting them with coal tar.

Sources of Benzo[a]pyrene

Benzo[a]pyrene is found in coal tar, in automobile exhaust fumes (especially from diesel engines), tobacco smoke, wood smoke, and in charbroiled food. Recent studies have revealed that levels of benzo[a]pyrene in burnt toast are significantly higher than once thought, although it is unproven whether burnt toast is itself carcinogenic.

Toxicity of Benzo[a]pyrene

A vast number of studies over the previous three decades have documented links between benzo[a]pyrene and cancers. It has been more difficult to link cancers to specific benzo[a]pyrene sources, especially in humans, and difficult to quantify risks posed by various methods of exposure (inhalation or ingestion). Researchers at Kansas State University recently discovered a link between vitamin A and emphysema in smokers. Benzo[a]pyrene was found to be the link to the deficiency, since it induces vitamin A deficiency in rats.

In 1996, a study was published that provided the clear molecular evidence conclusively linking components in tobacco smoke to lung cancer. Benzo[a]pyrene, found in tobacco smoke, was shown to cause genetic damage in lung cells that was identical to the damage observed in the DNA of most malignant lung tumours.

A 2001 National Cancer Institute study found levels of benzo[a]pyrene to be significantly higher in foods that were cooked well-done on the barbecue, particularly steaks, chicken with skin, and hamburgers. Japanese scientists showed that cooked beef contains mutagens, chemicals that are capable of altering the chemical structure of DNA. However, the foods themselves are not necessarily carcinogenic, even if they contain trace amounts of carcinogens, because the gastrointestinal tract protects itself against carcinomas by shedding its outer layer continuously. Furthermore, detoxification enzymes, such as cytochromes P450 have increased activities in the gut due to the normal requirement for protection from food-borne toxins. Thus in most cases small amounts of benzo[a]pyrene are metabolized by gut enzymes prior to being passed on to the blood. The lungs are not protected in either of these manners.

A recent study has found that cytochrome P450 1A1 (CYP1A1) and cytochrome P450 1B1 (CYP1B1) are both protective and, confusingly, necessary for benzo[a]pyrene toxicity. Experiments with strains of mice engineered to remove (knockout) CYP1A1and CYP1B1 reveal that CYP1A1 primarily acts to protect mammals from low doses of benzo[a]pyrene, and that removing this protection causes the biological accumulation of large concentrations of benzo[a]pyrene. Unless CYP1B1 is also knocked out, benzo[a]pyrene toxicity results from the bioactivation of benzo[a]pyrene to the ultimate toxic compound, benzo[a]pyrene-7,8-dihydrodiol-9,10-epoxide.

Interaction with DNA

Properly speaking, benzo[a]pyrene is a procarcinogen, meaning that the mechanism of carcinogensis of benzo[a]pyrene depends on enzymatic metabolism of benzo[a]pyrene to the ultimate mutagen, benzo[a]pyrene diol epoxide, pictured at right. This molecule intercalates in DNA, covalently bonding to the nucleophilic guanine nucleobases at the N2 position. X-ray crystallographic and nuclear magnetic resonance structure studies show that this binding distorts the DNA, inducing mutations by perturbing the double-helical DNA structure. This disrupts the normal process of copying DNA and induces mutations, which explains the occurrence of cancer after exposure. This mechanism of action is similar to that of aflatoxin which binds to the N7 position of guanine.

There are indications that specifically benzo[a]pyrene diol epoxide specifically targets the protective p53 gene. This gene is a transcription factor that regulates the cell cycle and hence functions as a tumor suppressor. By inducing G (guanine) to T (thymidine) transversions in transversion hotspots within p53, there is a probability that benzo[a]pyrene diol epoxide inactivates the tumor suppression ability in certain cells, leading to cancer.

Benzo[a]pyrene diol epoxide is the carcinogenic product of three enzymatic reactions:

1. Benzo[a]pyrene is first oxidized by cytochrome P4501A1 to form a variety of products, including (+)-benzo[a]pyrene 7,8-oxide.
2. This product is metabolized by epoxide hydrolase, opening up the epoxide ring to yield (-)-benzo[a]pyrene-7,8,dihydrodiol.
3. The ultimate carcinogen is formed after another reaction with cytochrome P4501A1 to yield the benzo[a]pyrene-7,8-dihydrodiol-9,10-epoxide. It is this diol epoxide that covalently binds to DNA.

Benzo[a]pyrene induces cytochrome P4501A1 (CYP1A1) by binding to the AHR (aryl hydrocarbon receptor) in the cytosol. Upon binding the transformed receptor translocates to the nucleus where it dimerises with ARNT (aryl hydrocarbon receptor nuclear translocator) and then binds xenobiotic response elements (XREs) in DNA located upstream of certain genes. This process increases transcription of certain genes, notably *CYP1A1*, followed by increased CYP1A1 protein production. This process is similar to induction of CYP1A1 by certain polychlorinated biphenyls and dioxins.

Recently, Benzo[a]pyrene has been found to activate a transposon, LINE1, in humans.

Physical Properties

Molecular Formula	$= C_{20}H_{14}O_3$
Formula Weight	= 302.32336
Composition	= C(79.46%) H(4.67%)O(15.88%)
Melting Point	= 179^0C
Boiling Point	= 495^0C
Molar Refractivity	= 92.08 cm^3
Molar Volume	= 192.5 cm^3
Parachor	= 582.7 cm^3
Index of Refraction	= 1.936
Surface Tension	= 83.8 dyne/cm
Density	= 1.569 g/cm^3
Polarizability	= 36.50 cm^3
Monoisotopic Mass	= 302.094294 Da
Nominal Mass	= 302 Da
Average Mass	= 302.3234 Da
Log P	= 3.82

Carcinogenic Compounds

It is well known that continued contact with certain tars may lead to the development of cancer, but only within the last few years this information has become available regarding the nature of the compound which may give rise to the disease. Progress in this field is large due the work of cool, which showed that 1, 2 benzpyrine a complex aromatic hydrocarbon present in tar possesses very strong carcinogenic properties. Another hydrocarbon of similar nature which has been synthesized by Cook is **1: 2: 5:6 – dibenzantracene**. Various other compounds of this kind have been prepared, the molecules of which all contain the compounds ring system of phenanthrene. The physiological effect is however, strongly dependent upon the molecular structure. Thus 4:5 – benzpyrine and 1:2 benzanthrene are almost inactive.

1:2 Benzpyrine

1:2:5:6 Benzpyrine

The most potent carcinogenic compounds so far discovered are methyl cholanthrene, which is a simple transformation product of deoxycholic acid of bile and 9:10 dimethyl 1:2 benzanthrene appear to result in the animal organism by the degradation of the sterols and bile acids, so the dehydrogenation of these compounds in a faulty mechanism may yield carcinogenic products.

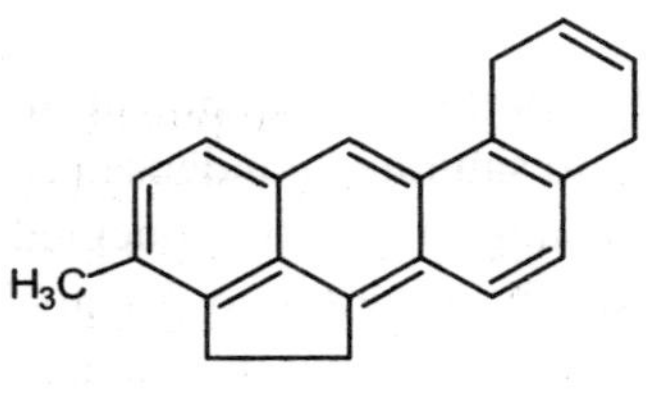

I

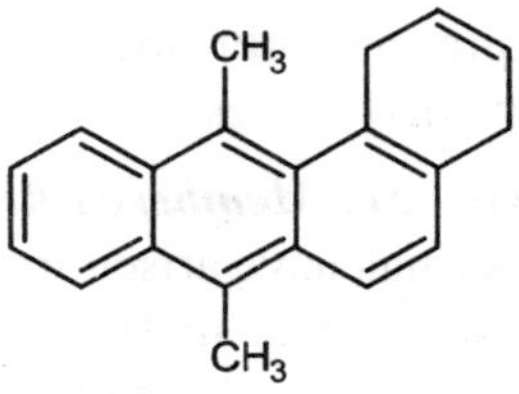

II

Stranles Ring and Condensed Ring Structures

For may years following the introduction of Bayer's Strain Theory*, ring structures were represented as being planer in configuration. Large rings containing seven or more ring atoms was supposed to be unstable and to exist in addition of strain, due to the distortion of valency bonds from the normal angle of 109, from 1905 onwards these ideas were modified by the work of Ruzita; unstable large ring compounds such as muscone and civetone and by Hückel's investigations on the decalins and hydrindanes.

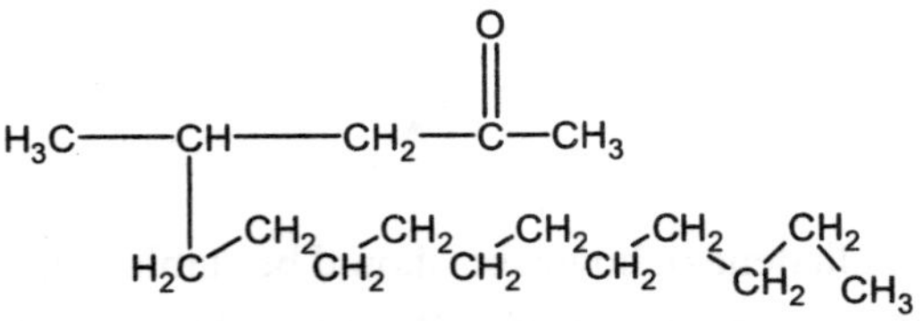

4-methylpentadecan-2-one
Muscone

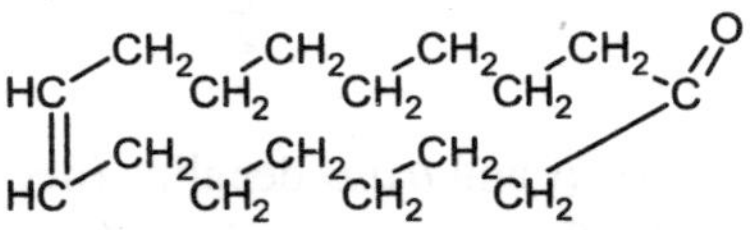

cycloheptadec-9-en-1-one
Civetone

A theory of stainless rings in which the tension is relieved by deviations from the plane model was put forward as early as 1892 in a mathematical paper by Sachse, and reached a more definite form in 1918 by Mohr. The theory may be summarized as follows. Rings containing 3, 4, and 5 carbon atoms must have a plane configuration, because this results in the least possible distortion from the normal valency angle. On the other hand, a ring containing 6 or more carbon may become completely starinless by assuming a folded form.

6 membered rings : Two models are possible on the tetrahedral basis as represented on following page, in the non – rigid *boat or trough* from and the semi rigid *chair* form. At first sight atoms 1 and 4

*we will read in the chapter of organic mechanism about this.

in the boat form appear to stand in special relationship to one another but this is not the case. An examination of the actual model, enabling the 1 – 4 position to pass successfully into the 2:5, 3 – 6 and so back to the original 1:4 arrangement. The chair model is semi – rigid and cannot merge into the boat form without definite strain being put upon the bonds. This strain, however, is considerably smaller than that required to maintain the six carbon atoms in one plane. Mohr concludes that in practice these two structures represent only one compound, in arrangement with our present knowledge about cyclohexane and its derivatives.

Fusion of Two – Six Membered Rings : The most important development of the Sachse – Mohr theory of starinless rings have arisen from its application to saturated condensed ring system saturated condensed ring system Mohr in 1918 it was the first to suggest that two buckled cyclohexane rings could be condensed together by either *cis* or *trans* linkage, yielding two decahydronaphthalenes. In addition each ring could exist theoretically in boat and chair forms, but in practice easily pass into one another.

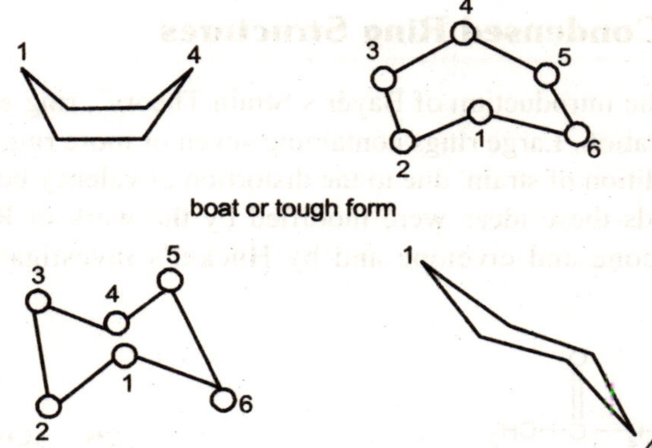

The two *cis* and *trans* decalins (decahydronaphthalenes), however, should be stable compounds as shown in the figure below, where the position of the hydrogen atoms attached to the common carbon atoms are indicated by bonds. On Bayer's theory each ring must exist in a plane form, leading to a *cis* decalin as the possible structure.

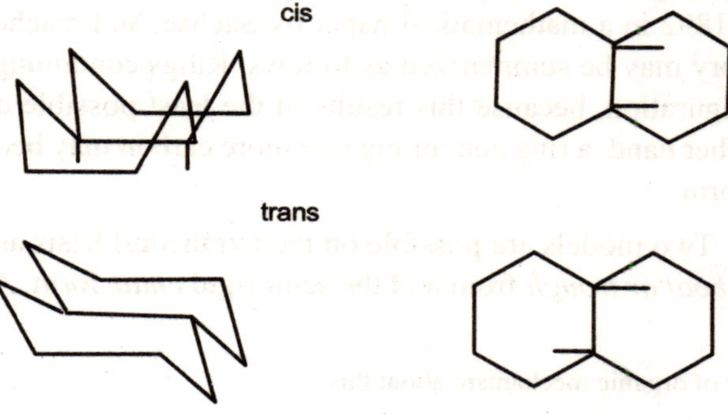

Seven years later, in 1925, the existence of two decalins was established experimentally by Hückel who started from the two known forms of β - decadol (decahydro - β - naphthol). These compounds had been reduction products of β - naphthols, although the nature of their isomerism was not known.

Hückel found that on oxidation with potassium dichromate and acetic acid, each β - decanol gave rise to a different β - decalone, a fact, which could be explained on Mohr's theory but not on the basis of a plane ring structure. Each decalone was reduced to a decalin by use of amalgamated zinc and hydrochloric acid (Memmensen) and each decalol was converted by oxidation with cold acid permanganate solution into a mixture of cyclohexane – I – 2 diacetic acid and cyclohexane – I – carboxy – 2 propionic acid. These two acids are obtained from β -decalol where the *cis*- configuration. This decalol and related decalone and decalin are therefore also of *cis* – structure.

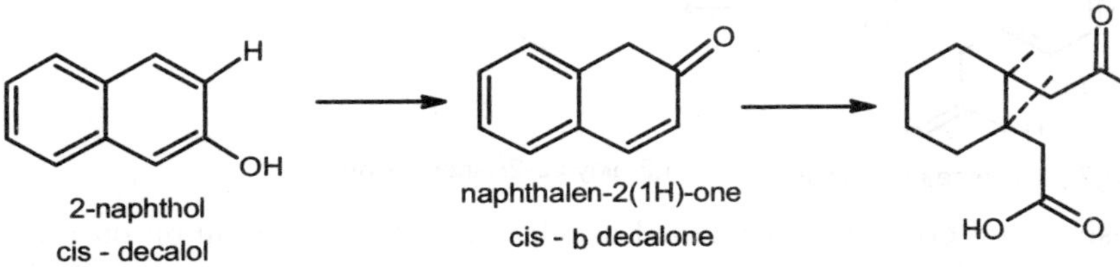

H OH

2-naphthol

cis - decalol

naphthalen-2(1H)-one

cis - b decalone

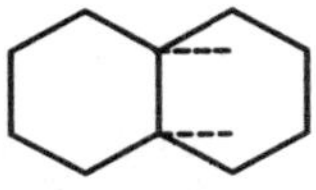

O OH HO O

cis - b decaline

The other series of compounds prepared from β - decalol where proved in a similar manner to possess *trans* structure. The physical properties of these are shown below

Name	Melting Point	Boiling Point	Density	Refractive Index
Trans - decalin	– 36°C	185°C	0.872	1.4713
Cis - decalin	– 51	193	0.898	1.413

The same two decalins where also isolated by careful fractionation of the mixture obtained from the direct hydrogenation of naphthalene. They are stable compounds although *cis* – decalin is slowly and almost quantitatively isomerised into *trans* decalin at room temperature in contact with aluminium chloride. The lower stability of the *cis* form under these conditions is in arrangement with it's somewhat heat of combustion.

It may be noted that β - decalol has three asymmetric carbon bonds an additional possibility of isomerism having been introduced owing to the *cis* and *trans* arrangement of the CHOH group which represent to the decalin residue. All four of the expected racemic* forms where isolated by Hückel by

*Constituting a compound or a mixture that is composed of equal amount of dextrorotatory and levorotatory forms of the same compound and is optically inactive. About this we will read in the chapter of organic mechanism.

reduction of the *cis* and *trans* β - decalones. Reduction of the oximes of the decalones resulted in the separations of four racemic β - amino – decalins.

Fusion of a Six with a Five Membered Ring Theoretically a plane five membered ring can only be attached to cyclohexane to form a strainless compound if the union is by *cis* – linkage. On the assumption of buckled rings, however, the tension arising from *trans* coupling is not great.

Hückel has made a careful study of the hydridane system and isolated both *cis* and *trans* hydrindanones. The structures of these compounds follow from their mode of formation from *cis* and *trans* cyclohexane diacetic acids by distillation with acetic anhydride.

2,2'-(1,2-phenylene)diacetic acid → 1,3-dihydro-2*H*-inden-2-one + CO_2 + H_2O

or from the corresponding esters by an internal acetoacetic ester condensation on the presence of sodium ethoxide, followed by hydrolysis with mineral acid and loss of carbon dioxide, thus with the *trans* ester.

Cyclohexane *trans* - diacetic ester → *trans* - hydrindanone α carboxylic ester → *trans* β hydrindanone

trans - β - hydrindanones is more stable than the *cis* – compound and has a lower heat of combustion. This is not in an agreement with theoretical conclusions, and led Hückel to suggest that the strain theory cannot be applied in a simple manner to a bicyclic system.

The above configurations where also confirmed by reducing the ketones to the hydrindanols. As was to be expected from the models, cis β - hydrindanols occurs in two geometrical isomerides each of which possesses a plane of symmetry containing the H and OH groups.

cis - form   *trans* - form

Only one *trans* - β hydrindanols exists and this is a racemic compound, which can be resolved into two optically active forms.

Fusion of Two Five membered rings : Theoretical considerations suggest that nay deviation from the plane model in either 5 – ring will result intension and that only fusion in the *cis* – position will give a stable compound. This is kept out by the work of Linstead, Meade and Cook on the bicycle – octanones. These compounds where prepared from the *cis* and *trans* Cyclopentane – diacetic acids, $C_5H_8\ (CH_2COOH)_2$. Of the *cis* and *trans* β - bicycloctanes (I and II), the former is the more readily obtained and is the more stable one.

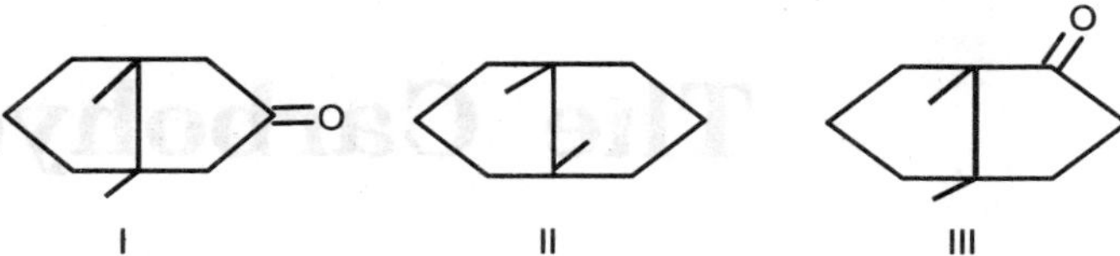

cis α - Bicyclo – octane (III) was the only a - ketone which could be isolated as the *trans* isomerides is apparently converted into the *cis* – form under the influence of heat.

Similar stability relationship exist among anhydrides of this type. Thus the anhydride of *cis* cylcopentane – I – carboxy – 2 – acetic acid is more stable that that of the *trans* compound.

CHAPTER

34

The Carbohydrates

Introduction

Carbohydrates, as the name implies believed to consist solely of the elements carbon and water in a ration 1:1 but our recent understanding is that a number of other atoms and functional groups can be present in a carbohydrate and the original formula of $C_xH_{2x}O_x$ is no long is no longer rigidly adhered to. Carbohydrates are traditionally considered to three broad groups :

1. Monosaccharide
2. Oligosaccharides
3. Polysaccharide

With monosaccharide being the simplest one, which cannot be hydrolyzed into smaller ones and from which the two other groups are obtained by formation of glycosidic linkages. This simple classification system defines oligosaccharides as simple polymers of monosachharides containing between two and 10 monosachharides unites or residues. Polysaccharides are higher molecular weight polymers of monosachharides which contain in excess of 10 residues, whilst the division between oligosaccharides and polysaccharides is arbitrarily set at 10 residues, nature has made the division much easier to observe since carbohydrates containing between five and 15 residues rarely exist in nature while only a few consist of 25 to 75 residues. The majority of naturally occurring polysaccharides contain 80 to 100 residues although a few negative cellulose comprising a series of polymers with an average molecular weight distribution equivalent to 3000 residues. Such weight but a range of macromolecules having a distribution of molecular weights. This phenomenon is known as mocroheterogneity.

Nomenclature

The members of these groups of compounds are denoted by the term – *ose*, the hydroxy – aldehyde (%CHO group) being called *aldoses* and the hydroxy ketones (–CO) are called *ketoses*. Whilst the number of carbon atoms in the compound is indicated by prefixing the great numerical to the syllable – ose. Thus glycolic aldehyde is an aldo – triose; where as dihydroxy – acetone is a keto – triose. The necessity for this system of nomenclature is apparent when dealing with the higher members of the series, each of which can exist, in several (glyceric aldehyde) 4 possible aldo – ketoses, 8 – possible aldopentoses and 16 possible aldo – hexoses. There are also heptoses, octoses, nonoses and decose with 7, 8, 9 and 10 atoms.

O=CH–CH(OH)–H_2C–OH

Aldotriose

HO–CH_2–C(=O)–CH_2–OH

keto triose

The existence of the large number of isomers in the example given above is due to the difference in space arrangement or configuration of the atoms, which is discussed in a later part of this chapter.

Classification of Carbohydrates

A. Monosaccharides (simple sugars)

1. **Hexoses**

   a. ***Glucose:*** honey, fruits, corn syrup, sweet grapes, sweet corn; hydrolysis of starch and cane sugar. Physiologically the most important sugar; the "sugar" carried by the blood and the principal one used for tissues.

   b. ***Fructose:*** honey, ripe fruits, some vegetables; hydrolysis of sucrose insulin. Can be changed to glucose in the liver and intestine and is an intermediate metabolite in glycogen breakdown.

   c. ***Galactose:*** not found free in nature; digestive end product of lactose hydrolysis. Can be changed to glucose in the liver; synthesized in body to make lactose and is a constituent of glycolipids.

   d. ***Mannose:*** found in legumes; hydrolysis of plant mannosans and gums. A constituent of polysaccharide of albumins, globulins, and mucoids.

2. **Pentoses**

   a. ***Arabinose:*** derived from gum arabic plus plum and cherry gums; not found free in nature. Has no known physiologic function in man; used in metabolism studies of bacteria.

   b. ***Ribose:*** derived from nucleic acid of meats and seafoods. Structural element of nucleic acids, ATP, and coenzymes (NAD and FAD).

   c. ***Ribulose:*** formed in metabolic processes. Intermediate in direct oxidative pathway of glucose breakdown.

   d. ***Xylose:*** wood gums, corncobs, and peanut shells; not found free in nature. Very poorly digested and has no known physiologic function; used medicinally as a diabetic food.

B. Oligosaccharides (2-10 sugar units)

1. **Disaccharides**

   a. ***Sucrose:*** cane and beet sugar, maple syrup, molasses, and sorghum. Hydrolyzed to glucose and fructose; a non-reducing sugar.

   b. ***Maltose:*** malted products and germinating cereals; an intermediate product of starch digestion. Hydrolyzed to two molecules of glucose; a reducing sugar; does not occur free in tissues.

   c. ***Lactose:*** milk and milk products; formed in the body from glucose nature. Hydrolyzed to glucose and galactose; may occur in urine during pregnancy; a reducing sugar.

2. **Trisaccharides**

   a. ***Raffinose:*** cottonseed meal, molasses, sugar beets and stems. Only partially digestible but can be hydrolyzed by enzymes of intestinal bacteria to glucose, fructose, and galactose.

   b. ***Melizitose:*** honey, poplars, and conifers. Composed of one fructose unit and two glucose units.

C. Polysaccharides (more than 10 sugar units)

1. **Digestible**

   a. ***Glycogen:*** meat products and seafoods; polysaccharides of the animal body, often called animal starch; storage form of carbohydrates in body, mainly in liver and muscle.

   b. ***Starch:*** cereal grains, unripe fruits, vegetables, legumes, and tubers. Most important food source of carbohydrates; storage form of carbohydrates in plants. Composed chiefly of amylose and amylopectin; hydrolyzed to glucose.

   c. ***Dextrin:*** toasted bread, intermediate product of starch digestion. Formed in course of hydrolytic breakdown of starch.

2. **Partially Digestible**

   a. ***Inulin:*** tubers and roots of dahlias, artichokes, dandelions, onions, and garlic. Hydrolizable to fructose; used in physiologic investigation for determination of glomerular filtration rate.

   b. ***Mannosan:*** legumes and plant gums. Hydrolyzable to mannose but digestion incomplete; further splitting by bacteria may occur in large bowel.

3. **Indigestible**

   a. ***Cellulose:*** skins of fruits, outercoverings of seeds, plus stalks and leaves of vegetables. Not subject to attack of digestive enzymes in man, thus an important source of "bulk" in diet; may be partially split to glucose by bacterial action in large bowel.

   b. ***Hemicellulose and pectin:*** woody fibers and leaves. Less polymerized than cellulose; may be digested to some extent by microbial enzymes, yielding xylose.

Indigestible is another name for indigestable polysaccharides is dietary fiber which is comprised of 2 groups:

c. ***Insoluble dietary fibers*** (cellulose, lignin, and cutin) which are the most abundant organic compounds in the world. They help prevent constipation, colon cancer, and diverticulosis, but not hypercholesterolemia.

d. ***Soluble dietary fibers*** (hemicellulose, pectins, gums, and algal polysaccharides) which are useful in decreasing serum cholesterol and in regulating blood glucose levels.

MONOSACCHARIDES

Introduction

The number of monosaccharides known in the neighbourhood of fifty, of which 10 occur in nature and the remaining are synthetic. The existence of such a large number of comnpounds is due to the presence of presence of asymmetric carbon atoms within the molecules. Aldohexoses, for example which includes glucose, a sugar of great theoretical and practical interest, contain no less than four asymmetric atoms, each of which may be present in either the *d* and *l* – configuration. It has already been shown that how rapidly stereo isomerides increases with each additional asymmetrical atom.

A list of best monocaccharides is given below in detail in the table

Name	Formula	Chemical name
Biose	$CH_2OH.CHO$	Glycolic aldehyde
Trioses	1. $CH_2OH.CHOH.CHO$	Glyceric aldehyde
	2. $CH_2OH.CO.\ CH_2OH$	Dihydroxy acetone
Tetrose	$CH_2OH.(CHOH)_2.CHO$	Erithrose
Pentose	$CH_2OH(CHOH)_3CHO$	Binose, xylose, ribose
Methyl pentose	$CH_3(CHOH)_4CHO$	Rhamnose
Hexose	1. $CH_2OH(CHOH)CHO$	Glucose, gulose, talose mannose
	2. $CH_2OH(CHOH)_4CHO$	Fructose , sorbose
Heptose	$CH_2OH(CHOH)_3CO\ CH_2OH$	Monoheptose, glucoheptose, galaheptose
Octose	$C_8H_{16}O_8$	
Nonose	$C_9H_{18}O_9$	

An explanation is required as to the way in which the optical isomerides are named. In accordance with a suggestion of Fischer, the structure of the sugar, where ever possible, referred back to that of the active glucoses. Any other monosaccharide standing in close structural relationship to one or other of the glucose is generally labelled with the corresponding letter *d* and *l* – irrespective of the actual sign of its rotation. Ordinarily fructose, or example, which is lævorotatory, but termed as *d* – fructose, owing to its spatial relationship to *d* – fructose. Naturally occurring arbinose, which is dexorotatory, is termed as l – arbinose to indicate its relationship to l – glucose. This is illustrated in the following formulæ, in which the terminal aldehydic or ketonic group is written uppermost links blinding the H and

OH addenda are assumed to be inclined from the chain of central carbon atoms towards the observer. According to the usual convention a CHOH group having the hydroxyl on the right of the formula employed.

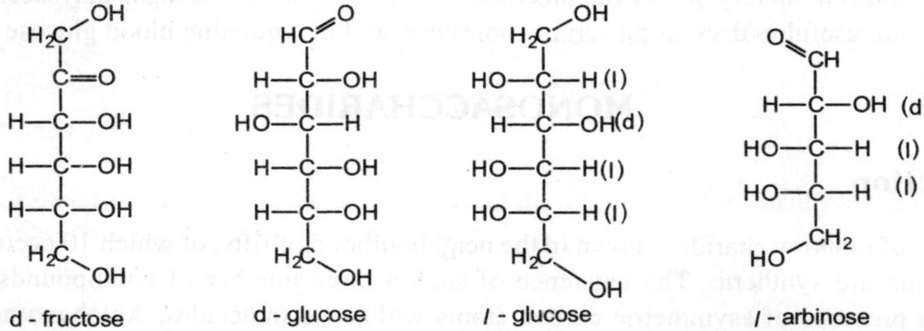

d - fructose   d - glucose   *l* - glucose   *l* - arbinose

In some cases the actual sign of the rotation given by a compound is indicated in the following pages by the use of the sign + and –, eg d(–) fructose for ordinary fructose. In addition, the prefix dl – indicates a racemic and mesoform.

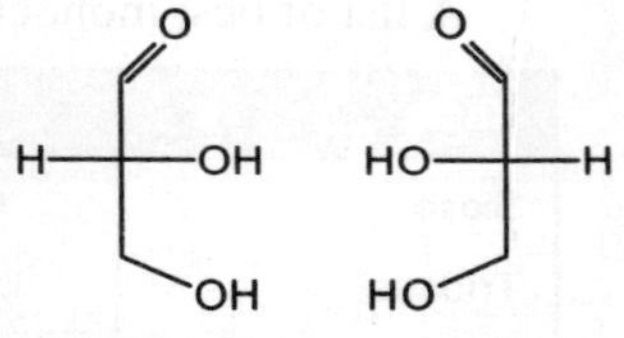

Wohl's work on glycerose now enables the family relationship to be carried back to the *d* and *l* glyceroses from which all other sugars can be derived by extending the molecule on the side of the aldehyde group by reaction with. The nomenclature fortunately remains unchanged as *d* – glycerose is *d* – rotatory and is genetically related to *d* – glucose. Hence it will be seen that the family of a sugar is determined in each case by the spatial arrangement of the CHOH group adjustment to the terminal CH_2OH, no matter what the disposition of the rest of the molecule may be. If this group has a dexo configuration, as indicated by writing it with OH to the right the sugar is classified as belonging to the *d* – family. The same system is applied to other derivatives of monosaccharides.

d glecerose   L glecerose

General Properties of Monosaccharides

The monosaccharides possess strong reducing properties, causing the separation of the silver form ammonia – silver nitrate solution and precipitating cuprous oxide from an alkaline solution of copper sulphate. They thus resembles aldehydes, moreover, like aldehydes and ketones, they form cyanohydrines with hydrocyanic acid, oximes with hydroxylamine, and phenylhydrazones with phenylhydrazones. The phenylhydrazones are as rule very soluble in water but by the further action of phenylhydrazine, they are converted into a yellow insoluble compound called osazones. The osazones are nearly insoluble in water, and readily separate from a solution of monosaccharide. They have, more over a definite melting point, and seen under microscope possess a characteristic appearance.

The reaction is therefore regarded as the most important process in detecting and identifying certain sugars. The monosaccharides are readily oxidized. The aldoses yield mono and di basic acids containing the same number of carbon atoms; the ketones break up into acids with a fewer carbon atoms. They all yield oxalic acid when warmed with strong nitric acid, with strong hydrochloric acid they form levulinic acid. Finally they undergo alcoholic fermentations with yeast. These reactions are going to be discussed under glucose.

Hexose

GLUCOSE

Introduction

Glucose a monosaccharide (or simple sugar), is an important carbohydrate in biology. The living cell uses it as a source of energy and metabolic intermediate. Glucose is one of the main products of photosynthesis and starts cellular respiration in both prokaryotes and eukaryotes. The name comes from the Greek word *glykys* (γλνκνζ), which means "sweet", plus the suffix "-ose" which denotes a sugar.

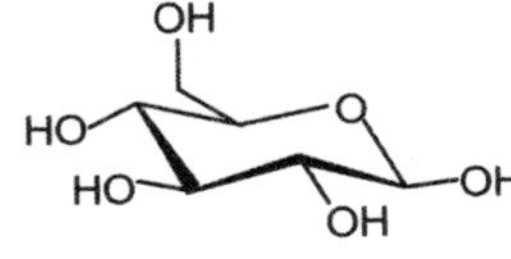

Two stereoisomers of the aldohexose sugars are known as glucose, only one of which (D-glucose) is biologically active. This form (D-glucose) is often referred to as **dextrose monohydrate**, or, especially in the food industry, simply **dextrose** (from *dextrorotatory glucose*. The mirror-image of the molecule, L-glucose, cannot be metabolized by cells in the biochemical process known as glycolysis. Glucose is commonly available in the form of a white substance or as a solid crystal. It can also be commonly found as an aqueous solution.

Synthesis

In small quantities pure glucose is most readily obtained from cane sugar. Cane sugar is dissolved in 80% alcohol and a little strong hydrochloric acid added. On gently warming the mixture, the cane sugar is hydrolysed, and breaks up into glucose and fructose

$$\underset{\text{Cane sugar}}{C_{12}H_{22}O_{11}} + H_2O \rightarrow \underset{\text{Glucose}}{C_6H_{12}O_6} + \underset{\text{Fructose}}{C_6H_{12}O_6}$$

Glucose, being less soluble in alcohol separates out as crystals.

Glucose is manufactured by boiling starch with dilute sulphuric acid; the starch is hydrolysed and converted into glucose. The liquid neutralized with soda ash and filtered and filtrate decolourised by filtration through animal charcoal. (See Fig. 34.1).

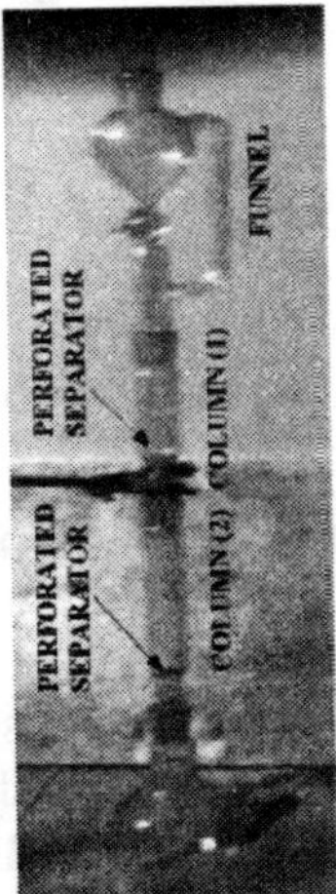

Fig. 34.1. Showing the picture of the column filter with animal charcoal used in Author's lab

The solution is evaporated to the requisite consistency in vacuum pans. The product solidifies on cooling and forms an amorphous looking mass, which always contains dextrin.

The vacuum pan (See Fig. 34.2)

It consists of an iron pan which is heated by vertical steam coils placed in the lower part of the vessel; two or three pans are connected so that the steam rising from the evaporation of the liquid in the first pan is utilized for heating the next and its vapour passes on to the third. Between the pans a small cylindrical vessel is interposed, which serves to collect any juice which "primes" or collects and carried over during boiling. The evaporation is continued until the liquid is so far concentrated that it shows "grains" or commences to crystallize. It is then run out and cooled and the un crystallized portion or molasses separate in a centrifugal extractor.

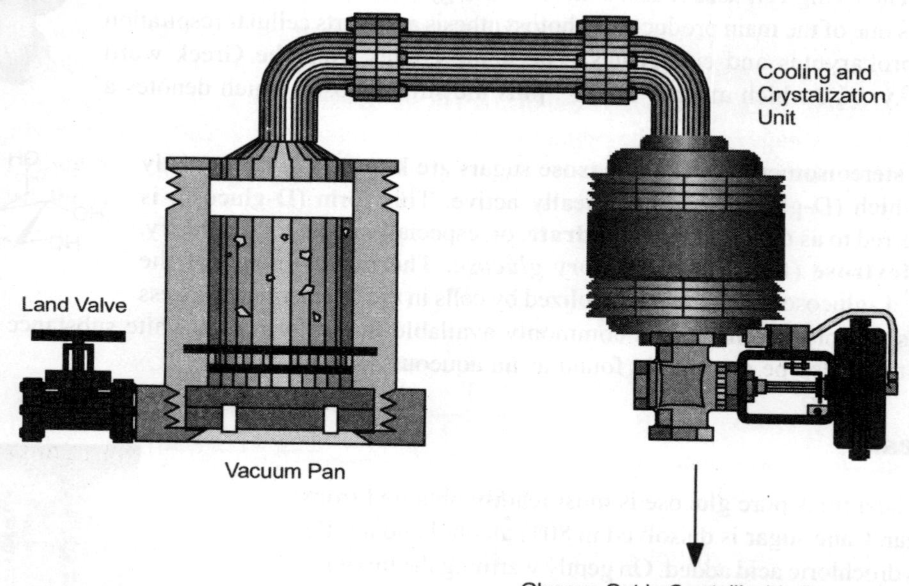

Fig. 34.2. Showing the Manufacturing process of Glucose from starch

Properties of Glucose

Pure glucose dissolves in 1.2 parts of water. It crystallizes from aqueous solution with 1 molecule of water, the crystal melts at 66°C, and whilst from alcohol the anhydrous compound separates, melting at 146°C. Glucose is dexorotatory in aqueous solution, when freshly dissolved in water it shows a specific rotation of +110.2°, which falls rapidly, and then gradually more and more slowly until after the lapse of about 6 hours it reaches a constant value of + 52.5°. This change of rotatory behaviour is called *muta rotation*, and this is due to structural changes which occur in solution. Many other sugars under go muta rotation. Another variety of glucose obtained by crystallization from alcohol has a specific rotation of + 19°, a solution of which on standing rises to + 52.5°. The two sugars are called α and β glucose (see constitution) and the value + 52.5°C is the specific rotatory power of an equilibrium mixture. In genral the α - sugars are characterized by a *fall* and the β - sugars a *rise* in the initial value of the rotation.

When lime or baryta solution is added to a solution of glucose and then alcohol, glucose, glucosates of calcium or barium are precipitated. These compounds are soluble in water and are decomposed by carbon dioxide into the original sugar and the carbonate of the metal. Calcium glucostate has the formula $C_6H_{12}O_6CaO$.

Some Detail Physical Properties of Glucose

Molecular Formula	$= C_6H_{12}O_6$
Formula Weight	= 180.15588
Composition	= C(40.00%) H(6.71%) O(53.29%)
Melting point	= 146°C
Molar Refractivity	$= 37.25 cm^3$
Molar Volume	$= 104.0\ cm^3$
Parachor	$= 312.7\ cm^3$
Index of Refraction	= 1.635
Surface Tension	= 81.7 dyne/cm
Density	$= 1.732\ g/cm^3$
Polarizability	$= 14.76\ 10^{-24} cm^3$
Monoisotopic Mass	= 180.063388 Da
Nominal Mass	= 180 Da
Average Mass	= 180.1559 Da

Structure of Glucose

Glucose forms a phenoacetyl derivative with acetic anhydride and therefore contains a hydroxyl groups. Each hydroxyl group probably attached to the different carbon atom, seeing that the arrangement of attachment of 2 hydroxyl groups to the same carbon atom would form a very unstable arrangement. The only point which is left uncertain chain. This point is determined by the reduction of glucose to sorbitol, and the conversion of the latter by means of hydriodic into normal secondary hexyl iodide.

It has also been shown that glucose combines with hydrocyanic acid and forms cyanhydrin which, on hydrolysis, yields an acid, By the reduction of this acid with hydriodic acid, normal heptylic acid is produced. These changes are represented by the following formulæ.

glucose → glucose cyanohydrin → hydroxy heptic acid → heptylic acid

The formula of glucose would therefore appear to be a penta hydroxy aldehyde

$$CH_2(OH).CH(OH).CH(OH).CH(OH).CH(OH).CHO$$

The formula does not, however strictly confirm to its behaviour for a study of the two methyl glucosides has revealed the fact that both glucose and the glucosides exist in cyclic form. Since the terminal (aldehyde) group of glucose is involved in the ring formation, this carbon atom becomes asymmetric, and in this way gives rise to the stereoisomers α and β - glucose and α and β glucosides. Further investigations has shown that the rings are 6 membered, with oxygen atom linked to the first and fifth group were present, it can presumably isomerise easily to an open chain structure. The α and β ring structure and the open chain formula for glucose are as follows.

It is convenient to be able to refer to the different carbon atoms by means of numbers, which are indicated on the right hand side. The two ring structures differ only in the disposition of the groups attached to No 1 carbon atom. The configuration of the other four asymmetric carbon atoms (no 2, 3, 4, and 6) has been accurately determined, and is here indicated by means of projection formulæ (see chap organic mechanism). Although these projection formulæ are quite satisfactory for open chain compounds, some confusion is at to arise whenever the chain is closed to a ring, so that Howorth has adopted the use of perspective formulæ, which reveal the ring as a hexagon the groups attached to the carbon atoms being shown either above or below the plane of the hexagon as the case may be.

α-glucopyranose β - glucopyranose

These are called pyranose forms. Other rings called furanose rings are formed with only four carbon atoms and one of oxygen. These terms are derived from *pyron* and *furon*.

Since every asymmetric atom gives rise to two stereoisomers every additional one of them must produce two more form each of these. Thus the aldopentoses with three asymmetric carbon atoms exist in eight active form, whilst the aldo hexoses with 4 asymmetric atoms give rise to 2 × 8 or 16, all of which are known, the most important of these are *d* – glucose, *d* – galactose and *d* – mannose.

Chemical Properties

In considering simple sugars or monosaccharides, it is useful to discuss in detain the chemical properties of the glucose. The fact is that it may be isolated from the acid – catalyzed hydrolysis of starch, suggests that the carbon atoms in glucose are attached to each other by carbon – carbon bonds rather than more labile ether bonds (although the existence) of the later is addition to the carbon – carbon bonds not excluded).

Quantitative Acylation of glucose (or equivalent benzoylation) suggests (in the equation below) that there are five hydroxy groups per molecule.

$$\text{(i) } C_6H_7O(OH)_5 + 5(CH_3CO)_2O \xrightarrow{Na^+OCOCH_3^-} C_6H_2O(OCOCH_3)_3 + 5CH_3CO_2H$$

Treatment of glucose with anhydrous methanol in presence of hydrogen chloride replaces only one hydroxy group with dilute aqueous acid, so that the product assumed be acetal.

$$C_6H_{11}O_8(OH) + CH_3OH \leftrightarrows C_6H_{11}O_5(OCH_3) + H_2O$$

All the hydroxy group in the sugar can be methylated by the treatment with methyl iodide and silver hydroxide. Of the methyl group the acetyl methyl is readily removed by the hydrolysis, others are not because they are linkages.

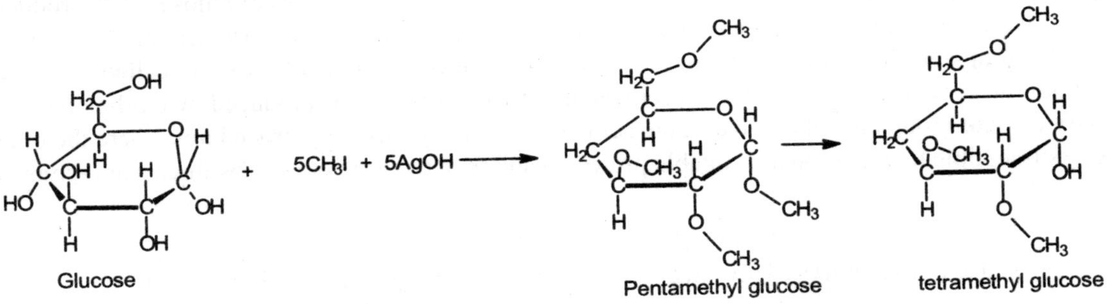

Glucose + $5CH_3I + 5AgOH \longrightarrow$ Pentamethyl glucose $\longrightarrow$ tetramethyl glucose

These results show that four of the five hydroxy groups are alcohol groups, while one is a hemiacetal group.

The presence of a carbonyl group (or equivalent) in glucose is demonstrable by treatment with hydroxylamine. The anticipated oxime is produced. However treatment of monosaccharide with phenylhydrazine produce an osazone reaction similar to that observed with benzoin. Oxidation occurs only at the hydroxy group adjustment to the original carboxyl group, so that a phenyl osazone result.

Glucose + phenylhydrazine → Glucose phenyl osazone + aniline + NH_3 (ammonia) + $2H_2O$ (water)

Both aldoses and ketoses are readily oxidized by Fehling's solution and Benedict's reagents these reagents are important for the determination of free available carboxyl group. Both consist of alkaline complexes of cupric ion. Fehling's solution is a mixture of citrate cupric ion and sodium carbonate. These reagents are reduced to cuprous oxide by α-hydroxy aldehydes and α- hydroxyketones. A positive test is indicated by complete disappearance of colour from the solution and the formation of a coral to red – brown precipitate.

One Possible Tatatocupriccomplex

Bromine and dilute acid are reagents useful for oxidation of aldehyde groups of aldoses. The products are termed aldonic acids. Oxidation of a aldoses with boiling concentrated nitric acid attack both the aldehyde group and the primary alcohol group to give a dicarboxylic acid formerly called saccharic acid, but now given the generic name glyceric acid. The aldehydic acids produced by oxidation of the primary alcohol group are called glycuronic acid. This oxidation involves procedures where the other groups susceptible to oxidation are suitable protected and then the protecting substituents are removed later.

$$RCO + Br_2 + H_2O \rightarrow RCO_2H + 2HBr$$

$$HCCH_2(CHOH)_nCHO + 6HNO_3 \rightarrow HOCO(CHOH)_nCO_2H + 6NO_2 + 4H_2O$$

Saccharic acid ← Gluconic acid ← Glucose → Glucuronic acid

Like similar aldehydes and ketones monosaccharides react with hydrogen producing cyanohydrines. This fact provides a tool effectual in the structural analysis of mono-saccharides as well as a means of synthesis of aldoses of one more carbon atom per molecule than the original aldose.

Interconversions: Heinrich Killani was first to apply cyanohydrin formation to the synthesis of higher aldoses. The reactions of his method (Killani synthesis) are summerised in outline. The new asymmetric center marked with an asterisk makes possible two product dia stereoisomers, which are formed in unequal amounts and differ in configuration at only one carbon atom (the 2 – position). These are called *epimers*[1].

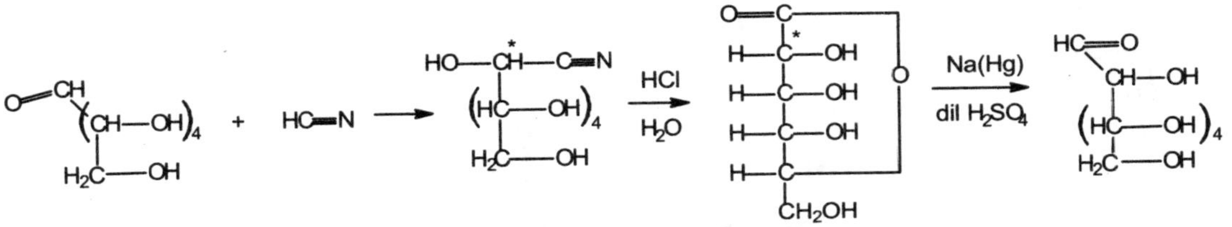

$$O{=}CH-(CH-OH)_4-H_2C-OH + HC{\equiv}N \longrightarrow HO-\overset{*}{C}H-C{\equiv}N,\ (HC-OH)_4,\ H_2C-OH \xrightarrow[H_2O]{HCl} \text{lactone} \xrightarrow[\text{dil } H_2SO_4]{Na(Hg)} HC{=}O,\ CH-OH,\ (HC-OH)_4,\ H_2C-OH$$

Vigorous reduction of the seven carbon lactone (I) with hydriodic acid results in the formation of heptanoic acid, This demonstrates the straight – chain nature of the carbon skeleton of glucose, as well as aldehyde (as opposed to keto) structure,

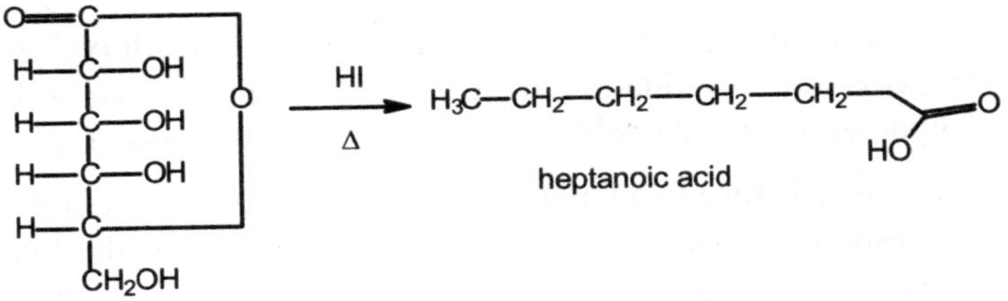

$$\text{lactone (I)} \xrightarrow[\Delta]{HI} H_3C-CH_2-CH_2-CH_2-CH_2-CH_2-C(=O)OH$$

heptanoic acid

---

[1]In chemistry , an **epimer** is a stereoisomer of another compound that has a different configuration at only one of several stereogenic centers. Stereoisomers include enantiomers and diastereomers, both which contain a stereogenic center (excluding geometric isomers, which is a class of diastereomers).

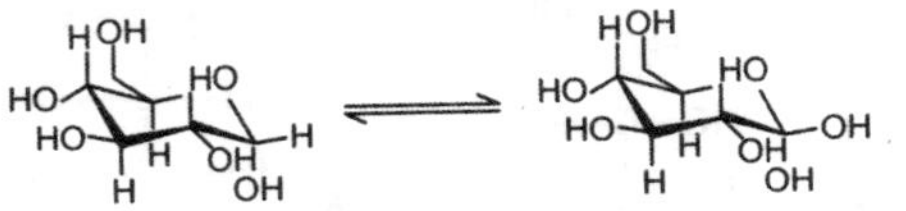

For example, the sugars α-glucose and β-glucose are epimers. In α-glucose, the -OH group on the first (anomeric) carbon is in the direction opposite the methylene group (in the axial position). In β-glucose, the -OH group is oriented in the same direction as the methylene group (in the equatorial position). These two molecules are both epimers and anomers

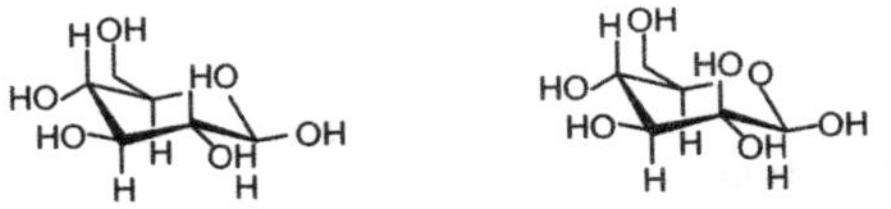

In this case, β-D-glucopyranose and β-D-mannopyranose are epimers because they differ only in the stereochemistry at the 2 position. The hydroxyl group in β-D-glucopyranose is equatorial (in the "plane" of the ring) while in β-D-mannopyranose the 2 hydroxyl group is axial (up from the "plane" of the ring). These two molecules are epimers but not anomers.In chemical nomenclature, one of the epimeric pairs is given the prefix **epi-** for example in quinine and *epi-quinine*. When the pairs are enantiomers, the prefix becomes **ent-**.

Degradation of aldoses to aldoses with one less carbon atom can be accomplished by either, in retrograde order of the Killani synthesis. This method is summerised in equation; the second degradation is the Ruff degradation.

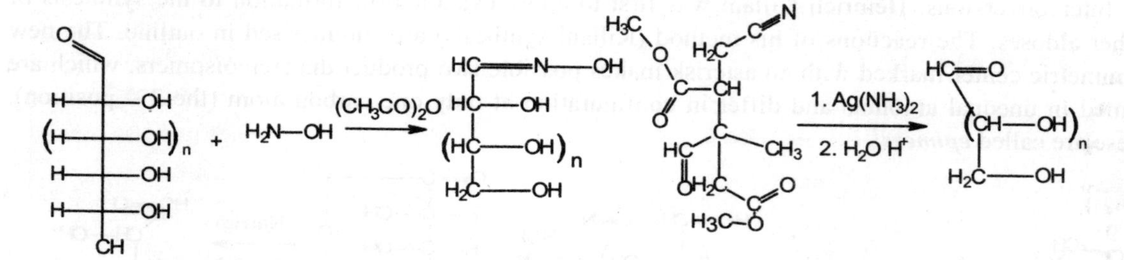

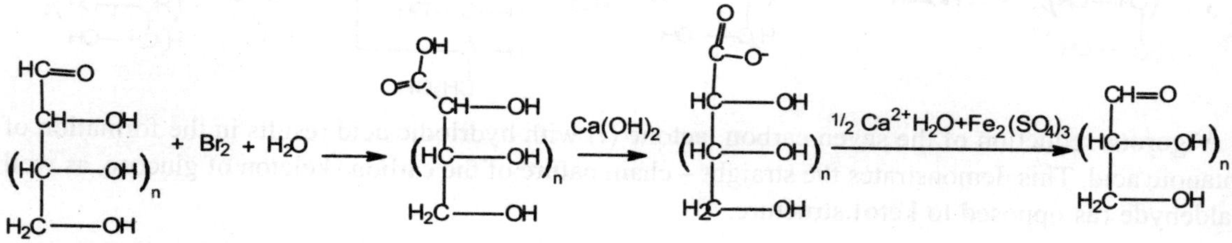

Conversion of aldose to a 2 ketose involves formation of the 1.2 – dicarboxyl compounds either by osazone synthesis and hydrolysis or by oxidation of the 2 – hydroxy group with hydrogen peroxide and ferric sulphate. The aldehyde group can be reduced preferentially with powdered zinc and acetic acid.

The reverse, conversion of a ketose to an aldose, is more equivocal. The first step is reduction of the ketose to the two corresponding stereo – isomeric polyols. Either of the two primary alcohols groups of a molecule of each polyol is then oxidized to each polyol is then oxidized to yield a lactone mixture. The lactone are then reduced to aldoses. Either the mixture of intermediate lactones or that of final aldoses must be separated by fractional crystallization.

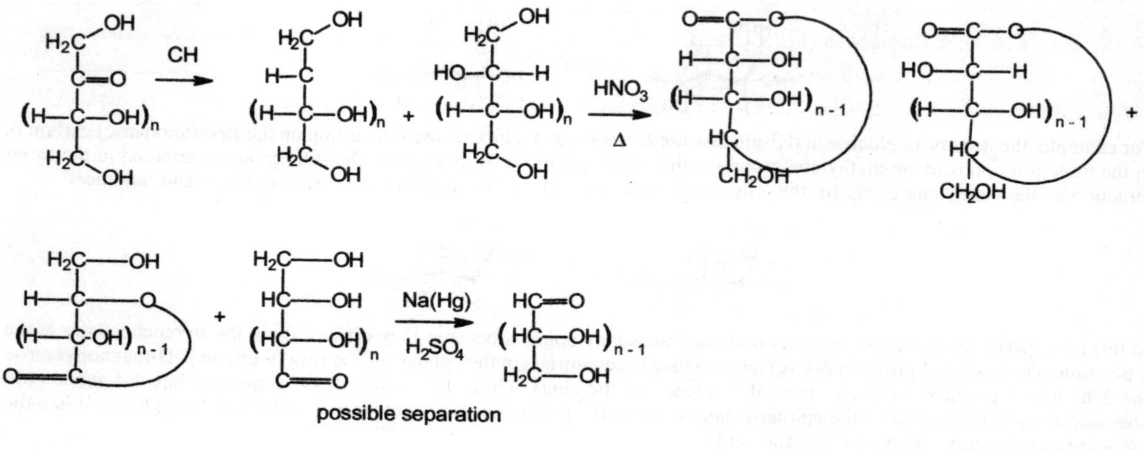

possible separation

Conversion of an aldose to its epimer at C_2 is readily accomplished. In practice, epimerization depends upon formation of equilibrium between the two epimeric aldonic acid in boiling pyridine. Consequently the same equilibrium results by starting with either epimer.

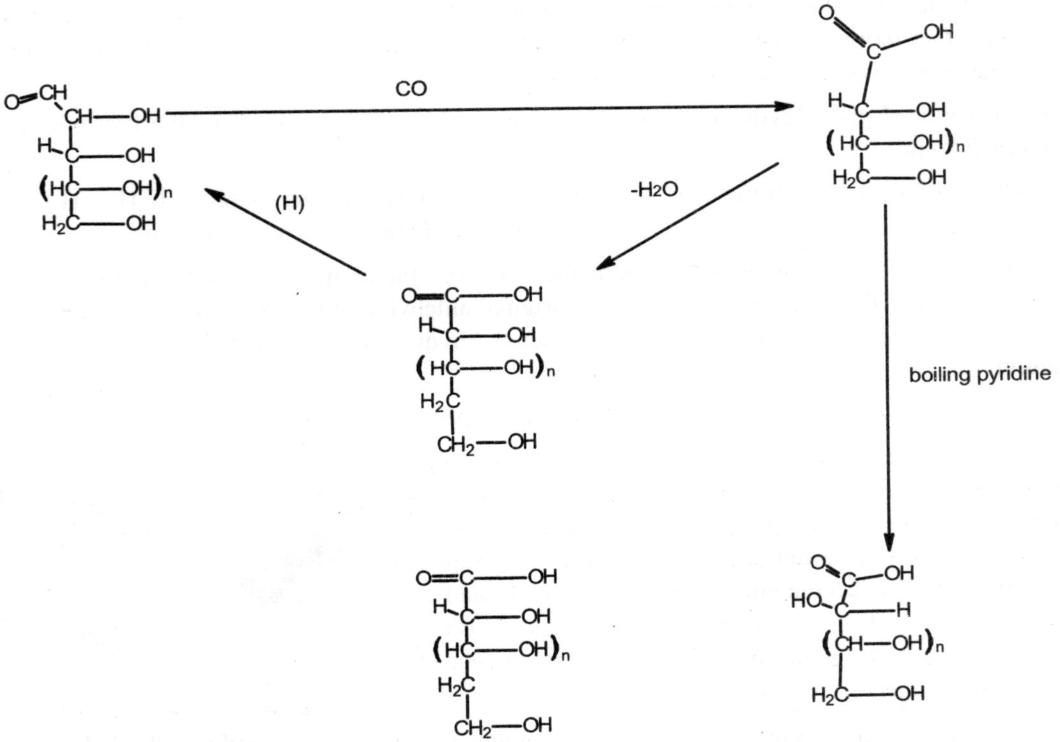

CO

-H2O

(H)

boiling pyridine

Another method of epimerization introduces the ketoses into the equilibrium. Treatment with dilute alkali operates through enedol forms to give the equilibrium mixture shown in the equation below.

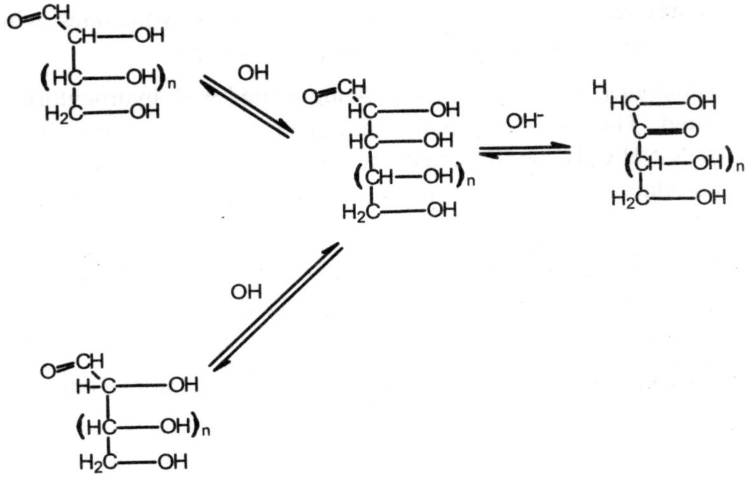

OH

OH^-

OH

Osazone and the Conversion of Aldoses and Ketoses

As aldehydes or ketones the glucose also reacts with hydrazines and hydroxylamine. Phenyl – hydrazine $C_6H_5NHNH_2$ has provided of the greatest value in the separation identification and interconversion of the various other monosaccharides, without the aid of this reagents the brilliant researches of Fischer in the sugar group would hardly have been possible.

When I mol of phenylhydrazine reacts with I mol of an aldose or ketoses, the first product is a normal hydrazone.

$$CH_2OH[CH.OH]_4.CHO{:}O + H_2N - NHC_6H_5 \rightarrow CH_2.OH\,[CH.OH]_4\,NH - N\,C_6H_5 + H_2O$$

Glucose phenyl hydrazone

On warming with excess of phenylhydrazine, however, the hydrazone first formed is oxidized in such way that the CHOH group adjacent to the original aldehydic or ketonic group is converted into a CO group. The latter than combines with more phenyl – hydrazine to give a di – hydrazone containing the group

I. These compounds are termed as osazones (see chemical properties).

Prior to Fischer's researches one of the greatest barriers to a wider knowledge of the monosachharides lay in the difficulty of spreading mixtures of these sugars by crystallization, owing to their high solubility in water and tendency to form syrups. The value depends on the fact that they are sparingly soluble and easily separable by crystallization, and in addition, from their characteristic melting – points and crystalline forms as seen under a microscope it is possible to identify.

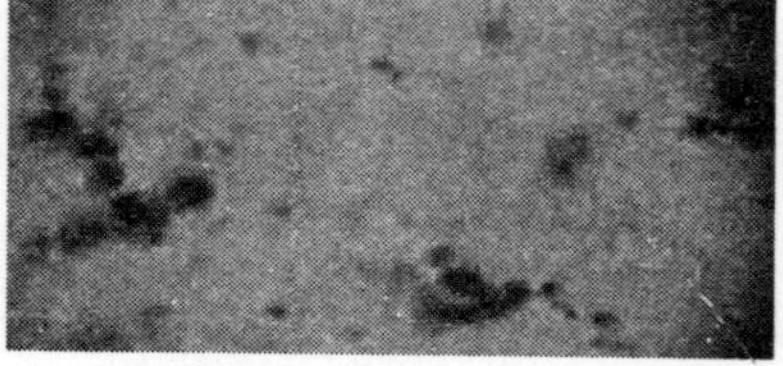

Fig. 34.3. Showing the Osazones under microscope.

The recognition and isolation of ketoses is a matter of some difficulty owing to their lack characteristic reactions, with secondary hydrazine, however, the ketoses give phenyl – methyl osazones by which they may be identified. Aldoses usually react with this base to form colourless hydrazones, which in all cases are readily distinguished or separated from the highly coloured osazones.

Osazones like hydrazones, are hydrolyzed on being heated with hydrochloric acid, when phenyl hydrazine is regenerated, The sugar originally employed, however, is not regained, as the group $C(:N.NH.C_6H_6) - CH(:N.NH.C_6H_5)$ is converted into the group – CO – CHO. The compound so formed is thus and oxidation product of the original sugar, and is termed as *osone* $CH_2OH.(CHOH)_3CO.CHO$ Glucosone.

On mild reduction of this compound with zinc dust and dilute acetic acid, the aldehydic group remaining unchanged. In this case, therefore the sugar finally obtained is fructose in place of the glucose used as starting material.

In these reactions we have a general method of transforming an aldose into a ketose, according to the scheme.

$$\text{Aldose} \xrightarrow[\text{hydrazine}]{\text{Phenyl}} \text{Osazone} \xrightarrow{\text{hydrolysis}} \text{Osone} \xrightarrow{\text{rediction}} \text{Ketose}$$

Further Structural Details of the Glucose Molecule

The data given by the idea that glucose is a mixture of tautomers involving equilibrium between free hydroxyaldehyde (2, 3, 4, 5 6 – pentahydroxyhexanal) and one or more internal hemiacetal forms.

Although there must be small amount of free aldehyde present, the principal substances in a solution of glucose are the hemiacetals and we must now enquire into their number and their structure. Experimentally, two hemiacetals α and β termed anomers are obtainable by special procedures. Thus, when glucose is crystallized from solution at room temperature, the α anomers is obtained. It has a specific rotation. + 111 when initially dissolved in water, but the rotation gradually falls to + 52.5.Crystallization of glucose at about 100 results in the β - anomers. The rotation of the solution of the β - anomers = + 19. The change in rotation with time, which is catalyzed by both acids and bases is called mutarotation and is caused by equilibrium mixture of the two containing 63% of β and 37% of an a anomers.

The mutarotation of glucose and the existence of two crystalline forms are accountable on the basis of the formation of cyclic hemiacetals (with introduction of an additional asymmetric center at carbon I) we have noted may times before that both five and six membered ring are readily formed and therefore sugar chemists had to understate a study of the ring size in sugars.

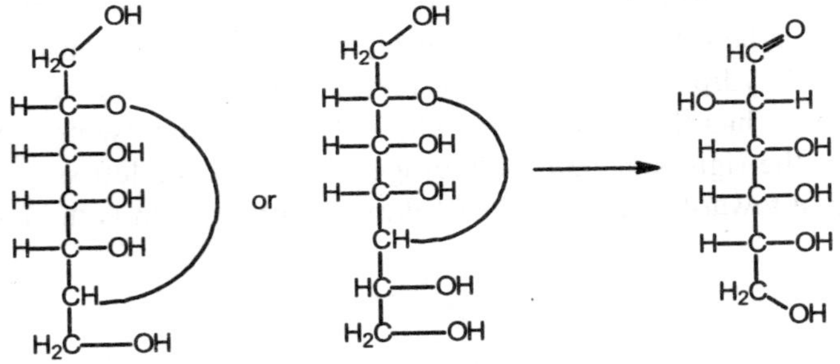

OH
H_2C
H—C—O
H—C—OH
H—C—OH
H—C—OH
CH
H_2C—OH

or

OH
H_2C
H—C—O
H—C—OH
H—C—OH
CH
HC—OH
H_2C—OH

→

HC=O
HO—C—H
H—C—OH
H—C—OH
H—C—OH
H_2C
OH

To ascertain the structure of the cyclic form of the glucose, the molecular configuration must be secured so that the molecule cannot change structure during a chemical reaction utilized form analysis. This apparently is achieved by complete methylation of glucose withy methyl sulphate and sodium hydroxide.

Hydrolysis of phentamethyl glucose in acid then frees the aldehyde group and the hydroxy group involved in the acetal linkage. Oxidation of the tetramethyl glucose occurs at the aldehyde group and the free hydroxy group. Upon oxidation the tetramethyl glucose forms a trimethoxy glutaric acid. These reactions are summerised in the above equation. Those sugar or sugar derivatives having a five membered ring lead to dimethoxy succcinic acids by the Haworth Procedure. Since the six membered

glucose — $(CH_3)SO_4$, NaOH → pentamethyl glucose → tetra methyl glucose ⇌ trimethyl glutaric acid —HNO_3→ 2,3,4-trihydroxy-pentanedioic acid

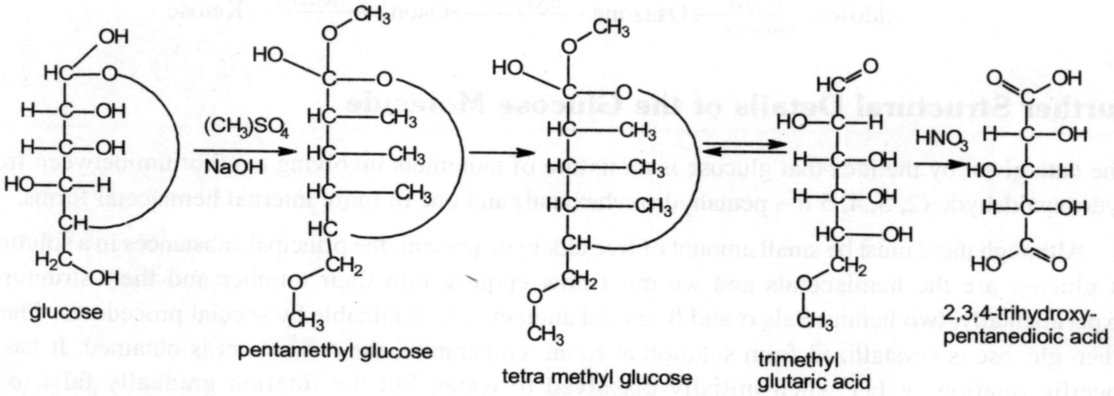

ring is structurally related to pyran a six membered acetal ring in a sugar is called a pyranose ring. A furanose ring because of its relation to furan.

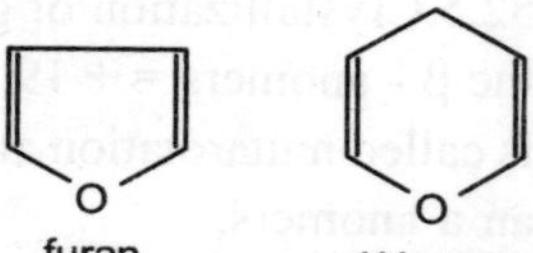

furan 4*H*-pyran

Structural analysis of the aldoses and the ketoses follows a pattern much like that outline d for glucose.

Cyclic Formulas for Glucose

After this work determining the relative configurations of the aldohesoses and aldopentoses. Emil Fischer devises a projection formation to represent the open(aldehydo) forms of these monosaccharides. In the Fischer projection, the carbon chain is represented vertically and the bonds from each carbon atom to its neighbouring carbon atoms are thought of a lying behind the carbon atom under consideration. The hydrogen atoms and hydroxy groups are considered to project up in front of the carbon atom to which they are attached. Thus the configurationally formula, II, represents a molecule in which all hydroxy groups are on the right side and all hydrogen atoms on the left side and the middle of the carbon chain loops upon towards the observer III the top carbon atom is the I position.

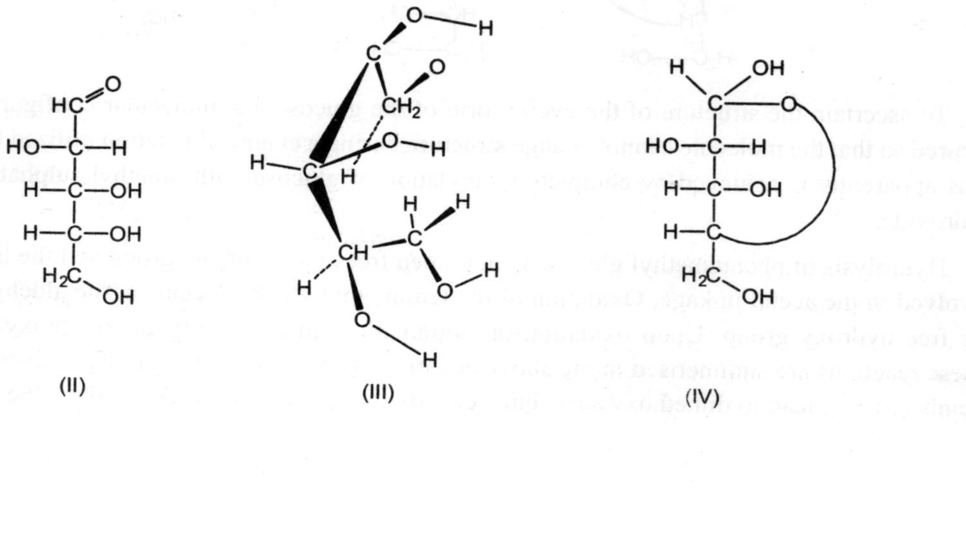

(II) (III) (IV)

Although the Fischer projection serves well enough for the open forms of the monosaccharides it is afterward when used to represent cyclic forms such as IV, The long bent " bond" between the oxygen atom and one of the carbon atoms does not represent the situation well; well hence it use is misleading in a formula intending to be a spatial representation of a molecule. To correct this, Haworth devised a modified projection formula which represent carbohydrates in conventionalized cyclic forms. The ring oxygen is usually placed at the upper right, with the carbon atoms placed in clockwise order of numbering after the oxygen atom. Hydroxy group placed to the right if Fischer projection of α - D fructofuranose and of the anomeric β – D fructofuranose.

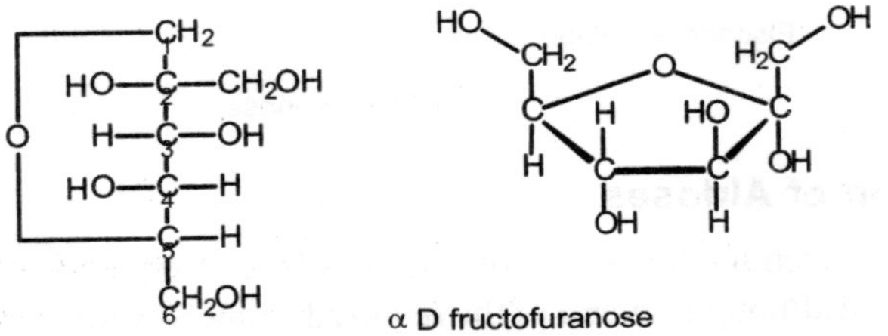

α D fructofuranose

In the oligosaccharides and polysaccharides the virtues of the Haworth projections are even more striking since, the links between units as well as the rings oxygens must be distorted in Fischer projections, where as Haworth projections cope well with this feature.

Sometimes the Haworth projection is inverted or reversed so as to avoid other obstacles to clear formulation. The projection is a picture of the ring with the top always farthest from the viewer.

Compare the formulas for β - D fructofuranose below with the given above.

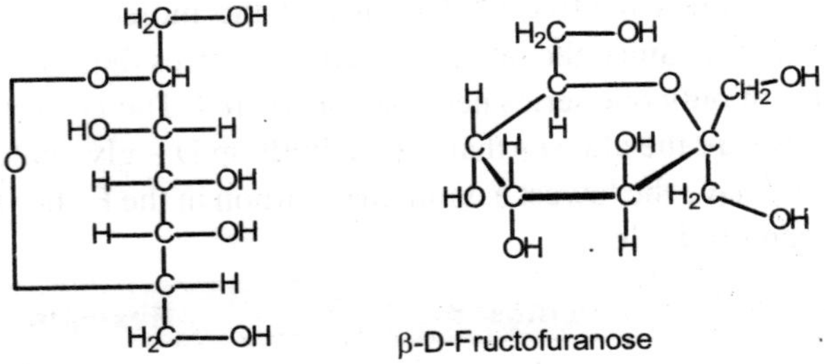

β-D-Fructofuranose

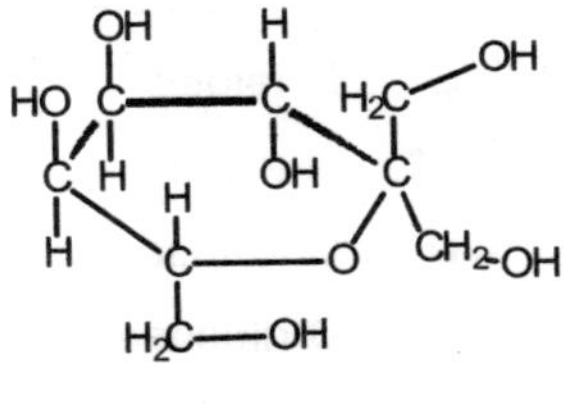

Inverted 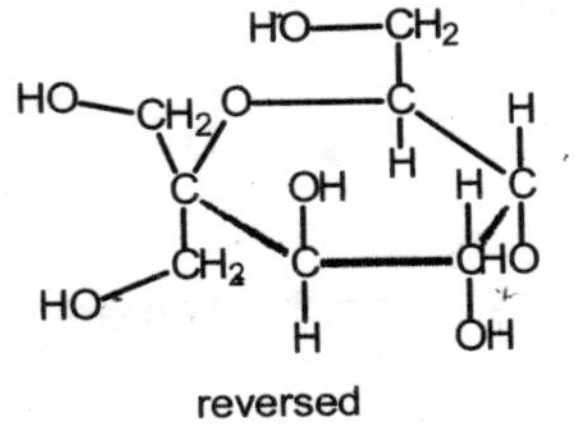 reversed 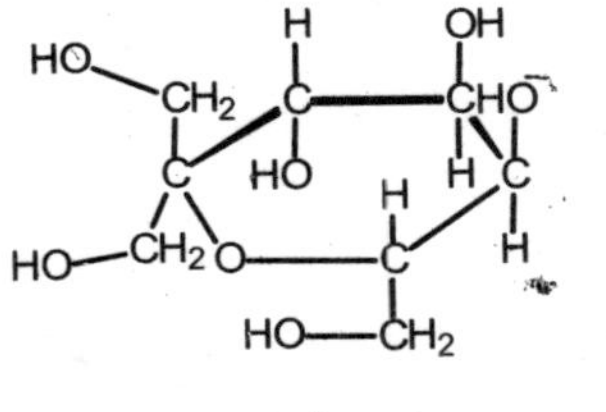 rotated

Still more realistic configuration at formula are often used in the consideration of spatial influences on reactive of equilibrium positions Compare the Haworth projection of α - D glucopyranose and the formational formula below.

Fischer projection

Haworth Projection

α-D-Glucopyranose

Optical Configuration of Aldoses

As late as 1885, although the tetrahedral theory of carbon valences required sixteen isomeric aldohexoses and eight aldopentoses, and although several of the required isomers were recognized, the configurations of the monosaccharides remained unknown. It was the great structural chemist, Emil Fischer, who basing much of his work on the studies of Killani, Wöhl and Ruff, solved the problem required.

The point of departure is the establishment of epimeric relationship and relationships among the aldoses via Killani synthesis and the Wöhl and Ruff degradation. These relationships are given in the table. Corresponding to the epimeric and series relationships are the configuration relationship of aldoses shown in the fig. 34.4. In this figure the molecular configuration are indicated by abbreviated Fisher projections using triangles to represent the orientation of hydroxy groups relative to the backbone of the molecule the orientation of hydroxy groups relative to the back of the molecule, which is the vertical line connecting the circle and triangle. Since the optically opposite series, but in minor image position, the configurational relationships of only one series need be considered. The D – series aldoses are aldoses discussed in the analysis. These are the aldoses that can be built up D – glyceraldehydes by successive Killani synthesis and, therefore, have the lowest asymmetric carbon in the Fisher convention attached to hydroxy on the right hydrogen on the left.

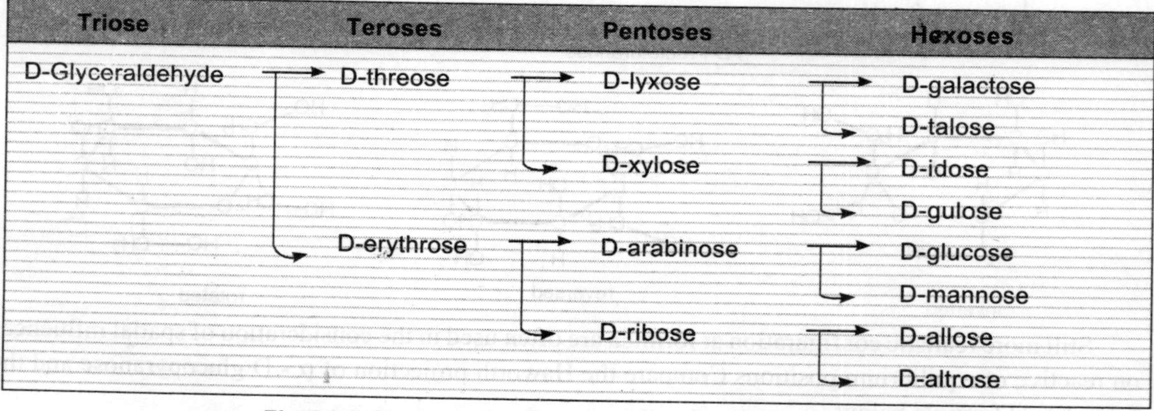

Triose	Teroses	Pentoses	Hexoses
D-Glyceraldehyde	D-threose	D-lyxose	D-galactose
			D-talose
		D-xylose	D-idose
			D-gulose
	D-erythrose	D-arabinose	D-glucose
			D-mannose
		D-ribose	D-allose
			D-altrose

Fig. 34.4. Showing the Table of Killani Synthesis series.

It should be apparent at first glance the whereas D – glyceraldehydes can be assigned by convention the configuration, the relationship between the other numbered formulæ and the aldoses listed in the table above are by no means self – evident. Tetrose 2 might be either D – threose or D – erithrose. To ascertain the configurational identity of the various aldoses. Fisher resorted to oxidations leading to glyceric (saccharic) acids. These yield structurally symmetrical compound of which a certain number should be meso acids. For example oxidation of ribose and xylose by nutric acid yields optically inactive hence, these two aldopentoses must be 4 and 6 (not necessary in order. To check this the carbonyl groups can be indicated by another sign like X and the configuration of the molecule checked for a plane of symmetry.

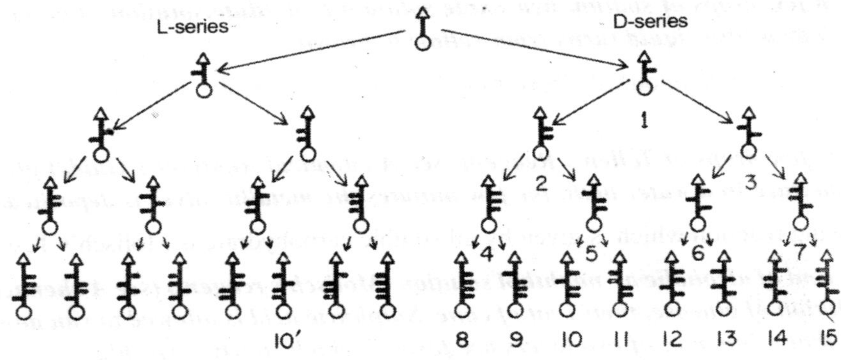

Fig. 34.5. The Configurational Relationship of Aldoses

Lyoxose and arbinose, which yield optically active glyceric acids, are 5 and 7 (not necessarily in order). Tallose and galactose are epimers derivable from lyxose glucose and mannose epimers derivable from arbinose. Of these hexode only galactose, yield a meso glyceric acid hence galactose is 14 and Tallose 15. Lyxose must by 7, arbinose 5 Xylose, the epimer of lyxode is 6 and ribose, the epimer of lyxose is 6 and ribose, the epimer of arbinose, is 4. Another inactive acid is produced by allose. Hence, allose is 8. Altrose which yields an optically active acid and is the epimer of allose is 9.

To establish the configuration of glucose recourse was made to Killani synthesis to the aldoheptoses. Oxidation of the two aldoheptoses yields one optically active glyceric acid and one meso acid. Consideration of formula 10 and 11, the epimeric hexoses related to arbinose, shows that the result substance which produces this result must have the configuration 10. Hence the glucose has the configuration 10. Mannose the epimer of glucose, which yields two optically active heptaglyceric acid is 11.

Upon oxidation, glucose gives glyceric acid specifically named v saccharic acid. Glucose yields enantimorph L – saccharic acid. Since D – glucose cannot belong to the L – series, this means that when the aldehyde group and the primary alcohol group lose their identify by transformation to carboxyl groups, the same result unuses as when L – glucose is oxidized. To see the relationship, one rotates a model of glucose 180°C so that its aldehyde group coincides with the primary alcohol group of L - glucose 10 and its primary alcohol group coincides with the aldehyde group of L glucose. The formula which fits this description is 12. The remaining hexode iodose is 13.

Occasionally one wishes to indicate the direction of optional rotation od an optically active rotation of an optically active material. This is done by placing (+) or (–) between the series indication and the remainder of the name, Thus D (+) glucose D(–) ribose D(+) glyceraldehydes and D(–) tartaric acid are examples.

Analysis of Glucose

Glucose gives the following series of reactions during analysis; caustic alkalies added to a solution of glucose and warmed, produce a brown solution.

Exp: Add a few drops of sodium hydroxide solution to a dilute solution of glucose, and warm gently. The colour of this liquid turns from yellow to brown.

Glucose reduces an Ammonical – Silver oxide solution to metallic silver which gets deposited on heating.

Exp: Add a few drops of Tollen's Reagent (see A chemical Analyser's Guide) glucose solution and heat the mixture in a water bath, for few minutes the metallic silver is deposited.

The following reaction which is given by all soluble carbohydrate is Molisch's Test

Exp: Add 1 ml of alcoholic a - naphthol solution (Molisch's reagent) (see A chemical Analyser's Guide)to a solution of glucose, then 1 ml of conc. Sulphuric acid is allowed to run down by the side of the tube so that a junction of two layers are formed which slowly turns blue.

An alkaline copper sulphate solution is reduced and cupric oxide is precipitated.

Exp: Add a few drops of copper sulphate solution to a solution of glucose and add a few drops of sodium hydroxide when the solution is heated the colour changes from blue to red and a red precipitate of cuprous oxide is produced.

Quantitative test of glucose

The above reaction is utilized for the quantitative estimation of glucose as well as other sugars. A standard solution of copper sulphate is prepared by dissolving 34.64 gm in 500ml (.434mol/L) A second solution is made by dissolving 60 gms of sodium hydroxide in 500ml water. This solution when mixed and run into a flask and boiled in a measured amount, and the sugar solution is added gradually from a burette into the hot solution mixture until all the cuprous oxide is precipitated. The quantity of sugar solution taken is a measure of the amount of glucose present 10cc of the alkaline copper solution corresponds to 0.05 gm of glucose.

Use

We can speculate on the reasons why glucose, and not another monosaccharide such as fructose (Fru), is so widely used in evolution, the ecosystem, and metabolism. Glucose can form from formaldehyde under abiotic conditions, so it may well have been available to primitive biochemical systems. Probably more important to advanced life is the low tendency of glucose, by comparison to other hexose sugars, to non-specifically react with the amino groups of proteins. This reaction (glycation) reduces or destroys the function of many enzymes. The low rate of glycation is due to glucose's preference for the less reactive cyclic isomer. Nevertheless, many of the long-term complications of diabetes (e.g., blindness, kidney failure, and peripheral neuropathy) are probably due to the glycation of proteins or lipids. In contrast, enzyme-regulated addition of glucose to proteins by glycosylation is often essential to their function.

As an Energy Source

Glucose is a ubiquitous fuel in biology. It is used as an energy source in most organisms, from bacteria to humans. Use of glucose may be by either aerobic or anaerobic respiration (fermentation). Carbohydrates are the human body's key source of energy, through aerobic respiration, providing approximately 3.75 kilocalories (16 kilojoules) of food energy per gram.[4] Breakdown of carbohydrates (e.g. starch) yields mono- and disaccharides, most of which is glucose. Through glycolysis and later in the reactions of the Citric acid cycle (TCAC), glucose is oxidized to eventually form CO_2 and water, yielding energy, mostly in the form of ATP. The insulin reaction, and other mechanisms, regulate the concentration of glucose in the blood. A high fasting blood sugar level is an indication of prediabetic and diabetic conditions.Glucose is a primary source of energy for the brain, and hence its availability influences psychological processes. When glucose is low, psychological processes requiring mental effort (e.g., self-control) are impaired.

FRUCTOSE

Introduction

Fructose (also **levulose** or **laevulose**) is a simple reducing sugar (monosaccharide) found in many foods and is one of the three most important blood sugars along with glucose and galactose. Honey, tree fruits, berries, melons, and some root vegetables, such as beets, sweet potatoes, parsnips, and onions, contain fructose, usually in combination with sucrose and glucose. Fructose is also derived from the digestion of sucrose, a disaccharide consisting of glucose and fructose that is broken down by glycoside hydrolase enzymes during digestion. Fructose is the sweetest naturally occurring sugar, estimated to be twice as sweet as sucrose. Fructose is often recommended for, and consumed by, people with diabetes mellitus or hyperglycemia, because it has a very low glycemic index (GI) relative to cane sugar (sucrose). However, this benefit is tempered by concern that fructose may have an adverse effect on plasma lipid and uric acid levels, and the resulting higher blood levels of fructose can be damaging to proteins (see below). The low GI is due to the unique and lengthy metabolic pathway of fructose, which involves phosphorylation and a multi-step enzymatic process in the liver. See health effects and glycation for further information.

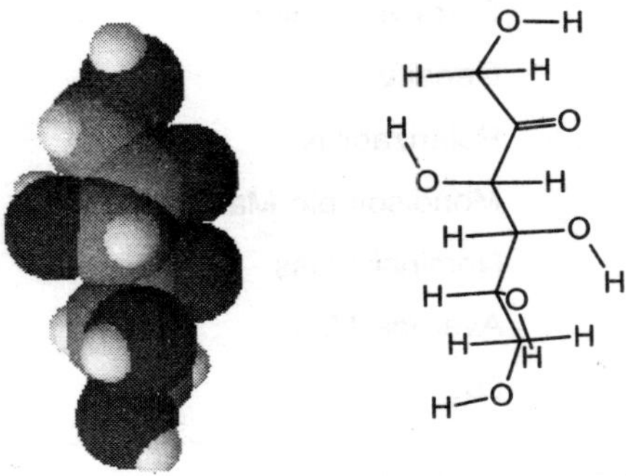

Synthesis

Fructose can be obtained from cane sugar by hydrolysis with sulphuric acid in dilute state. The acid id removed by the precipitation with barium carbonate and the filtrate is concentrated: Milk of lime is then added, when the lime compound or the calcium fructosate (corresponding to calcium glucosate which is only slightly soluble separates out and is filtered and washed. The calcium compound is then suspended in water and decomposed by carbon dioxide. The solution is again filtered from calcium carbonate and evaporated. On introducing a crystal of fructose into the syrup, the later crystallizes.

Fructose is also prepared from inulin which turns to fructose inulin which turns to fructose on hydrolysis with sulphuric acid. After removing the acid the liquid is evaporated preferably under reduced pressure. When a syrup is left which solidifies if a crystal of the solid is added.

Physical Properties

Molecular Formula	$= C_6H_{12}O_6$
Formula Weight	= 180.15588
Composition	= C(40.00%) H(6.71%) O(53.29%)
Melting point	= D-fructose: 103°C
Molar Refractivity	= 37.42 cm^3
Molar Volume	= 113.3 cm^3
Parachor	= 351.7 cm^3
Index of Refraction	= 1.574
Surface Tension	= 92.6 dyne/cm
Density	= 1.589 g/cm^3
Polarizability	= 14.83 $10^{-24} cm^3$
Monoisotopic Mass	= 180.063388 Da
Nominal Mass	= 180 Da
Average Mass	= 180.1559 Da
Log P	= -1.63

Chemical Properties and Constitution

Fructose is more soluble than glucose but crystallizes from rhombic prisms which melt at 103°C. it has a sweet taste and gives many of the reactions of glucose. Although fructose is not an aldehyde but a ketone, it nevertheless reduces alkaline copper solution. This is due to the presence of the easily oxidisable group – CO_2 – CH_2(OH) with phenylhydrazine, fructose yields an osazone, which is identical with glucosazone. Fructose also undergoes fermentation with yeast, through less readily than glucose,

glucose being first removed when a solution of the two sugars is fermented. Like glucose fructose undergoes mutarotation in solution the specific rotation rising from %135.5 to %92.0°.

Fructose forms a pentacetyl derivative like glucose. On reduction it is converted into a mixture of sorbitol and mannitol, since an additional asymmetric combination has been introduced. On oxidation it does not like glucose form an acid with same number of carbon atom but breaks up into formic acid and trihydroxybutaric acid. This decomposition points to the presence of a ketone group in the molecule, which is further confirmed by the following reactions

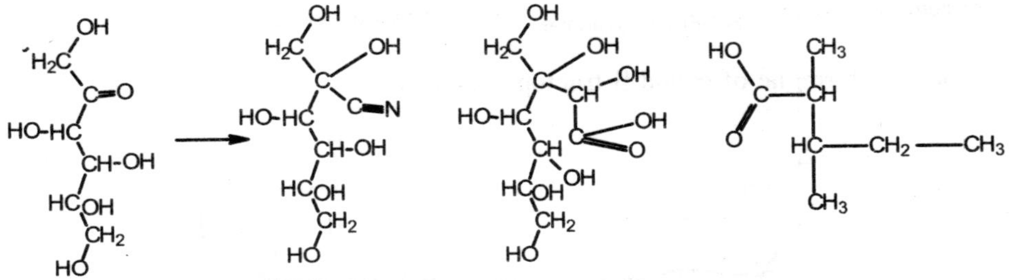

fructose Cyanohydrin of fructose hydroxy acid

The above formula for fructose agrees moreover, with the synthesis of inactive fructose from the mixture glyceric aldehyde and dihydroacetone and with the production of the same osazone as that obtained from glucose, a reaction in which the two end carbon atoms of the chain are involved.

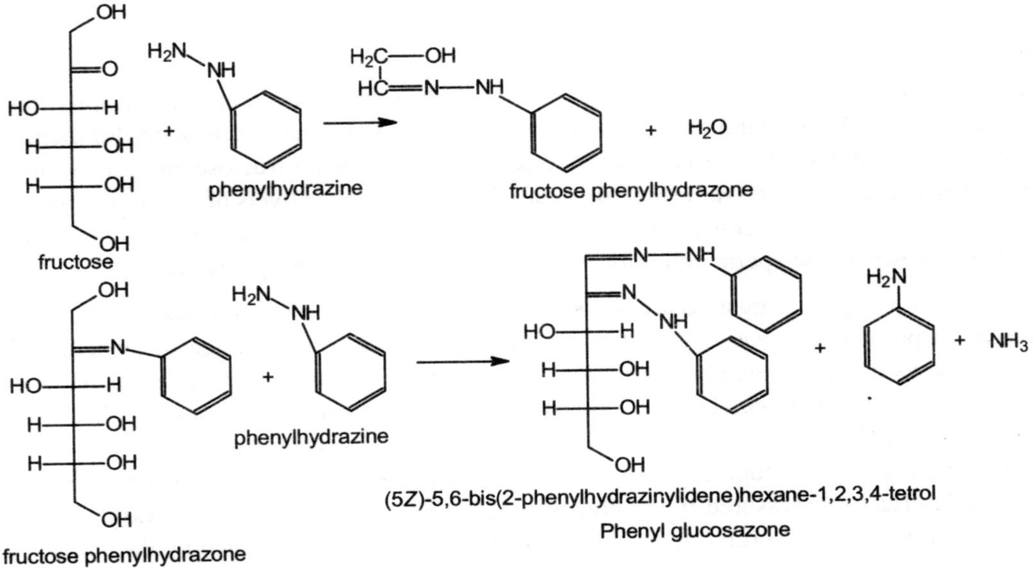

fructose + phenylhydrazine → fructose phenylhydrazone + H_2O

fructose phenylhydrazone + phenylhydrazine → (5Z)-5,6-bis(2-phenylhydrazinylidene)hexane-1,2,3,4-tetrol + H_2N–Ph + NH_3

(5Z)-5,6-bis(2-phenylhydrazinylidene)hexane-1,2,3,4-tetrol

Phenyl glucosazone

Although fructose in many of its reactions a ketone its exists like glucose in α and β stereoisomeric pyranose forms and also in g or furanose rings

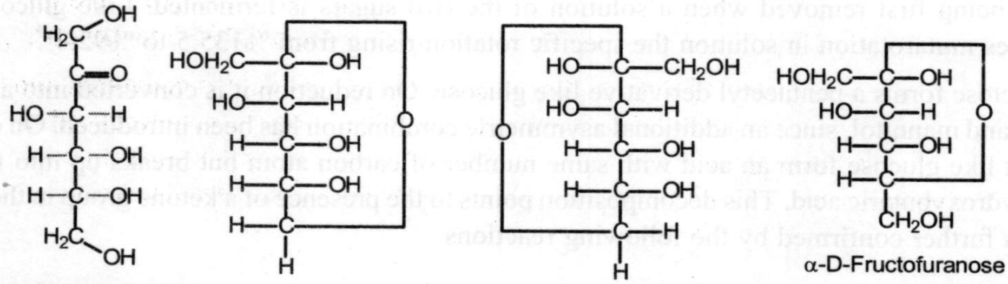

Keto fiormula fructose α-D-Fructopyranose β-D-Fructopyranose α-D-Fructofuranose

The perspective formulæ of α and β fructopyranose are

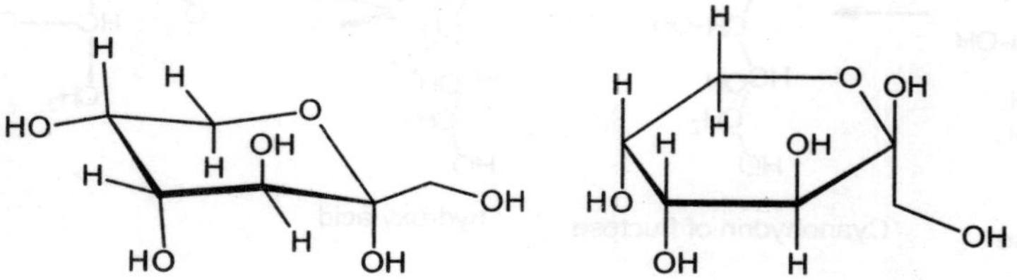

α-D-Fructofuranose β-D-Fructopyranose

γ - fructose is a constituent of cane sugar but when liberated from its union with glucose passes into the pyranose form.

Uses

Fructose is often recommended for diabetics due to its glycemic index being significantly lower than both glucose, sucrose and starches. The medical profession thinks fructose is better for diabetics than sugar, but every cell in the body can metabolize glucose. However, all fructose must be metabolized in the liver. The livers of the rats on the high fructose diet looked like the livers of alcoholics, plugged with fat and cirrhotic This is not entirely true as certain other tissues do use fructose directly, notably the cells of the intestine, and sperm cells (for which fructose is the main energy source).

Fructose is a reducing sugar, as are all monosaccharides. The spontaneous addition of single sugar molecules to proteins, known as glycation, is a significant cause of damage in diabetics. Fructose appears to be as dangerous as glucose in this regard and so does not seem to be a better answer for diabetes for this reason alone. This may be an important contribution to senescence and many age-related chronic diseases. Fructose is used as a substitute for sucrose (composed of one unit each of fructose and glucose linked together with a relatively weak glycosidic bond) because it is less expensive and has little effect on measured blood glucose levels. Often, fructose is consumed as high fructose corn syrup, which is corn syrup (glucose) that has been enzymatically treated by the enzyme glucose isomerase. This enzyme converts a portion of the glucose into fructose thus making it sweeter. This is done to such a degree as to yield corn syrup with an equivalent sweetness to sucrose by weight. While most carbohydrates have around the same amount of calories, fructose is sweeter and manufacturers can use less of it to get the same result. The free fructose present in fruits, their juice, and honey is responsible for the greater sweetness of these natural sugar sources. Some studies point to fructose as key factors in hyperactivity and tooth decay in children.

GALACTOSE

Galactose (Gal) (also called **brain sugar** is a type of sugar which is less sweet than glucose and not very water-soluble. It is considered a nutritive sweetener because it has food energy. Galactan is a polymer of the sugar galactose. It is found in hemicellulose and can be converted to galactose by hydrolysis. It is found in dairy products, in sugar beets and other gums and mucilages. It is also synthesized by the body, where it forms part of glycolipids and glycoproteins in several tissues.

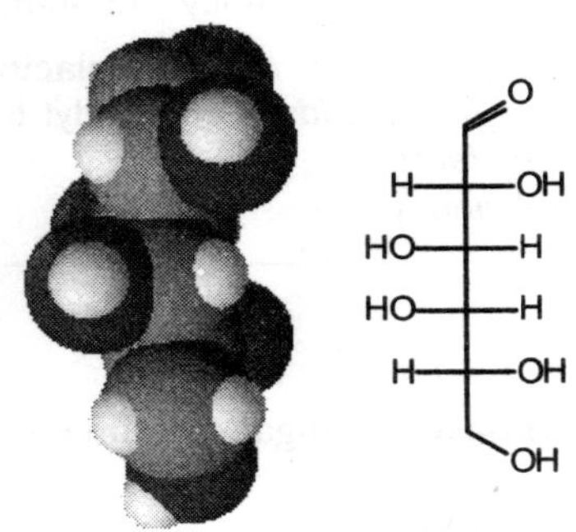

Relationship to Lactose

Galactose is a monosaccharide constituent, together with glucose, of the disaccharide lactose. The hydrolysis of lactose to glucose and galactose is catalyzed by the enzyme lactase, a β-galactosidase. In the human body, glucose is changed into galactose in order to enable the mammary glands to secrete lactose. Galactose and glucose are produced by hydrolysis of lactose by β-galactosidase. This enzyme is produced by the *lac* operon in *Escherichia coli* (*E. coli*).

Clinical Significance

Two studies have suggested a possible link between galactose in milk and ovarian cancer. Other studies show no correlation, even in the presence of defective galactose metabolism. More recently, pooled analysis done by the Harvard School of Public Health showed no specific correlation between lactose containing foods and ovarian cancer, and showed statistically insignificant increases in risk for consumption of lactose at > 30 g/d. More research is necessary to ascertain possible risks. There are some ongoing studies which suggest that galactose may have a role in treatment of focal segmental glomerulosclerosis (a kidney disease resulting in kidney failure and proteinuria). This effect is likely to be a result of binding of galactose to FSGS factor.

Structure and Isomerism

The first and last -OH groups point the same way and the second and third -OH groups point the other way. D-Galactose has the same configuration at its penultimate carbon as D-glyceraldehyde. Galactose is a diastereomer of glucose.

Metabolic Disorders

There are 3 important disorders involving galactose:

Name	Enzyme	Description
Galactosemia	Galactokinase	Causes cataracts and mental retardation. If a galactose-free diet starts sufficiently early, the cataracts will regress without complications however neurological damage is permanent.
UDPgalactose-4-epimerase deficiency	UDPgalactose-4-epimerase	Is extremely rare (only 2 reported cases). It causes nerve deafness.
Galactose-1-phosphate uridyl transferase deficiency	Galactose-1-phosphate uridyl transferase	Is the most problematic, as galactose-free diets do not have considerable long-term effects.

MANNOSE

Mannose is a sugar monomer of the hexose series of carbohydrates.

Formation

Mannose can be formed by the oxidation of mannitol.

It can also be formed from D-glucose in the Lobry-de Bruyn-van Ekenstein transformation

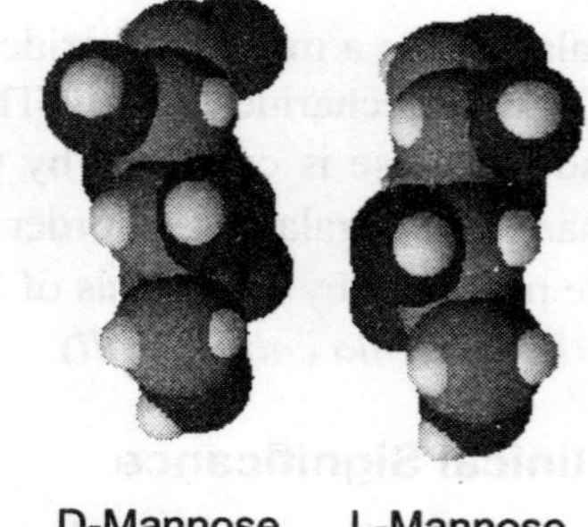

D-Mannose	L-Mannose
HC=O	HC=O
HO—C—H	H—C—OH
HO—C—H	H—C—OH
H—C—OH	HO—C—H
H—C—OH	HO—C—H
H_2C—OH	H_2C—OH

Etymology

The root of both "mannose" and "manitol" is manna, which the Bible records as the food supplied to the Israelites during their journey through the Sinai Peninsula. Manna is a sweet secretion of several trees and shrubs, such as *Fraxinus ornus*.

Cranberry juice

D-Mannose, which appears in some fruits including cranberry, has been postulated to prevent the adhesion of bacteria to the epithelium of the urinary tract and Urinary bladder through a mechanism presumed to be competitive in nature with the polysaccharide coating of the cystic epithelium. While no studies have objectively verified this mechanism, anecdotal reports from patients show some improvement in symptoms.

Configuration

The fact that D-mannose has the same configuration at its penultimate carbon as D-glyceraldehyde is unsurprising as that is what defines the dextro classification. However, mannose differs from D-glucose by inversion of the C2 chiral centre. This apparently simple change leads to the drastically different chemistry of the two hexoses, as it does the remaining six hexoses.

Metabolism

Mannose enters the carbohydrate metabolism stream in two steps:

- First it undergoes phosphorylation to mannose-6-phosphate by hexokinase.
- Then, it is converted to fructose-6-phosphate by mannose phosphate isomerase.

Mannose is present in numerous glycoconjugates including N-linked glycosylation of proteins. C-mannosylation is also abundant and can be found in collage-like regions. Mannose is a C2 epimer of glucose and displays a ^2C_4 pucker in the solution ring form.

Pentoses

ARBINOSE

Arabinose is an aldopentose — a monosaccharide containing five carbon atoms, and including an aldehyde (CHO) functional group. It has chemical formula $C_5H_{10}O_5$ and a molar mass of 150.13 g/mol.

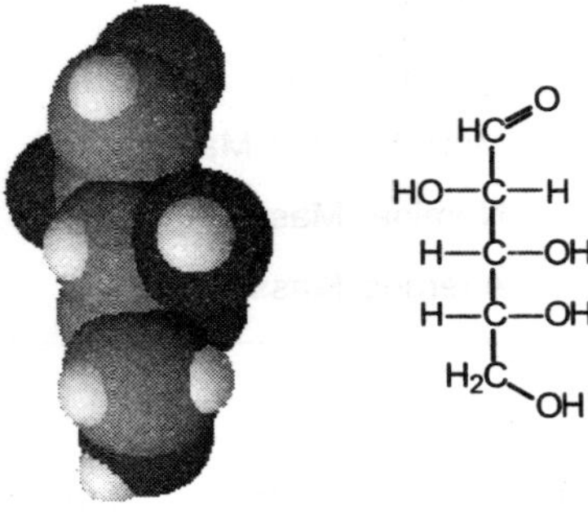

For biosynthetic reasons, saccharides are almost always more abundant in nature as the "D" form, or structurally analogous to D-(+)-glyceraldehyde. However, L-arabinose is in fact more common than D-arabinose in nature and is found in nature as a component of biopolymers such as hemicellulose and pectin. The L-arabinose operon is a very important operon in molecular biology and bioengineering.

A classic method for the organic synthesis of arabinose from glucose is the Wohl degradation

RIBOSE

Ribose, primarily seen as **D-ribose**, is an aldopentose — a monosaccharide containing five carbon atoms, and including an aldehyde functional group in its linear form. It has the chemical formula $C_5H_{10}O_5$, and was discovered in 1905 by Phoebus Levene.As a component of the RNA that is used for genetic transcription, ribose is critical to living creatures. It is related to deoxyribose, which is a component of DNA. It is also a component of ATP, NADH, and several other chemicals that are critical to metabolism.Refer to the article on deoxyribose for more information on both sugars, how they relate to each other, and how they relate to genetic material.

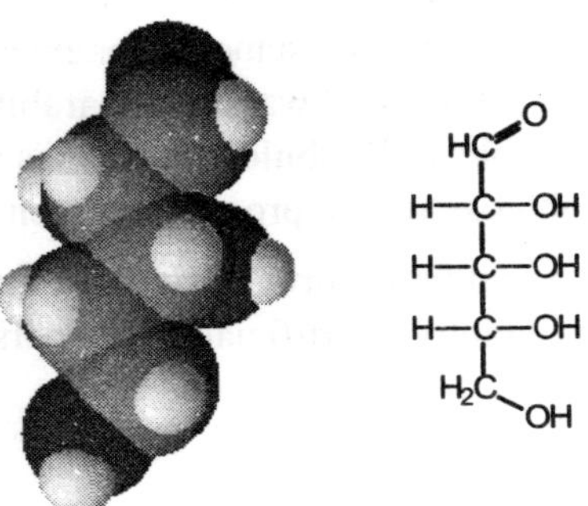

Physical Properties

Molecular Formula	$= C_5H_{10}O_5$
Formula Weight	= 150.1299
Composition	= C(40.00%) H(6.71%) O(53.29%)
Melting Point	= 99 °C
Molar Refractivity	$= 31.41\ cm^3$
Molar Volume	$= 99.5\ cm^3$
Parachor	$= 299.0\ cm^3$
Index of Refraction	= 1.543
Surface Tension	= 81.4 dyne/cm
Density	$= 1.508\ g/cm^3$
Polarizability	$= 12.45\ 10^{-24} cm^3$
Monoisotopic Mass	= 150.052823 Da
Nominal Mass	= 150 Da
Average Mass	= 150.1299 Da

RIBULOSE

Ribulose is a ketopentose — a monosaccharide containing five carbon atoms, and including a ketone functional group. It has chemical formula $C_5H_{10}O_5$. Two enantiomers are possible, D-ribulose (D-erythro-pentulose) and L-ribulose (L-erythro-pentulose). D-Ribulose is the diastereomer of D-xylulose.Ribulose sugars are composed in the pentose phosphate pathway. They are important in the formation of many bioactive substances. For example, D-ribulose is an intermediate in the fungal pathway for D-arabitol production. Also, as the 1,5-bisphosphate, D-ribulose combines with carbon dioxide at the start of the photosynthetic process in green plants (carbon dioxide trap).

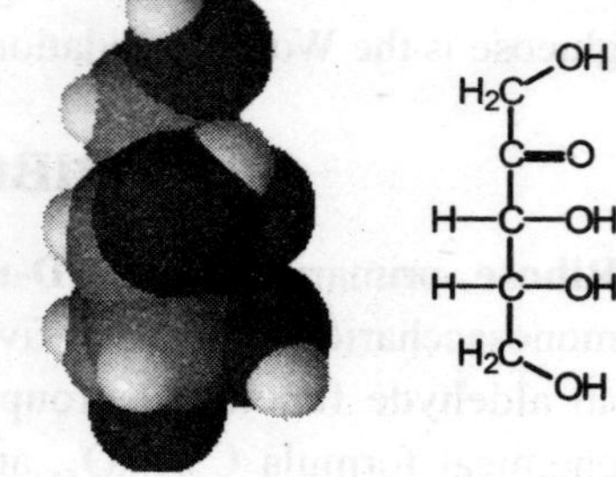

A synthetic form of ribulose known as sucroribulose is found in many brands of artificial sweeteners.

XYLOSE

Xylose or wood sugar is an aldopentose — a monosaccharide containing five carbon atoms and including an aldehyde functional group. It has chemical formula $C_5H_{10}O_5$. Xylose is found in the embryos of most edible plants.

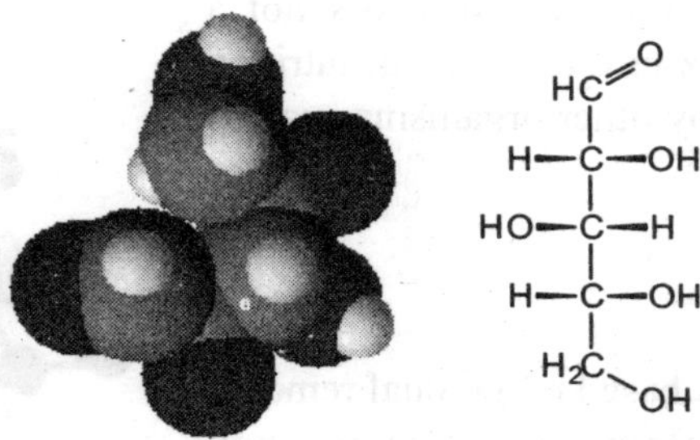

In animal medicine, xylose is used to test for malabsorption by administering to the patient in water after fasting. If xylose is detected in blood and/or urine within the next few hours, it has been absorbed by the intestines.

Physcial Properties

Molecular Formula	$= C_5H_{10}O_5$
Formula Weight	= 150.1299
Composition	= C(40.00%) H(6.71%) O(53.29%)
Melting Point	=144-145 °C
Molar Refractivity	= 31.41 cm^3
Molar Volume	= 99.5 cm^3
Parachor	= 299.0 cm^3
Index of Refraction	= 1.543
Surface Tension	= 81.4 dyne/cm
Density	= 1.508 g/cm^3
Polarizability	= 12.45 $10^{-24} cm^3$
Monoisotopic Mass	= 150.052823 Da
Nominal Mass	= 150 Da
Average Mass	= 150.1299 Da

Oligosaccharides

Disaccharides

SUCROSE

Sucrose (common name: **table sugar**, also called **saccharose**) is a disaccharide (glucose + fructose) with the molecular formula $C_{12}H_{22}O_{11}$. Its systematic name is α-D-glucopyranosyl-(↔2)-β-Dfructofuranoside (ending in "oside", because it's not a reducing sugar). It is best known for its role in human nutrition and is formed by plants but not by other organisms such as animals.

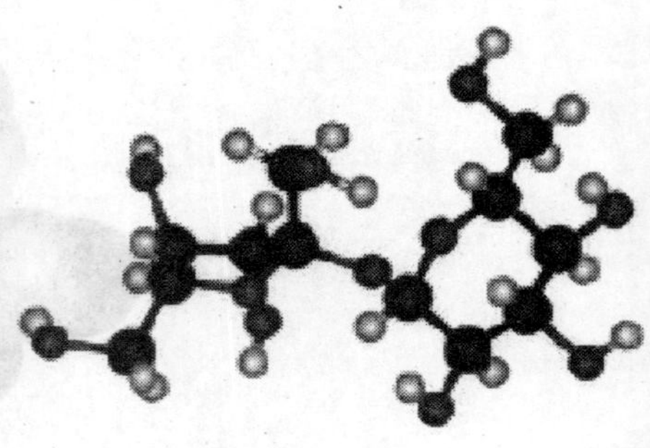

Synthesis of Sucrose

The manufacture of cane sugar is based on gradual removal of fibre, non-sugars, water, colouring matter and non crystallisable sugars.

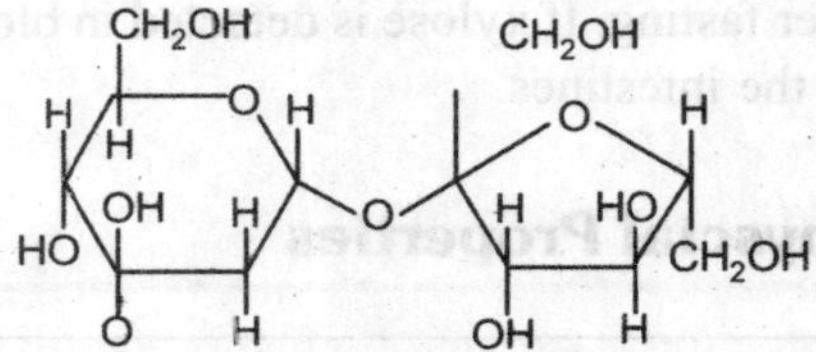

(a) Fibre is removed by crushing and milling;

(b) Non-sugars are removed by clarification;

(c) Water is removed by evaporation and centrifuging;

(d) Colouring matter is removed by bleaching *or* absorption by bone charcoal *or* activated carbon;

(e) Non-crystallisable sugars are eliminated during crystallisation. White sugar is produced either in two steps *or* in one step.

(a) In two steps method, first raw *or* un-refined crystalline sugar is manufactured, and then white sugar is manufactured by refining raw sugar.

(*b*) In one step process in which chrification is done by sulphitation *or* double carbonation method. In this process the syrup is bleached by sulphur dioxide and the thin film *of* molasses on the crystals is finally removed by double centrifuging.

Extraction of the Juice or Separation of the Fibre from the Juice (Fig. 34.6)

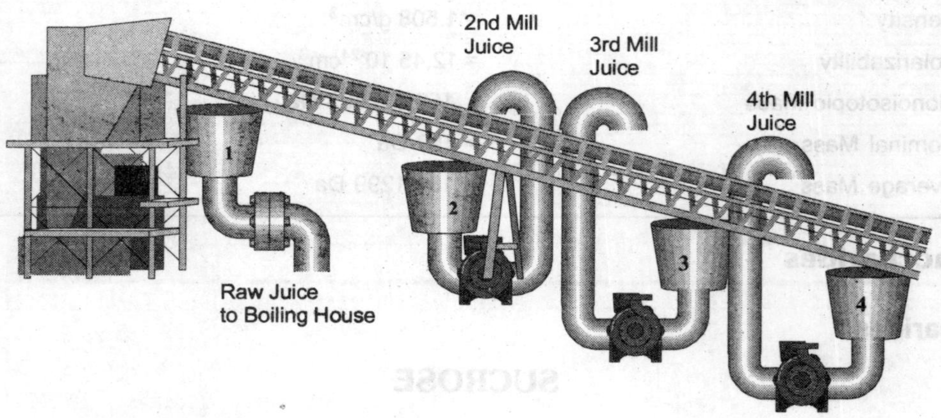

Fig. 34.6. Showing the Pressing of Sugar cane

The process of extraction consists of :

(a) Cutting of the stalk and leaves.

(b) Cutting of the canes by means of sharp rapidly revolving knives (550-600 r.p.m.) fitted on a horizontal shaft 2" inches apart. The canes move on a carrier under the knives which make diagonal cuts, converting the canes into chips in which the soft portion is exposed to the maximum.

(c) The chips are then brought to the extractor consisting of one two rolled toothed crusher and four three-rolled grooved mills. (In India, the crusher and the mills are mostly made of grey cast iron; in other places it is made *of* cast steel)

The three rollers of the mill are placed at the crusher of a triangle; only the top roller of the crusher or of the mill moves, the bottom rollers in the milling train remain stationery during operation. A pressure of 3 to 6 tons per sq. in. is maintained on each of the top roller. There are intermediate carriers between the crusher and the first mill and between the two subsequent mills to carry the bagasse to the next mill or out of the milling train. The crusher and the first mill are provided with vibrating screens and arrangement for transferring the juice to the raw juice tank. The 2nd, 3rd and 4th mills are provided with containers to hold unstrained juice which is used for combined imbibition process (Fig. 34.2).

When the chips are fed into the crusher, the chips during their passage through the rollers are pressed to extract about 50 % of the juice in the cane. The bagasse (the fibre) containing 50% of the juice is fed into the first mill. The juice from these two are strained on the strainer below and disposed as mentioned .above. The bagasse still containing some juice pass successively through -the second, third and fourth mills and then to the storage or to the kiln in which it is fed for steam raising. Even in recent times, most of the Indian mills used bagasse as fuel, but the present tendency is to replace it by coal or coke.

Compound Imbibition Process

After the crusher, the baggasse has to be, in all cases, moistened for efficient pressing. Had it been done by water the sugar extracted would have been less; hence the maximum amount of sugar is pressed out by combined imbibition process. In this process the juice from the second mill is sprayed all the baggasse as it passes from' the crusher and the first 'milt. The juice from the third mill is sprayed between first and second mill, juice from the fourth mill is sprayed between the second and thud mills, while the baggasse passing from the third to the fourth mill is sprayed with hot water. About 25 % of the weight of the cane is used in the imbibition process.

Clarification

Why clarification is done?

Cane juice contains

78 -86 % water
10% -20 % sucrose
0.5- 1 % organic matter
0.5- 2.5 % reducing sugar
0.3- 0.77% ash

Besides these the cane juice contains finely divided pith and fibre and also some colloidal matters. These substances together with gums, pectin's, proteins and other non-sugars definitely hamper crystallisation of sucrose. Some of these are also responsible for the colouring of the crystallized sugar.

The process of clarification by which the above impurities are eliminated is, therefore, an important step in the production of maximum quantity of white crystalline sugar. In simple defecation by lime the impurities are not completely eliminated; hence by crystallising the syrup (the concentrated form of juice) a coloured crystaline sugar known as raw sugar is obtained. This raw sugar is then converted into white crystalline sugar by a refining process.

In most of the countries, white sugar is produced by the above-mentioned process. In India"Java and some other parts of the world white sucrose is produced by one step. In these places clarification is done either by sulphitation or carbonation process. In the sulphitation and sometimes in the carbonation process also the concentrated juice is treated with sulphur dioxide. In carbonation process, more. efficient clarification is done by double carbonation process. Hence these processes are known as double sulphitation and double carbonation processes.

Lime Defecation Process: The cane juice is acid having pH from 5.1 to 5.7. This acidity helps the inversion of sucrose to reducing non-crystallisable sugars. For this reason the juice after extraction should be made alkaline without delay. If acidity is removed by liming, i.e. by adding adequate amount of milk of lime, then insoluble salts of organic acids such as oxalic and' propionic are precipitated. Lime reacts with phosphoric' acid or insoluble, phosphates to form precipitate of. tricalcium phosphate Now 'when the alkalinity is increased, gums, pectins, and proteins, etc. which are in colloidalform are precipitated due to coagulation in presence of alkali. These colloidal matters are also precipitated by being entrained in the precipitates of phosphates etc. Along with colloidal matter finely divided pith and fibre and also clay are precipitated.

The raw cane juice is heated in juice heaters to 60°C and then limed to a pH of about 8. The mixing is helped by stirring. The limed juice is then passed through brass tubular heater, where it is heated to boiling. The hot limed juice is then pumped into settling tank for subsidization of the scum and the mud. By heating the protein matter is coagulated and settling of the precipitates is accelerated. In the settling tank the hot limed juice separates into three layers." The dear juice in the middle lies between the top layer of scum (consisting of fine fibre particles) and the bottom layer of mud (consisting of the heavy precipitates. The scum is skimmed of and the clear juice is allowed to drain off through a pipe which is gradually lowered to a level above the mud layer. The clear juice is sent to the boiling house for evaporation to syrup. The mud and the scum are filtered in a filter press. The cakes are washed during filtering (the wash water is sent back to the clarifier), and then dried. These tilter cakes contain sucrose and phosphate. Hence, the added cakes are used as manure after rotting for a year. This removes the sucrose which attracts white ants.

Clarification by Sulphitation : To prepare white sugar by One step clarification is done in India by passing sulphur dioxide gas through limed juice. The production of white is further helped by bleaching the syrup by sulphur dioxide. Better clarification takes place due to two causes:

(1) By adding more lime, more collidal matters and non-sugars are precipitated, because increased alkalinity increases the coagulation of colloids;

(2) By addition of precipitates of calcium sulphite formed by the action of sulphur dioxide on lime as well as on the ttricalcium sucrate ($C_{11}H_{22}O_{11}$, 3ao.-3so$_2$ = $3CaSO_3$—$C_{12}H_{22}O_{11}$) there takes place the precipitation of more colloidal matter by entrainment. The heavy precipitate of calcium sulphite also brings about quick and better subsidization. Sulphur dioxide also helps to bleach the juice.

Sulphur dioxide is passed from the bottom of a tank (with conical bottom) fitted with perforated baffles (at different levels) down which heated (temp. 140°F) limed cane juice comes down in a shower. The gas is passed until the pH of the resultant solution is '7 (neutral). The discharge from the sulphitation tank is heatf;d almost to boiling point in a second juice heater. The amount of lime added is more than that added in simple defecation process (2-4 lbs of lime per ton of cane). From the second heater the hot fluid containing suspended precipitates is sent to the settling tank, where, after subsidization, the discharge separates into dear juice and mud. The subsequent operations have been already mentioned.

Clarification by Carbonation: In India a *few* factories have substituted the sulphitation process by double carbonation process. In this process, the limed juice is treated with carbon dioxide gas to precipitate the excess of lime as $CaCO_3$. This $CaCO_3$ subsequently acts like $CaSO_3$in the sulphitation process. CaCOs is formed by the action of CO: on $Ca(OH)_2$ as well as on $C_{12}H_{22}O_{11}$ 3CaO (tricalcium sucrate). In this process much more lime is added than in sulphitation process. (1-1.5 p.c. of lime on the weight of cane). This amount of lime works out to be 16-24 Ibs of lime per ton of cane. Due to the large alkalinity prodliced by such excessive amount of lime, the removal of non-sugar and colouring matter become more complete. Moreover, due to the formation of much larger quantity of insoluble precipitate in the juice, the subsidization becomes quicker and clarification becomes more effective. Due to removal of soluble impurities more completely the resultant syrup becomes more mobile and more sucrose crystallises cut. Due to the mobility of the syrup the movement of the syrup up and down the calendria in the striking pan becomes easier. This ensures uniform heating and better crystallization. Had sulphur dioxide been used for the CO_2, very large amount of valuable sulphur would have been necessary. Hence, the tenuency in Indian mills is to take to carbonation process. In carbonation process, the excessive amount of lime precipitates the gums, pectins, etc. much more completely when liming takes place below 60°C. But when the alkalinity is neutralised by CO2 ,these-precipitates pass into solution. Also these precipitates become soluble near boiling point of the juice. Hence, instead of single carbonation, double carbonation is adopted. (Described below).

First, cane juice is heated to 55°C and limed by adding milk of lime to the heated juice. Temperature during liming is not allowed to rise above 55°C. Keeping the limed juice at this temperature in ajacketed tank carbon dioxide obtained by heating lime stone in a kiln, is pumped.through the limed juice until the alkalinity is represented by 0.05 p.c. of lime.

The juice along with the precipitate is filtered en-masse (without settling). The juice is then raised to boiling point, saturated with CO_2, again boiled to break the bicarbonate formed, and then filtered en-masse. The filter cake is washed; washings may be added to the juice or sent back for clarification. The juice is sent to the boiling house to make syrup.

Evaporation to make Syrup (Fig. 34.7)

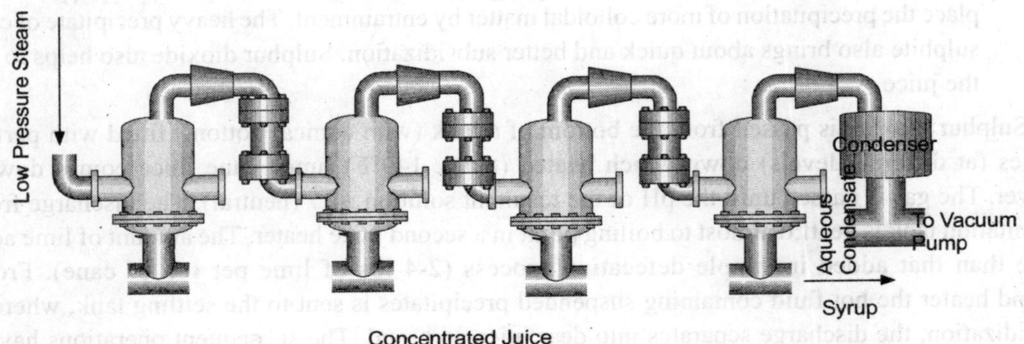

Fig. 34.7. Showing the Quadruple Effect Vacuum Evaporator

Crystalline sugar is obtained from clarified cane juice in two steps. First, the juke is concentrated to syrup containing 50% water; next the syrup is concentrated to massecuite A, containing 9-11 % water. Crystalline sugar—coloured or white-is obtained from the massecuite by removing the molasses.

Before the introduction of the present day method of evaporation under vacuum, the juice was concentrated in open cauldron 'heated by solid fuels. In India, 'Gur' (crystalline sugar mixed with molasses) is still made by this process. Formerly, crystalline sugar was manufactured from 'gur; by removing the molasses by an indigenous process. By this process much sucrose is lost as caramel, and portion of sucrose is converted into glucose'. Even concentration by steam coils under atmospheric pressure produces loss. Hence, the present-day method of evaporation at a lower temperature under reduced pressure has been adopted.

The clarified juice obtained from any of the above methods is concentrated into the thick syrup by concentrating in multiple-effect evaporators. Usually quadruple-effect or tripple-effect evaporators are used.

An evaporator, is a tall cylindrical vessel made of riveted mild steel plates or cast in sections from cast iron. The top is bottle-necked and the bottom is spheroidal. The evaporators are placed vertically and in a battery of three or four. The cylinder is divided into three chambers by fitting two parallel plates a few feet arart at the lower part of the cylinder; the chambers above the upper plate and below the lower plate are the juice chambers, and the chamber between the two plates form the steam chamber or calendria.

'Several narrow vertical pipes join the upper plate to the lower plate (the pipes open in, the two juice chambers). A few wider pipes also vertically connect the upper and the lower pipe. The juice circulates inside the tubes and the steam circulates outside the tubes. There is a steam inlet pipe to each of the calendrias. The bottle-neck is closed at the top, but is fitted with a vapour outlet pipe at one side. The outlet pipe of the first evaporator is joined to the calcndria of the second evaporator, and so on up to the last calendria. Arrester baffle plates are fitted to the neck to prevent any juice or syrup to pass through the vapour outlet pipe. A narrow pipe known as "Ammonia pipe' (by analogy with beet juice evaporators) joins the calendria to the main body to remove air and other uncondensible gases accumulated in the calendria of the body, and ultimately delivers these gases to the vacuum and condensers attached to the tast body.

Heating of the juice in the first evaporator is done by passing steam exhaust or line through the calendria. The juice submerges the calendria, and during heating the juice which is admitted from the bottom by a feed pipe circulates up through the narrow tubes, and down through the wider tubes. In this way by rapid circulation and heating at a vacuum, the water is rapidly evaporated. The vapour from the first evaporator enters into the calendria of the second evaporator. This stem, although of the lower temperature, helps rapid evaporation of the partially thickened syrup which passed from the first evaporator to the second by the opening of the valve of the bottom feed pipe: being actuated by difference of pressure. The low temperature steam helps rapid evaporation because of a higher vacuum maintained in the second evaporator. In quadruple-effect evaporators, the vacuum maintained in the 3rd body is greater than in the second, and that maintained in the last body is greater than in the 3rd. The vacuum pump together with a condenser is attached to the vapour outlet pipe of the 4th evaporator. This pump gradually reduces the pressure from the first to the last body. The vacuum and corresponding temperature maintained in each evaporator along with the steam temperature is as follows:

	Evaporators			
	I	II	III	IV
Temp of Heating Steam	125°C	99°C	86°C	76°C
Temp of boiling Juice	99°C	86°C	76°C	60°C
Vacuum maintained (in cm)	2.744	30.48	50.8	62.03

From the data mentioned above, it is found that boiling of the juice take place from 210°F to 140°F in the last evaporator. This means that most of the water has been removed at comparatively lower temperature and in absence of air, to ensure least loss of sugar by evaporation. The evaporation and subsequent transfer of fluid is so adjusted as to keep the level of the liquid in the evaporators above the level of the calendria.

From the last evaporator the pale yellow thick syrup (50-66 % solid matter) is discharged into the syrup tank. From this tank it is pumped to the top of the sulphitation tank wherefrom it is allowed to percolate down the cylinder over the' perforated baffle plates, in a shower, to meet an ascending stream of sulphur dioxide.

The sulphur dioxide bleaches the syrup and reacts with residual calcium ions. This process is omitted when manufacture of raw sugar is aimed at. This process is, therefore, followed by the Indian factories a few factories where clarification is done by carbonation, bleaching by sulphur dioxide may be omitted if double carbonation has been properly carried out.

Crystallization (Fig. 34.8)

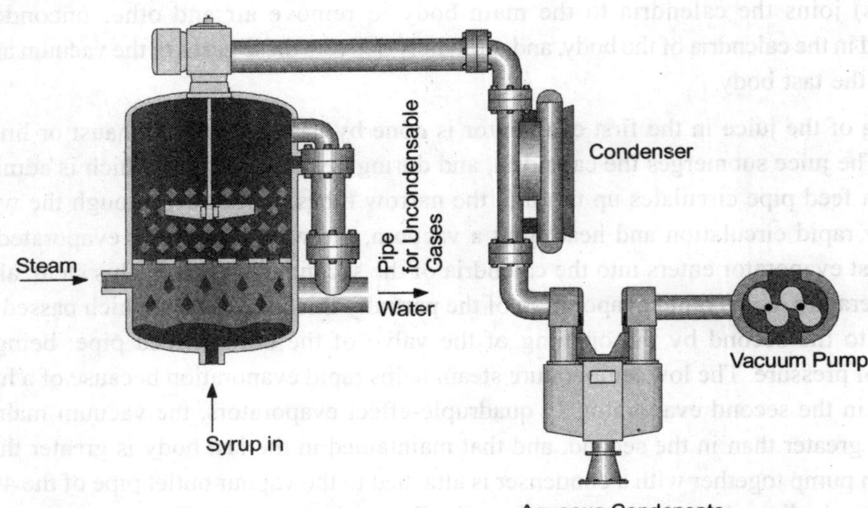

Fig. 34.8. Showing the Single Effect Vacuum Evaporators

The syrup is next treated for forming crystalline sugar. This is done by removing most of the water by heating in a single-effect vacuum evaporator known as pan. The vacuum pan is either fitted with calendria or steam coils. The calendria fitted pan is similar to the evaporator used for boiling the juice. But it is larger in diameter and the tubes of its calendria are wider. It is fitted with vacuum gauge and a condenser, a thermometer, a sampling stick and safety device to prevent the boiling mass getting into the vapour) outlet pipe and outlet pipe at the bottom to dump the massecuite. There is the usual feed pipe at the bottom.

The crystallization takes place in two steps, viz.

(a) Formation of seed crystals 'by graining;

(b) Allowing the seed crystals to grow into fully developed crystals.

(a) The syrup is introduced into the pan until the steam chest is under it. Then by heating under vacuum the syrup is concentrated until, it is supersaturated with sucrose (83.5-84.5%.sucrose at 80°C). Then the operator adopts either waiting method or shock method to arrive at the graining point, i.e., induce the formation of seed crystals by breaking the supersaturation. In the waiting method the operator gradually cools down to break the supersaturation. In the shock method which is usually adopted the supersaturation is broken by sudden chilling which is, done either by increasing the vacuum or by decreasing the rate of steam flow or by suddenly introducing large 'amount of cold syrup'. Circulation. of the syrup is necessary for producing proper seed crystal

(b) After. graining point is reached more and more syrup is introduced in instalments. By proper circulation of the very thick syrup up and down the tubes of the calendria which is helped by injecting dry steam through, the bottom even, boiling is carried out at high vacuum (28.5 inches) to force out

sucrose out of the concentrated syrup which is deposited as crystals on the nuclei formed at the graining point. As more and more syrup is introduced the crystals already formed grow more and more, until a semi-solid of mass fills up the pan. When this point is reached the flow of syrup is stopped and Concentration continued up to 92° - 93° Brix (9 to 11 % water) at a temperature of 65°C and vacuum of 28.7 inches. The semi-solid mass consisting of sugar crystals arid molasses-called massecute A is removed through the bottom outlet in crystalliser tanks. During the formation of crystals in the pan, false grains are formed which hamper the growth of crystals. False grains consist of minute crystals formed between the proper crystals. The object of the operator is to allow the grains to grow and to prevent the formation of fresh grains.

Physical Properties

Molecular Formula	$= C_{12}H_{22}O_{11}$
Formula Weight	= 342.29648
Composition	= C(42.11%) H(6.48%) O(51.42%)
Melting Point	=186 °C
Solubility in water	= 211.5 g/100 ml (20 °C)
Molar Refractivity	$= 71.04\ cm^3$
Molar Volume	$= 189.2\ cm^3$
Parachor	$= 583.9\ cm^3$
Index of Refraction	= 1.674
Surface Tension	= 90.6 dyne/cm
Density	$= 1.808\ g/cm^3$
Polarizability	$= 28.16\ 10^{-24} cm^3$
Monoisotopic Mass	= 342.116212 Da
Nominal Mass	= 342 Da
Average Mass	= 342.2965 Da
Log P	= –1.82

Chemical Properties

Pure sucrose is most often prepared as a fine, white, odorless crystalline powder with a pleasing, sweet taste; the common table sugar. Large crystals are sometimes precipitated from water solutions of sucrose onto a string (or other nucleation surface) to form rock candy, a confection. Like other carbohydrates, sucrose has a hydrogen to oxygen ratio of 2:1. It consists of two monosaccharides, α-glucose and fructose, joined by a glycosidic bond between carbon atom 1 of the glucose unit and carbon atom 2 of the fructose unit. What is notable about sucrose is that unlike most polysaccharides, the glycosidic bond is formed between the reducing ends of both glucose and fructose, and not between the reducing end of one and the non-reducing end of the other. The effect of this inhibits further bonding to other saccharide units. Since it contains no free anomeric carbon atom, it is classified as a non-reducing sugar. Sucrose melts and decomposes at 186 °C to form caramel, and when combusted produces carbon, carbon dioxide, and water. Water breaks down sucrose by hydrolysis, however the process is so gradual that it could sit in solution for years with negligible change. If the enzyme sucrase is added however, the reaction will proceed rapidly. Reacting sucrose with sulfuric acid dehydrates the sucrose and forms the element carbon, as demonstrated in the following equation:

$$C_{12}H_{22}O_{11} + H_2SO_4 \text{ catalyst} \rightarrow 12\ C + 11\ H_2O$$

Sugar as a Macronutrient

In mammals, sucrose is very readily digested in the stomach into its component sugars, by acidic hydrolysis. This step is performed by a glycoside hydrolase, which catalyzes the hydrolysis of sucrose to the monosaccharides glucose and fructose. Glucose and fructose are rapidly absorbed into the bloodstream in the small intestine. Undigested sucrose passing into the intestine is also broken down by sucrase or isomaltase glycoside hydrolases, which are located in the membrane of the microvilli lining the duodenum. These products are also transferred rapidly into the bloodstream.

Sucrose is digested by the enzyme invertase in bacteria and some animals. Acidic hydrolysis can be used in laboratories to achieve the hydrolysis of sucrose into glucose and fructose.

Use in Human Nutrition

Sucrose is an easily assimilated macronutrient that provides a quick source of energy to the body, provoking a rapid rise in blood glucose upon ingestion. However, pure sucrose is not normally part of a human diet balanced for good nutrition, although it may be included sparingly to make certain foods more palatable. Overconsumption of sucrose has been linked with some adverse health effects. The most common is dental caries or tooth decay, in which oral bacteria convert sugars (including sucrose) from food into acids that attack tooth enamel. Sucrose, as a pure carbohydrate, has an energy content of 3.94 kilocalories per gram (or 17 kilojoules per gram). When a large amount of foods that contain a high percentage of sucrose is consumed, beneficial nutrients can be displaced from the diet, which can contribute to an increased risk for chronic disease. It has been suggested that sucrose-containing drinks may be linked to the development of obesity and insulin resistance.[1] However, most soft drinks in the USA are now made with high-fructose corn syrup, not sucrose. The rapidity with which sucrose raises blood glucose can cause problems for people suffering from defects in glucose metabolism, such as persons with hypoglycemia or diabetes mellitus. Sucrose can contribute to development of the metabolic syndrome. In an experiment with rats that were fed a diet one-third of which was sucrose, the sucrose first elevated blood levels of triglycerides, which induced visceral fat and ultimately resulted in insulin resistance. Another study found that rats fed sucrose-rich diets developed high triglycerides, hyperglycemia, and insulin resistance.[4] Extension of average lifespan can be achieved by good diet, exercise and avoidance of hazards such as smoking and excessive eating of sugar-containing foods.

MALTOSE

Maltose, or malt sugar, is a disaccharide formed from two units of glucose joined with an $\alpha(1\rightarrow4)$ linkage. It is the second member of an important biochemical series of glucose chains. The addition of another glucose unit yields maltotriose; further additions will produce dextrins (also called maltodextrins) and eventually starch.

Maltose can be broken down into two glucose molecules by hydrolysis. In living organisms, the enzyme maltase can achieve this very rapidly. In the laboratory, heating with a strong acid for several minutes will produce the same result.

The production of maltose from germinating cereals, such as barley, is an important part of the brewing process. When barley is malted, it is brought into a condition in which the concentration of maltose-producing amylases has been maximized. Mashing is the process by which these amylases convert the cereal's starches into maltose. Metabolism of maltose by yeast during fermentation then leads to the production of ethanol and carbon dioxide.

Physical Properties

Molecular Formula	$= C_{12}H_{22}O_{11}$
Formula Weight	= 342.29648
Composition	= C(42.11%) H(6.48%) O(51.42%)
Melting Point	= 102-103 °C
Molar Refractivity	= 71.13 cm^3
Molar Volume	= 190.8 cm^3
Parachor	= 579.5 cm^3
Index of Refraction	= 1.668
Surface Tension	= 85.0 dyne/cm
Density	= 1.793 g/cm^3
Polarizability	= 28.20 $10^{-24}cm^3$
Monoisotopic Mass	= 342.116212 Da
Nominal Mass	= 342 Da
Average Mass	= 342.2965 Da

LACTOSE

Lactose (also referred to as *milk sugar*) is a sugar which is found most notably in milk. Lactose makes up around 2–8% of milk (by weight). The name comes from the Latin word for milk, plus the -ose ending used to name sugars. Its systematic name is β-D-galactopyranosyl-(1 ↔ 4)α-D-glucopyranose.

Chemistry

Lactose is a disaccharide that consists of β-D-galactose and β-D-glucose molecules bonded through a β1-4 glycosidic linkage.

Solubility

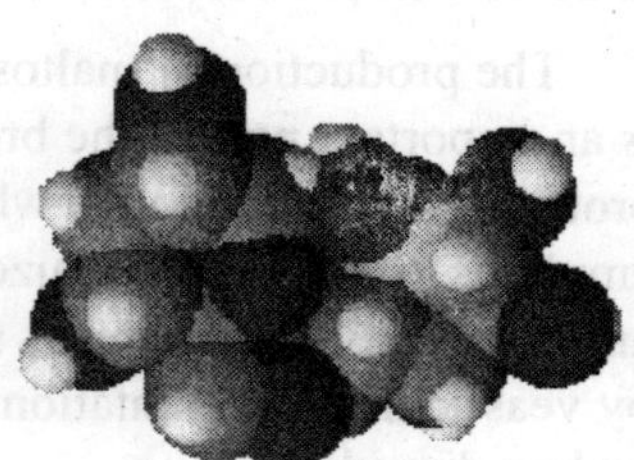

Lactose has a solubility of 1 in 4.63 measured %w/v. This translates to 0.216 g of lactose dissolving readily in 1 mL of water.

The solubility of lactose in water is 18.9049 g at 25°C, 25.1484 g at 40°C and 37.2149 g at 60°C per 100 g solution. Its solubility in ethanol is 0.0111 g at 40°C and 0.0270 g at 60°C per 100 g solution.

Digestion of lactose

Infant mammals are fed on milk by their mothers. To digest it an enzyme called lactase (β-D-galactosidase) is secreted by the intestinal villi, and this enzyme cleaves the molecule into its two subunits glucose and galactose for absorption.Since lactose occurs mostly in milk, in most mammals the production of lactase gradually decreases with maturity.Many people with ancestry in Europe, the Middle East, India, or parts of East Africa, maintain normal lactase production into adulthood. In many of these cultures, mammals such as cattle, goats, and sheep are milked for food. Hence, it was in these regions that genes for lifelong lactase production first evolved.

Trisaccharides

RAFFINOSE

Raffinose is a trisaccharide composed of galactose, fructose, and glucose. It can be found in beans, cabbage, brussels sprouts, broccoli, asparagus, other vegetables, and whole grains. Raffinose can be hydrolyzed to D-galactose and sucrose by the enzyme α-galactosidase (α-GAL), an enzyme not found in humans. α-GAL also hydrolyzes other α-galactosides such as stachyose, verbascose, and galactinol, if present. The enzyme does not cleave β-linked galactose, as in lactose.

The raffinose family of oligosaccharides (RFOs) are alpha-galactosyl derivatives of sucrose, and the most common are the trisaccharide raffinose, the tetrasaccharide stachyose, and the pentasaccharide verbascose. RFOs are almost ubiquitous in the plant kingdom, being found in a large variety of seeds from many different families, and they rank second only to sucrose in abundance as soluble carbohydrates.

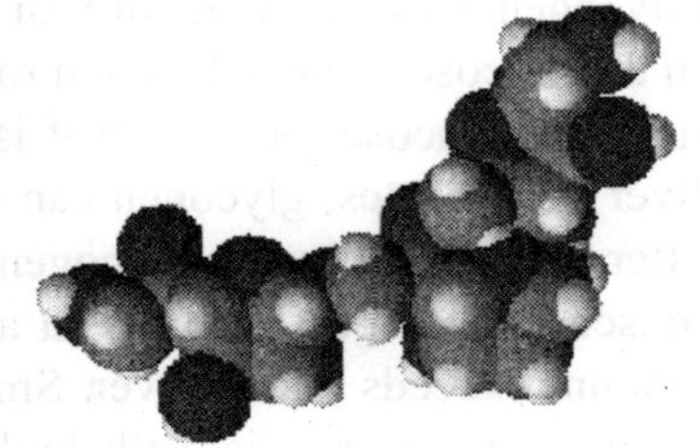

Humans and other monogastric animals (pigs and poultry) do not possess the α-GAL enzyme to break down RFOs and these oligosaccharides pass undigested through the stomach and upper intestine. In the lower intestine, they are fermented by gas-producing bacteria which do posses the α-GAL enzyme and make carbon dioxide, methane, and/or hydrogen — leading to the flatulence commonly associated with eating beans and other vegetables. α-GAL is present in digestive aids such as the product Beano.

Procedures concerning cryopreservation have utilized raffinose to provide hypertonicity for cell desiccation prior to freezing. Either raffinose or sucrose is used as a base substance for sucralose.

MELEZITOSE

Melezitose, also spelled **melicitose**, is a nonreducing trisaccharide sugar that is produced from lice such as *Cinara pilicornis* by an enzyme reaction. This is beneficial to the insects, as it reduces the stress of osmosis by reducing their own water potential. The melezitose is part of the honeydew of the honey produced by bees, and acts as an attractant for ants. This is useful to the lice as they have a symbiotic relationship with ants. Melezitose can be partially hydrolyzed to glucose and turanose the latter of which is an isomer of sucrose.

Polysachharide

Digestible

GLYCOGEN

Glycogen is a polysaccharide of glucose (Glc) which functions as the primary short term energy storage in animal cells. It is made primarily by the liver and the muscles, but can also be made by the brain, uterus, and the vagina. Glycogen is the analogue of starch, a less branched glucose polymer in plants, and is commonly referred to as **animal starch**, having a similar structure to amylopectin.

Glycogen is found in the form of granules in the cytosol in many cell types, and plays an important role in the glucose cycle. Glycogen forms an energy reserve that can be quickly mobilized to meet a sudden need for glucose, but one that is less compact than the energy reserves of triglycerides (fat). In the liver hepatocytes, glycogen can compose up to 8% of the fresh weight (100–120 g in an adult) soon after a meal. Only the glycogen stored in the liver can be made accessible to other organs. In the muscles, glycogen is found in a much lower concentration (1% of the muscle mass), but the total amount exceeds that in liver. Small amounts of glycogen are found in the kidneys, and even smaller amounts in certain glial cells in the brain and white blood cells. The uterus also stores glycogen during pregnancy to nourish the embryo.

Synthesis

Glycogen synthesis differs from glycogen breakdown. Unlike breakdown, synthesis is endergonic, meaning that glycogen is not synthesized without the input of energy. Energy for glycogen synthesis comes from UTP, which reacts with glucose-1-phosphate, forming UDP-glucose, in reaction catalysed by UDP-glucose pyrophosphorylase. Glycogen is synthesized from monomers of UDP-glucose by the enzyme Glycogen synthase, which progressively lengthens the glycogen chain. As glycogen synthase can only lengthen an existing chain, the protein glycogenin is needed to initiate the synthesis of glycogen.

Breakdown

Glycogen is cleaved from the non-reducing ends of the chain by the enzyme glycogen phosphorylase to produce monomers of glucose-1-phosphate that is then converted to Glucose 6-phosphate. A special debranching enzyme is needed to remove the alpha(1-6) branches in branched glycogen and reshape the chain into linear polymer. The G6P monomers produced have three possible fates:

- G6P can continue on the glycolysis pathway and be used as fuel.
- G6P can enter the pentose phosphate pathway via the enzyme Glucose-6-phosphate dehydrogenase to produce NADPH and 5-carbon sugars.
- In the liver and kidney, G6P can be dephosphorylated back to Glucose by the enzyme Glucose 6-phosphatase. This is the final step in the gluconeogenesis pathway.

Structure and Biochemistry

Glycogen is a highly branched polymer that is better described as a dendrimer of about 60,000 glucose residues and has a molecular weight between 10^6 and 10^7 daltons (~4.8 million) Most of Glc units are linked by α-1,4 glycosidic bonds, approximately 1 in 12 Glc residues also makes -1,6 glycosidic bond with a **second Glc**, which results in the creation of a branch. Glycogen does not possess a reducing end: the 'reducing end' glucose residue is not free but is covalently bound to a protein termed glycogenin as a beta-linkage to a surface tyrosine residue. Glycogenin is a glycosyltransferase and occurs as a dimer in the core of glycogen. The glycogen granules contain both glycogen and the enzymes of glycogen synthesis (glycogenesis) and degradation (glycogenolysis). The enzymes are nested between the outer branches of the glycogen molecules and act on the non-reducing ends. Therefore, the many non-reducing end-branches of glycogen facilitate its rapid synthesis and catabolism.

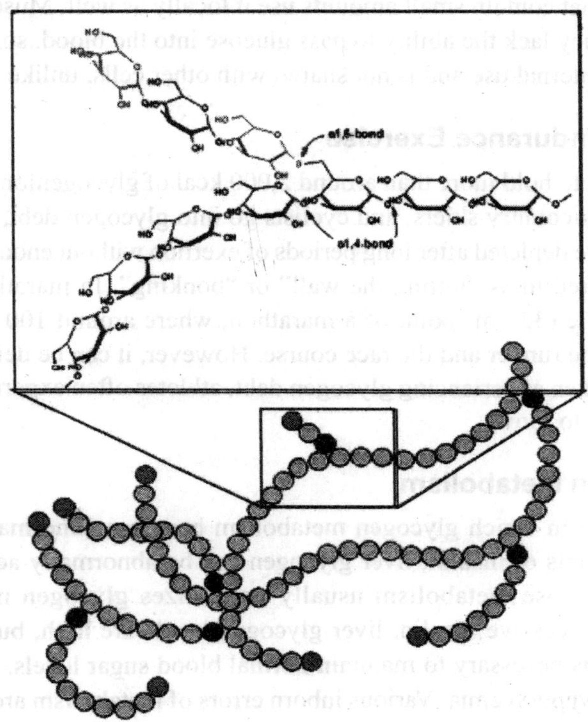

Fig. 34.9. Showing the Glycogen structure

Function and Regulation of Liver Glycogen

As a meal containing carbohydrates is eaten and digested, blood glucose levels rise, and the pancreas secretes insulin. Glucose from the portal vein enters the liver cells (hepatocytes). Insulin acts on the hepatocytes to stimulate the action of several enzymes, including glycogen synthase. Glucose molecules are added to the chains of glycogen as long as both insulin and glucose remain plentiful. In this postprandial or "fed" state, the liver takes in more glucose from the blood than it releases. After a meal has been digested and glucose levels begin to fall, insulin secretion is reduced, and glycogen synthesis stops. About four hours after a mealglycogen begins to be broken down and converted again to glucose. Glycogen phosphorylase is the primary enzyme of glycogen breakdown. For the next 8–12 hours, glucose derived from liver glycogen will be the primary source of blood glucose to be used by the rest of the body for fuel. Glucagon is another hormone produced by the pancreas, which in many respects serves as a counter-signal to insulin. When the blood sugar begins to fall below normal, glucagon is secreted in increasing amounts. It stimulates glycogen breakdown into glucose even when insulin levels are abnormally high.

In Muscle and Other Cells

Muscle cell glycogen appears to function as an immediate reserve source of available glucose for muscle cells. Other cells that contain small amounts use it locally as well. Muscle cells lack glucose-6-phosphatase enzyme, so they lack the ability to pass glucose into the blood, so the glycogen they store internally is destined for internal use and is not shared with other cells, unlike liver cells.

Glycogen Debt and Endurance Exercise

Due to the body's inability to hold more than around 2,000 kcal of glycogenlong-distance athletes such as marathon runners, cross-country skiers, and cyclists go into glycogen debt, where almost all of the athlete's glycogen stores are depleted after long periods of exertion without enough energy consumption. This phenomenon is referred to as "hitting the wall" or "bonking". In marathon runners it normally happens around the 20 mile (32 km) point of a marathon, where around 100 kcal are spent per mile depending on the size of the runner and the race course. However, it can be delayed by a carbohydrate loading before the task. When experiencing glycogen debt, athletes often experience extreme fatigue to the point that it is difficult to move.

Disorders of Glycogen Metabolism

The most common disease in which glycogen metabolism becomes abnormal is diabetes, in which, because of abnormal amounts of insulin, liver glycogen can be abnormally accumulated or depleted. Restoration of normal glucose metabolism usually normalizes glycogen metabolism as well. In hypoglycemia caused by excessive insulin, liver glycogen levels are high, but the high insulin level prevents the glycogenolysis necessary to maintain normal blood sugar levels. Glucagon is a common treatment for this type of hypoglycemia. Various inborn errors of metabolism are caused by deficiencies of enzymes necessary for glycogen synthesis or breakdown. These are collectively referred to as glycogen storage diseases.

STARCH

Introduction

These are both complex carbohydrate polymers of glucose (chemical formula of glucose $C_6H_{12}O_6$), making starch a glucose polymer as well, as seen by the chemical formula for starch, regardless of the ratio of amylose:amylopectin. The word is derived from Middle English *sterchen*, meaning to stiffen, which is appropriate since it can be used as a thickening agent when dissolved in water and heated.

Synthesis

The starch is located in the crown and glutinous regions of the endosperm. Large, round granules are loosely held in the crown region, associated with a relatively small amount of protein. The separation of these granules in the milling process offers no serious problem. Smaller, polygonally shaped granules are tightly held in a water- insoluble protein matrix in the glutinous region. The hydration and disintegration

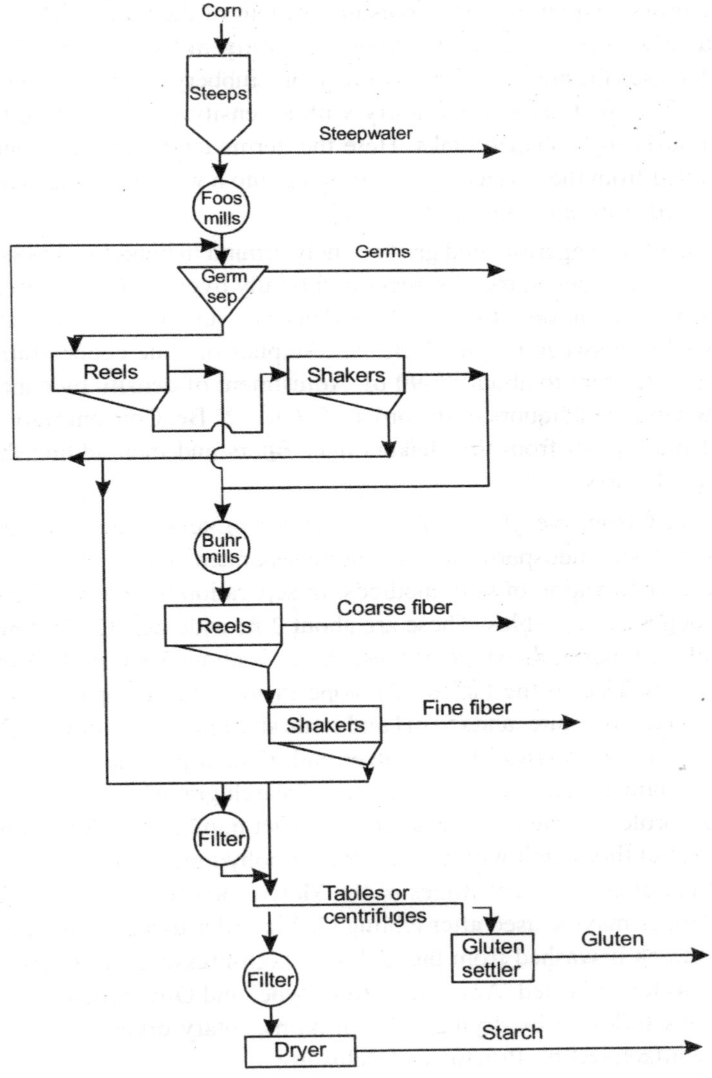

Fig. 34.10. Showing the Industrial Preparation of Starch

of this protein in order to loosen the starch, the separation of the starch from the finely dispersed protein, and the accomplishment of these ends without a serious deterioration of the more fragile and less resistant granules in the crown region constitute the chief problems in milling. The steps of the *wet milling process* are indicated diagrammatically in Fig. 34.10.

In order to facilitate the separation of the constituents of the grain, the corn is hydrated by soaking *(steeping)* in water at about 50°C for 35-45 hours. Usually, about 0.15 – 0.20% of sulfur dioxide is added to the water (pH about 4.0) in order to control biological activity (19) and to facilitate the disintegration of the protein structures. The moisture content of the corn increases during steeping to about 45%. The steeped corn is passed to coarse grinding mills (such as the Foos mill) called degenninators, which loosen the hulls and separate the tough, rubbery germs with very little disintegration of these constituents. The grist, as a water slurry with a density corresponding to 8-10.5°Be. at 90-95°F., is fed continuously to V-shaped tanks. Here the germs tend to float because of their high oil content and are removed from the surface by overflowing into a weir box. The heavier part of the grist is continuously removed from the bottom of the tank.

The coarse portion of the degerminated grist is finely ground in buhrstone mills. Hulls and fiber are removed by passage over a series of metal screens and finally over a series of screens made of bolting silk. The resulting slurry contains only finely dispersed protein, starch, and some water-soluble material. The density is adjusted to between 6 and 12°Be., the sulphur dioxide concentration to between 0.06 and 0.08%, and the temperature to about 85-90 of. Adjustment of density by concentration is usually required in mills with table feed liquors of the order of 10 to 12° Be. Concentration may be affected by passing a portion of the liquors from the shakers over filters and then adding the filter cake to the balance of the unfiltered slurry.

Starch is *sedimented* from the gluten (water-insoluble proteins, zein and zeanin, together with minor amounts of corn fiber, endosperm lipides, and unseparated starch) either by gravity, by the use of centrifuges, or by a combination of both methods. In separation by gravity, the slurry is allowed to run down inclined troughs called tables. These are about 2 ft. wide and 120 ft. long. For feed rates of 1.5-2.0 gaL/minute of mill liquor, the slope of these tables is about 3>16 in.jft. Variables affecting the separation are: (1) rate of feed to the tables, (2) slope of the table compared to its dimensions, (3) density of the slurry, (4) temperature, and (5) pH and ionic strength of the liquor (12). Starch sediments on the tables, while the gluten passes off the lower end. Gluten particles which are not sufficiently hydrated and which contain a relatively large number of starch granules will obviously settle with the starch and increase the protein content of the latter. A smaller number of starch granules may not cause the gluten to sediment, but this starch will increase the carbohydrate content of the gluten fraction and result in a reduced yield of starch. Centrifuges of the Merco (bowl) type may be used to separate the starch, or these centrifuges may be used after tabling (38) in order further to purify either the starch or the gluten fractions. Starch is washed from the tables by high-pressure water jets. After filtration, the starch is washed with water and dried. American (drum-type) and Oliver (disk-type) filters are usually used in these operations followed by drying either in kilns, rotary dryers, or preferably endless belt dryers of the type manufactured by Proctor and Schwartz

The wet milling of corn employs the countercurrent principle throughout in the use of water. Aside from the water which enters the system as moisture in the corn grain, the balance, making a total of

10-12 gal. per bushel of corn processed, is added in the final washing of the starch before filtration. This wash water is then used to flush the starch from the tables and is used in turn in various cycles of the milling process, until as gluten overflow water from the tables it contains about 1-2 grams of dissolved material per 100 ml. After sedimentation of the gluten in tanks called gluten settlers, the water is finally used in the steeping operation. After steeping and before final dewatering, the pH of the starch is usually adjusted to a range between 5 and 7. The water is removed from the corn and evaporated to recover the extracted corn solids. Steep water may contain as much as 8 grams of nonvolatile solids per 100 ml. and has proved to be a practical commercial source of meso-inositol, which is of interest as a member of the vitamin-B complex. The various by-product constituents of the grain such as the germs, the hulls, and fiber are in turn washed free of adhering starch by the process waters in the system and are then further processed.

Physical Properties

Molecular Formula	$= C_{48}H_{76}O_{38}$
Formula Weight	= 1261.09424
Composition	= C(45.72%) H(6.07%) O(48.21%)
Molar Refractivity	= 262.69 cm^3
Molar Volume	= 774.8 cm^3
Parachor	= 2243.7 cm^3
Index of Refraction	= 1.593
Surface Tension	= 70.3 dyne/cm
Density	= 1.627 g/cm^3
Polarizability	= 104.14 $10^{-24}cm^3$
Monoisotopic Mass	= 1260.401458 Da
Nominal Mass	= 1260 Da
Average Mass	= 1261.0942 Da

Chemical Properties

Most varieties of starch contain two types of polymers which differ from each other in molecular weight and in chemical structure. The linear polymer, **amylose,** consists of 200-1000 glucopyranose units joined to each other through a-l,4-glucosidic linkages, whereas the branched or ramified polymer, **amylopectin,** is made up of 1500 or more glucopyranose units (6,10,15,20) (see Fig. 34.11). In addition to the normal or predominating a-l,4 linkages, an anomalous a-l,6 Iinkage is present in the ramified structure at the origin or point of branching in a ratio of about 1 :25.

The linear polymer and the longer branches of the nonlinear polymer show a pronounced tendency to orient and associate with other linear members. This property is characteristic of many thread-like molecules which contain a large number of hydroxyl or other hydrogen-bond-forming groups distributed along their chain length. Starch molecules are highly associated in the granule.

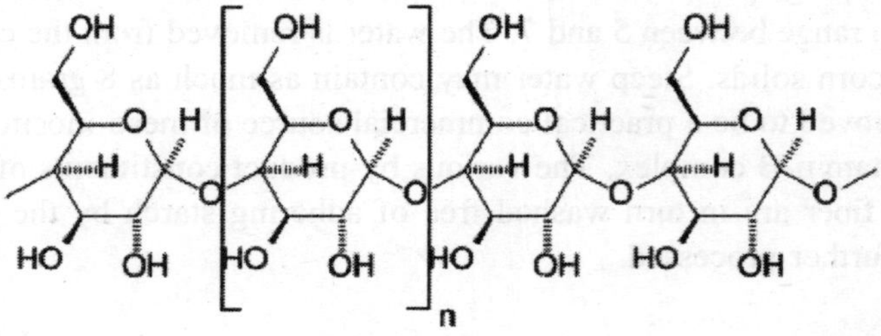

AMYLOSE MOLECULE

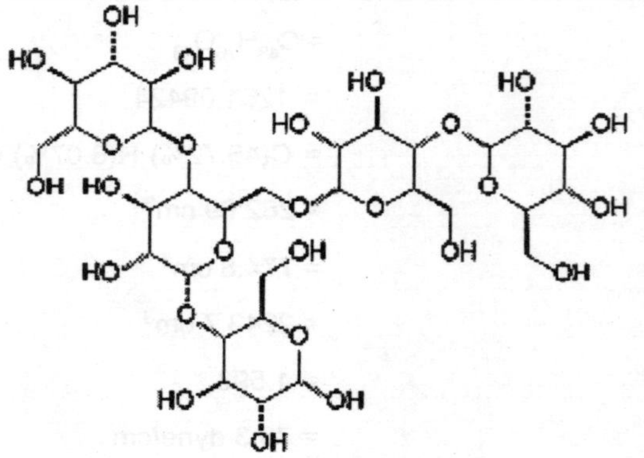

AMYLOPECTINE MOLECULE

When starch is dispersed in water, the sol may be fractionated into its polymeric components by the addition of polar substances, particularly the higher alcohols. The amylose forms an insoluble complex with butyl alcohol which may be sedimented by centrifugation. The amount of amylose obtained varies with the origin of the starch. Starches from so-called waxy grains, such as waxy corn and sorghum, contain no amylose; in contrast, the starch from certain varieties of peas contains 75% by weight of amylose. The common industrial starches contain about 15-30% by weight of amylose.

The chemical property of starch which is at present of greatest industrial importance is the comparative ease with which the plymeric bond is hydrolyzed by both enzymes and acids. When starch is gelatinized by heating in water or when it is dissolved by suitable solvating agent~, for example, sodium hydroxide, followed by neutralization, its glucoside linkages (the polymeric bonds) are readily hydrolyzed by **amylases.** These enzymes are obtained from a large number of sources such as bacteria, fungi, vegetables, and animals. The amylases present in these preparations are grouped into three broad classifications or types since the course of the hydrolysis varies in a characteristic manner with the type of amylase employed.

α-Amylase splits the high-polymer starch molecules into molecules of progressively smaller size. Eventually, sugar molecules are produced. Characteristic of this reaction is the rapid decrease in viscosity of the starch medium with very little production of reducing sugars. Accordingly, the reaction is used industrially to produce low-viscosity sizes and adhesives from native starch. *{3-Amylase* splits the disaccharide *maltose* directly from starch molecules. Amylose molecules are almost quantitatively hydrolyzed to maltose in this manner. Amylopectin molecules are hydrolyzed 50-60%. The residue which remains ("beta-amylase limit. dextrin") is of comparatively high molecular weight and still contains all of the anomalous glucoside linkages which were present in the original starch molecules. A third type of amylase, called *amyloglucosidase*, splits starch molecules of both structures directly to the monosaccharide, D-glucose (dextrose). The saccharifying reactions are used by fermentation industries employing starch as substrate. In as much as both the maltose- and glucose-producing enzymes are believed to act only on the non-reducing terminal sugar groups of the starch polymer, it follows that one of the principal functions of a-amylase, as it occurs normally in nature in conjunction with saccharifying enzymes, is to facilitate the action of the latter by creating more terminal groups per unit weight. Thus, the complete hydrolysis of starch to sugars is accelerated.

Although amylases of anyone class act on starch in qualitatively the same manner, members of a class, depending on source, may show quantitative differences in their mode of action. The relatively complex chemistry involved in the action of amylases has been summarized by Kerr and Gehman.

All starches are hydrolyzed by acid to D-glucose, as in the commercial production at high temperatures of dextrose *(q.v.)* and starch sirups (mixtures consisting of dextrose, maltose, and higher polysaccharides generally called dextrins). In common with the action of enzymes, acid at first causes a reduction in paste viscosity. In contrast to enzymes, however, acid readily penetrates at least the amorphous regions of the native un swollen granule, so that hydrolysis may take place even at temperatures materially lower than those required for gelatinization. This fact is utilized in the manufacture of modified, "thin-boiling" starches (18). When these acid-modified starches are gelatinized, the consistency of their pastes at a given concentration is less than that of the native starch. The intermediate product of acid degradation is known as **amylodextrin.**

The alcohol groups in starch may be oxidized to form carbonyl or carboxyl groups. The course of oxidation varies with the oxidant and conditions employed. Specific oxidants such as nitrogen dioxide act principally to convert primary alcoholic groups to carboxyls, whereas others such as periodate act on the two adjacent secondary hydroxyl groups, converting them first into ketone groups and subsequently into aldehydes with a rupture of the bond between carbons 2 and 3. Supplementary oxidants such as bromine water may then convert them further to diacids. Less specific agents used industrially, such as hypochlorite, very likely create many carboxyl groups along the starch molecule, thus increasing its solubility in water particularly in neutralized or dilute alkaline solution. The alcohol groups in starch react with aldehydes, under acidic conditions, by a condensation reaction, in which cross-linkages are very likely established between two adjacent starch molecules through the carbonyl carbon of the aldehyde. The end product is completely nondispersible in water. Starch also reacts to form a product which is nondispersible in water with agents such as epichlorohydrin, dibasic acids or their acid chlorides, potassium pyroantimonate, and allyl bromide. Where the reaction involves the uniting of two starch molecules, it may be viewed as a polymerization. Starch, or more particularly the linear polymer of starch, reacts characteristically with iodine to form a blue-colored complex. The iodine test is used in food analysis as a test for unconverted starch.

The aldehyde group in the starch molecule is reducible. Polyhydric alcohols such as sorbitol are formed by hydrogenation (involving also hydrolysis) and hydrogenolysis

Tests

Iodine solution is used to test for Starch. A bluish-black colour indicates the presence of iodine in the starch solution. It is thought that the iodine fits inside the coils of amylose.[5] A 0.3% w/w solution is the standard concentration for a dilute starch indicator solution. It is made by adding 4 grams of soluble starch to 1 litre of heated water; the solution is cooled before use (starch-iodine complex becomes unstable at temperatures above 35 °C). This complex is often used in redox titrations: in presence of an oxidizing agent the solution turns blue, in the presence of reducing agent, the blue color disappears because triiodide (I_3^-) ions break up into three iodide ions, disassembling the complex.

Under the microscope, starch grains show a distinctive Maltese cross effect (also known as 'extinction cross' and birefringence) under polarized light.

Uses

Starch in Food

In terms of human nutrition, starch is by far the most consumed polysaccharide in the human diet. It constitutes more than half of the carbohydrates even in many affluent diets, and much more in poorer diets. Traditional staple foods such as cereals, roots and tubers are the main source of dietary starch.

Starch (in particular cornstarch) is used in cooking for thickening foods such as sauces. In industry, it is used in the manufacturing of adhesives, paper, textiles and as a mold in the manufacture of sweets such as wine gums and jelly beans. It is a white powder, and depending on the source, may be tasteless and odorless.

Fruit, seeds, rhizomes or tubers of plants and is the major source of energy in these food items. The major resources for starch production and consumption worldwide are rice, wheat, corn, and potatoes. Cooked foods containing starches include boiled rice, various forms of bread and noodles (including pasta).

As an additive for food processing, arrowroot and tapioca are commonly used as well. Commonly used starches around the world are: arracacha, buckwheat, banana, barley, cassava, kudzu, oca, sago, sorghum, regular household potatoes, sweet potato, taro and yams. Edible beans, such as favas, lentils and peas, are also rich in starch.

When a starch is pre-cooked, it can then be used to thicken cold foods. This is referred to as a pregelatinized starch. Otherwise starch requires heat to thicken, or "gelatinize."[vague]The actual temperature depends on the type of starch.

A modified food starch undergoes one or more chemical modifications, which allow it to function properly under high heat and/or shear frequently encountered during food processing. Food starches are typically used as thickeners and stabilizers in foods such as puddings, custards, soups, sauces, gravies, pie fillings, and salad dressings, but have many other uses.

Resistant starch is starch that escapes digestion in the small intestine of healthy individuals.

Plants use starch as a way to store excess glucose, and thus also use starch as food during mitochondrial oxidative phosphorylation.

Commercial Applications

Papermaking is the largest non-food application for starches globally, consuming millions of metric tons annually. In a typical sheet of copy paper for instance, the starch content may be as high as 8%. Both chemically modified and unmodified starches are used in papermaking. In the wet part of the papermaking process, generally called the "wet-end", starches are chemically modified to contain a cationic or positive charge bound to the starch polymer, and are utilized to associate with the anionic or negatively charged paper fibers and inorganic fillers. Starch also helps get out cleaning stains from dirty washing.

These cationic starches impart the necessary strength properties for the paper web to be formed in the papermaking process (wet strength), and to provide strength to the final paper sheet (dry strength). In the dry end of the papermaking process the paper web is rewetted with a solution of starch paste that has been chemically, or enzymatically depolymerized. The starch paste solutions are applied to the paper web by means of various mechanical presses (size press). The dry end starches impart additional strength to the paper web and additionally provide water hold out or "size" for superior printing properties.

Corrugating glues are the next largest consumer of non-food starches globally. These glues are used in the production of corrugated fiberboard (sometimes called corrugated cardboard), and generally contain a mixture of chemically modified and unmodified starches that have been partially gelatinized to form an opaque paste. This paste is applied to the flute tips of the interior fluted paper to glue the fluted paper to the outside paper in the construction of cardboard boxes. This is then dried under high heat, which provides the box board strength and rigidity.

Another large non-food starch application is in the construction industry where starch is used in the or wall board manufacturing process. Chemically modified or unmodified starches are added to the rock mud containing primarily gypsum. Top and bottom heavyweight sheets of paper are applied to the mud formulation and the process is allowed to heat and cure to form the eventual rigid wall board. The starches act as a glue for the cured gypsum rock with the paper covering and also provide rigidity to the board.

Clothing starch or **laundry starch:** is a liquid that is prepared by mixing a vegetable starch in water (earlier preparations also had to be boiled), and is used in the laundering of clothes. Starch was widely used in Europe in the 16th and 17th centuries to stiffen the wide collars and ruffs of fine linen which surrounded the necks of the well-to-do. During the 19th century and early 20th century, it was stylish to stiffen the collars and sleeves of men's shirts and the ruffles of girls' petticoats by applying starch to them as the clean clothes were being ironed.

Aside from the smooth, crisp edges it gave to clothing, it served practical purposes as well. Dirt and sweat from a person's neck and wrists would stick to the starch rather than fibers of the clothing, and would easily wash away along with the starch. After each laundering, the starch would be reapplied.

Starch is also used to make some packing peanuts, and some dropped ceiling tiles.

Printing industry - in the printing industry food grade starch is used in the manufacture of anti-set-off spray powder used to separate printed sheets of paper to avoid wet ink being set off. Starch is also used in the manufacture of glues for book-binding.

Hydrogen production - Starch can be used to produce Hydrogen.

Oil exploration - starch is used as to adjust the viscosity of drilling fluid which is used to lubricate the drill head in (mineral) oil extraction.

Use as a mold: -Gummed sweets such as jelly beans and wine gums are not manufactured using a mold in the conventional sense. A tray is filled with starch and levelled. A positive mold is then pressed into the starch leaving an impression of 1000 or so jelly beans. The mix is then poured into the impressions and then put into a stove to set. This method greatly reduces the number of molds that must be manufactured.

Starch can be modified by addition of some chemical forms to be a hard glue for paper work , some of those forms are Borax , Soda Ash , which mixed with the starch solution at 50-70C to gain a very good adhesive, Sodium Silicate can be added to reinforce this formula.

DEXTRIN

Dextrins are a group of low-molecular-weight carbohydrates produced by the hydrolysis of starch. Dextrins are mixtures of linear α-(1,4)-linked D-glucose polymers starting with an α-(1,6) bond. Because branched amylopectin and glycogen also contain α-(1,6) bonds, which α-amylase cannot hydrolyze in humans, the digest resulting from this action contains a mixture of dextrins. They have the same general formula as carbohydrates but are of shorter chain length. Industrial production is, in general, performed by acidic hydrolysis of potato starch. Dextrins are water-soluble, white to slightly yellow solids that are optically active. Under analysis, dextrins can be detected with iodine solution, giving a red coloration.

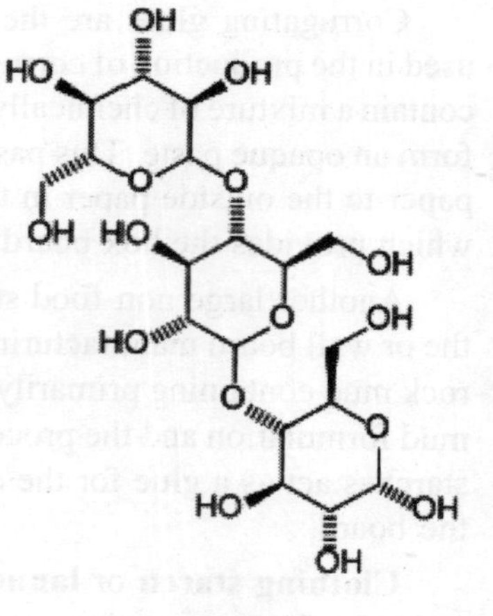

The cyclical dextrins are known as cyclodextrins. They are formed by enzymatic degradation of starch by certain bacteria, for example, Bacillus macerans. Cyclodextrins have toroidal structures formed by 6-8 glucose residues.

Dextrins find widespread use in industry, due to their non-toxicity and their low price. They are used as water-soluble glues, as thickening agents in food processing, and as binding agent in pharmaceuticals. In pyrotechnics, they are added to fire formulas, allowing them to solidify as pellets or "stars." Cyclodextrins find additional use in analytical chemistry as a matrix for the separation of hydrophobic substances, and as excipients in pharmaceutical formulations. Not all forms of dextrin are digestible, and indigestible dextrin is sometimes used in fiber supplements.

For example, **maltodextrin** either can be moderately sweet or have hardly any flavour at all. Maltodextrin is a polysaccharide that is used as a food additive. It is produced from starch and is usually found as a creamy-white hygroscopic powder. Maltodextrin is easily digestible, being absorbed as rapidly as glucose.

Maltodextrin can be derived from any starch. In the US, this starch is usually rice, corn or potato; elsewhere, such as in Europe, it is commonly wheat. This is important for coeliacs, since the wheat-derived maltodextrin can contain traces of gluten. There have been recent reports of coeliac reaction to maltodextrin in the United States. This might be a consequence of the shift of corn to ethanol production and its replacement with wheat in the formulation.

Foods containing maltodextrin may contain traces of amino acids, including glutamic acid as a manufacturing by-product. The amino acids traces would be too small to have any dietary significance.

Partly Digestible

INULIN

Inulins are a group of naturally occurring polysaccharides (several simple sugars linked together) produced by many types of plants.[1] They belong to a class of fibers known as fructans. Inulin is used by some plants as a means of storing energy and is typically found in roots or rhizomes. Most plants which synthesize and store inulin do not store other materials such as starch.

Biochemistry

Inulins are polymers mainly comprised of fructose units and typically have a terminal glucose. The fructose units in inulins are joined by a beta-(2-1) glycosidic bond. Plant inulins generally contain between 20 to several thousand fructose units. Smaller compounds are called fructooligosaccharides, the simplest of these is 1-ketose, which has 2 fructose units and 1 glucose unit.

Inulins are named in the following manner, where **n** is the number of fructose residues and **py** is the abbreviation for pyranosyl:

- Inulins with a terminal glucose are known as *alpha-D-glucopyranosyl-[beta-D-fructofuranosyl](n-1)-D-fructofuranosides*, abbreviated as **GpyFn**.
- Inulins without glucose are *beta-D-fructopyranosyl-[D-fructofuranosyl](n-1)-D-fructofuranosides*, abbreviated as **FpyFn**.

Hydrolysis of inulins may yield fructooligosaccharides, which are oligomers with a degree of polymerization (DP) of <= 10.

Uses

Processed Foods

Inulin is used increasingly in foods because it has unusual nutritional characteristics. It ranges from completely bland to subtly sweet and can be used to replace sugar, fat, and flour. This is particularly advantageous because inulin contains a third to a quarter of the food energy of sugar or other carbohydrates and a sixth to a ninth of the food energy of fat. It also increases calcium absorption[2] and possibly magnesium absorption,[3] while promoting intestinal bacteria. Nutritionally, it is considered a form of soluble fiber, and it is important to note that consuming large quantities (particularly for sensitive and/or unaccustomed individuals) can lead to gas and bloating. Inulin has a minimal impact on blood sugar, and—in sharp contrast to fructose—is not insulemic and does not raise triglycerides,[4] making it generally considered suitable for diabetics and potentially helpful in managing blood sugar-related illnesses.

Industrial Use

Nonhydrolyzed inulin can also be directly converted to ethanol in a simultaneous saccharification and fermentation process which may have great potential for converting crops high in inulin into ethanol for fuel.

Medical

Inulin is used to help measure kidney function by determining the Glomerular filtration rate (**GFR**). GFR is the volume of fluid filtered from the renal (kidney) glomerular capillaries into the Bowman's capsule per unit time.

Calculation of Glomerular Filtration Rate (GFR)

Inulin is uniquely treated by nephrons in that it is completely filtered at the glomerulus but neither secreted nor reabsorbed by the tubules. This property of inulin allows the clearance of inulin to be used clinically as a highly accurate measure of Glomerular filtration Rate (GFR)—the percentage of plasma from the afferent arteriole that is filtered into Bowman's capsule.

It is useful to contrast the properties of inulin with those of para-aminohippuric acid (PAH). PAH is completely filtered from plasma at the glomerulus and not reabsorbed by the tubules, in a manner identical to inulin. PAH is different from inulin in that the fraction of PAH that bypasses the glomerulus and enters the nephron's tubular cells (via the Peritubular capillaries) is completely secreted. Renal Clearance of PAH is thus useful in calculation of renal plasma flow (RPF), which empirically is (1-Hematocrit) times renal blood flow. Of note, the clearance of PAH is reflective only of RPF to portions of the kidney that deal with urine formation, and thus underestimates actually RPF by about 10%.[7]

The measurement of GFR by inulin is still considered the gold-standard. Practically, however, it has now been largely replaced by other, simpler measures that are approximations of GFR. These measures, which involve clearance of such substrates as EDTA and creatinine, have had their utility confirmed in large cohorts of patients with chronic kidney disease.

Health Effects

Inulin is indigestible by the human enzymes ptyalin and amylase, which are adapted to digest starch. As a result, inulin passes through much of the digestive system intact. It is only in the colon that bacteria metabolise inulin, with the release of significant quantities of carbon dioxide, hydrogen and/or methane.

Inulin-containing foods can be rather gassy, particularly for those unaccustomed to inulin, and these foods should be consumed in moderation at first.

There are two types of dietary fiber, soluble and insoluble. Insoluble fiber increases the movement of materials through the digestive system and increases stool bulk; it is especially helpful for those suffering from constipation or stool irregularity. Soluble fiber dissolves in water to form a gelatinous material. Some soluble fibres may help lower blood cholesterol and glucose levels. Inulin is considered a soluble fiber. Because normal digestion does not break inulin down into monosaccharides, it does not elevate blood sugar levels and may therefore be helpful in the management of diabetes. Inulin also stimulates the growth of bacteria in the gut.[4] Inulin passes through the stomach and duodenum undigested and is highly available to the gut bacterial flora. This contrasts with proprietary probiotic formulations based on Lactic acid bacteria (LAB) in which the bacteria have to survive very challenging conditions through the gastrointestinal tract before they are able to colonize the gut.

Some traditional diets contain up to 20g per day of inulin or fructo-oligosaccharides. Many foods naturally high in inulin or fructo-oligosaccharides, such as chicory, garlic, and leek, have been seen as "stimulants of good health" for centuries. Inulin is also used in medical tests to measure the total amount of extracellular volume and determine the function of the kidneys.

Inulin is generally recognized as safe (GRAS) by the FDA. Contrary to the health benefits as mentioned above, allergic reactions to inulin in foods have been reported in a letter[11] to the New England Journal of Medicine. Between about 30-40% of the population suffers from fructose malabsorption. Since inulin is a fructan, it is problematic for people with fructose malabsorption. It is recommended that fructan intake for people with fructose malabsorption be kept to less than 0.5 grams/serving.

Natural Sources of Inulin

Plants that contain high concentrations of inulin include:

- Elecampane (*Inula helenium*)
- Dandelion (*Taraxacum officinale*)
- Wild Yam (*Dioscorea* spp.)
- Jerusalem artichokes (*Helianthus tuberosus*)
- Chicory (*Cichorium intybus*)
- Jicama (*Pachyrhizus erosus*)
- Burdock (*Arctium lappa*)
- Onion (*Allium cepa*)
- Garlic (*Allium sativum*)
- Agave (*Agave* spp.)
- Yacon (*Smallanthus sonchifolius* spp.)

INDIGESTIBLE

CELLULOSE

History

Cellulose was discovered in 1838 by the French chemist Anselme Payen, who isolated it from plant matter and determined its chemical formula. Cellulose was used to produce the first successful thermoplastic polymer, celluloid, by Hyatt Manufacturing Company in 1870. Hermann Staudinger determined the polymer structure of cellulose in 1920. The compound was first chemically synthesized (without the use of any biologically-derived enzymes) in 1992, by Kobayashi and Shoda.

Cellulose is an organic compound with the formula $(C_6H_{10}O_5)_n$, a polysaccharide consisting of a linear chain of several hundred to over ten thousand β(1→4) linked D-glucose units.

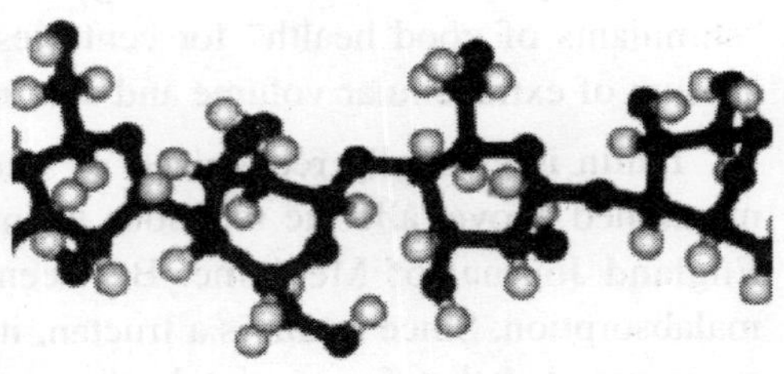

Cellulose is the structural component of the primary cell wall of green plants, many forms of algae and the oomycetes. Some species of bacteria secrete it to form biofilms. Cellulose is the most common organic compound on Earth. About 33 per cent of all plant matter is cellulose (the cellulose content of cotton is 90 per cent and that of wood is 50 per cent).

For industrial use, cellulose is mainly obtained from wood pulp and cotton. It is mainly used to produce cardboard and paper; to a smaller extent it is converted into a wide variety of derivative products such as cellophane and rayon.

Some animals, particularly ruminants and termites, can digest cellulose with the help of symbiotic micro-organisms that live in their guts. Cellulose is not digestible by humans and is often referred to as 'dietary fiber' or 'roughage', acting as a hydrophilic bulking agent for feces.

Structure and Properties

Cellulose is taste- and odorless, hydrophilic, not soluble in water or most organic solvents, chiral and biodegradable.

Cellulose is derived from D-glucose units, which condense through β(1→ 4)-glycosidic bonds. This linkage motif contrasts with that for α(1→4)-glycosidic bonds present in starch, glycogen, and other carbohydrates. Cellulose is a straight chain polymer: unlike starch, no coiling occurs, and the molecule adopts an extended and rather stiff rod-like conformation. The multiple hydroxyl groups on the glucose residues from one chain form hydrogen bonds with oxygen molecules on another chain, holding the chains firmly together side-by-side and forming *microfibrils* with high tensile strength. This strength is important in cell walls, where they are meshed into a carbohydrate *matrix*, conferring rigidity to plant cells.

Compared to starch, cellulose is also much more crystalline. Whereas starch undergoes a crystalline to amorphous transition when heated beyond 60-70°C in water (as in cooking), cellulose requires a temperature of 320°C and pressure of 25 MPa to become amorphous in water.

Chemically, cellulose can be broken down into its glucose units by treating it with concentrated acids at high temperature.

Many properties of cellulose depend on its degree of polymerization or chain length, the number of glucose units that make up one polymer molecule. Cellulose from wood pulp has typical chain lengths between 300 and 1700 units; cotton and other plant fibers as well as bacterial celluloses have chain lengths ranging from 800 to 10,000 units.[5] Molecules with very small chain length resulting from the break down of cellulose are known as cellodextrins; in contrast to long-chain cellulose, cellodextrins are typically soluble in water and organic solvents.

Plant-derived cellulose is usually contaminated with hemicellulose, lignin, pectin and other substances, while microbial cellulose is quite pure, has a much higher water content, and consists of long chains.

Assaying Cellulose

Given a cellulose-containing material, the portion that does not dissolve in a 17.5% solution of sodium hydroxide at 20 °C is *á cellulose*, which is true cellulose. Acidification of the extract precipitates β *cellulose*. The portion that dissolves in base but does not precipitate with acid is γ *cellulose.*

Cellulose can be assayed using a method described by Updegraff in 1969, where the fiber is dissolved in acetic and nitric acid, and allowed to react with anthrone in sulfuric acid. The resulting coloured compound is assayed spectrophotometrically at a wavelength of approximately 635 nm.

In addition, cellulose is represented by the difference between acid detergent fiber (ADF) and acid detergent lignin(ADL).

Biosynthesis

In vascular plants cellulose is synthesized at the plasma membrane by rosette terminal complexes (RTC's). The RTC's are hexameric protein structures, approximately 25 nm in diameter, that contain the cellulose synthase enzymes that synthesise the individual cellulose chains.[8] Each RTC floats in the cell's plasma membrane and "spins" a microfibril into the cell wall.

The RTC's contain at least three different cellulose syntheses, encoded by *CesA* genes, in an unknown stoichiometry.[9] Separate sets of *CesA* genes are involved in primary and secondary cell wall biosynthesis. Cellulose synthase utilizes UDP-D-glucose precursors to generate microcrystalline cellulose. Cellulose synthesis requires chain initiation and elongation, and the two processes are separate. *CesA* glucosyltransferase initiates cellulose polymerization using a steroid primer, 'sitosterol-beta-glucoside' and UDP-glucose.[10] A cellulase may function to cleave the primer from the mature chain.

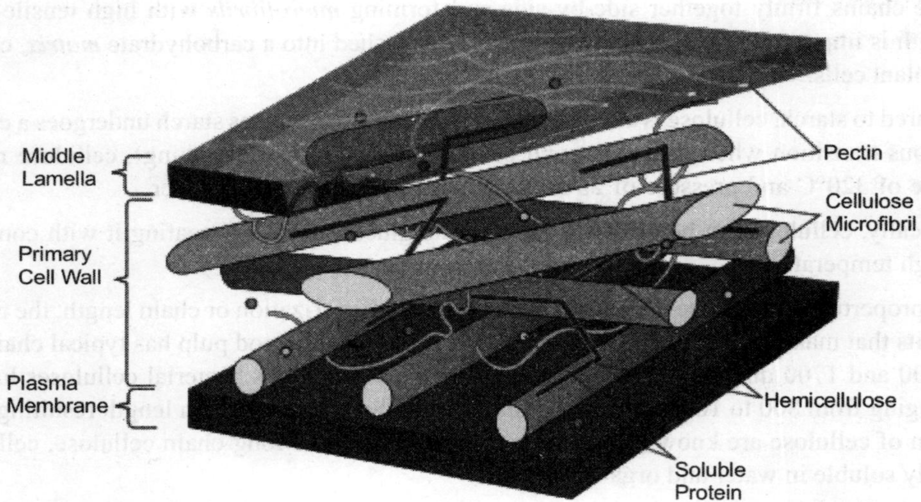

Fig. 34.11. Location and arrangement of cellulose microfibrils in the plant cell wall.

Breakdown (Cellulolysis)

Cellulolysis is the process of breaking down cellulose into smaller polysaccharides called cellodextrins or completely into glucose units; this is a hydrolysis reaction. Because cellulose molecules bind strongly to each other, cellulolysis is relatively difficult compared to the break down of other polysaccharides.[11]

Mammals do not have the ability to break down cellulose directly. Some ruminants like cows and sheep contain certain symbiotic anaerobic bacteria (like *Cellulomonas*) in the flora of the gut wall, and these bacteria produce enzymes to break down cellulose; the break down products are then used by the mammal. Similarly, lower termites contain in their hindguts certain flagellate protozoa which produce such enzymes; higher termites contain bacteria for the job. Fungi, which in nature are responsible for recycling of nutrients, are also able to break down cellulose.

The enzymes utilized to cleave the glycosidic linkage in cellulose are glycoside hydrolases including endo-acting cellulases and exo-acting glucosidases. Such enzymes are usually secreted as part of multienzyme complexes that may include dockerins and cellulose binding modules; these complexes are in some cases referred to as cellulosomes.

Hemicellulose

Hemicellulose is a polysaccharide related to cellulose that comprises ca. 20% of the biomass of most plants. In contrast to cellulose, hemicellulose is derived from several sugars in addition to glucose, including especially xylose but also mannose, galactose, rhamnose, and arabinose. Hemicellulose consists of shorter chains - around 200 sugar units as opposed to 7,000 - 15,000 glucose molecules in the average cellulose polymer. Furthermore, hemicellulose is branched, whereas cellulose is unbranched.

Commercial products

Cellulose is the major constituent of paper and cardboard and of textiles made from cotton, linen, and other plant fibers.

Cellulose can be converted into cellophane, a thin transparent film, and into rayon, an important fiber that has been used for textiles since the beginning of the 20th century. Both cellophane and rayon are known as "regenerated cellulose fibers"; they are identical to cellulose in chemical structure and are usually made from viscose, a viscous solution made from cellulose. A more recent and environmentally friendly method to produce rayon is the Lyocell process. Cellulose is used in the laboratory as the stationary phase for thin layer chromatography. It is the raw material in the manufacture of nitrocellulose (cellulose nitrate) which was historically used in smokeless gunpowder and as the base material for celluloid used for photographic and movie films until the mid 1930s. Cellulose is used to make hydrophilic and highly absorbent sponges and water-soluble adhesives and binders such as methyl cellulose and carboxymethyl cellulose which are used in wallpaper paste. Microcrystalline cellulose (E460i) and powdered cellulose (E460ii) are used as an inactive filler in tablets and as thickeners and stabilizers in processed foods.

Cellulose Source and Energy Crops

The major combustible component of non-food energy crops is cellulose, with lignin second. Non-food energy crops are more efficient than edible energy crops (which have a large starch component), but still compete with food crops for agricultural land and water resources.[6] Typical non-food energy crops include industrial hemp, switchgrass, *Miscanthus*, *Salix* (willow), and *Populus* (poplar) species.

Some bacteria can convert cellulose into ethanol which can then be used as a fuel; see cellulosic ethanol.

Nitrocellulose (also: **cellulose nitrate**, **flash paper**) is a highly flammable compound formed by nitrating cellulose through exposure to nitric acid or another powerful nitrating agent. When used as a propellant or low-order explosive, it is also known as **guncotton**.

Guncotton and Gunpowder

Various types of smokeless powder, consisting primarily of nitrocellulose

Henri Braconnot discovered in 1832 that nitric acid, when combined with starch or wood fibers, would produce a lightweight combustible explosive material which he named *xyloïdine*. A few years later in 1838 another French chemist Théophile-Jules Pelouze (teacher of Ascanio Sobrero and Alfred Nobel) treated paper and cardboard in the same way. He obtained a similar material he called *nitramidine*. Both of these substances were highly unstable, and were not practical explosives.

However, Christian Friedrich Schönbein, a German-Swiss chemist, discovered a more practical solution around 1846. As he was working in the kitchen of his home in Basle, he spilled a bottle of concentrated nitric acid on the kitchen table. He reached for the nearest cloth, a cotton apron, and wiped it up. He hung the apron on the stove door to dry, and as soon as it was dry there was a flash as the apron exploded. His preparation method was the first to be widely imitated — one part of fine cotton wool to be immersed in fifteen parts of an equal blend of sulfuric and nitric acids. After two minutes the cotton was removed and washed in cold water to set the esterification level and remove all acid residue. It was then slowly dried at a temperature below 100°C. Schönbein collaborated with the Frankfurt professor Rudolf Böttger, who had discovered the process independently in the same year. By a strange coincidence there was even a third chemist, the Braunschweig professor F. J. Otto, who had also produced guncotton in 1846 and was the first to publish the process, much to the disenchantment of Schönbein and Böttger. (Itzehoer Wochenblatt, 29 October 1846, columns 1626 f.)

The process uses the nitric acid to convert the cellulose into cellulose nitrite and water:

$$2HNO_3 + C_6H_{10}O_5 \rightarrow C_6H_8(NO_2)_2O_5 + 2H_2O$$

The sulfuric acid is present to prevent the water produced in the reaction from diluting the concentrated nitric acid.

The power of guncotton made it suitable for blasting. As a projectile driver, it has around six times the gas generation of an equal volume of black powder and produces less smoke and less heating. However the sensitivity of the material during production led the British, Prussians and French to discontinue manufacture within a year.

Jules Verne viewed the development of guncotton with optimism. He referred to the substance several times in his novels. His adventurers carried firearms employing this substance. Most notably, in his *From the Earth to the Moon*, guncotton was used to launch a projectile into space.

Further research indicated that the key was the very careful preparation of the cotton: unless it was very well cleaned and dried, it was likely to explode spontaneously. The British, led by Frederick Augustus Abel, also developed a much lengthier manufacturing process at the Waltham Abbey Royal Gunpowder Mills, patented in 1865, with the washing and drying times each extended to 48 hours and repeated eight times over. The acid mixture was also changed to two parts sulfuric acid to one part nitric.

Guncotton remained useful only for limited applications. For firearms, a more stable and slower burning mixture would be needed. Guncotton-like preparations were eventually prepared for this role, known at the time as smokeless powder.

Guncotton, dissolved at approximately 25% in acetone, forms a lacquer used in preliminary stages of wood finishing to develop a hard finish with a deep luster. It is normally the first coat applied, sanded, and followed by other coatings that bond to it.

Nitrate Film

Nitrocellulose was used as the first flexible film base, beginning with Eastman Kodak products in August, 1889. Camphor is used as plasticizer for nitrocellulose film. It was used until 1933 for X-ray films (where its flammability hazard was most acute) and for motion picture film until 1951. It was replaced by safety film with an acetate base. German WW2 newsreel film (circa 1940) believed to be nitrocellulose, made by AGFA The use of nitrocellulose film for motion pictures led to a widespread requirement for fireproof projection rooms with wall coverings made of asbestos. Famously, the US Navy shot a training film for projectionists which included footage of a controlled ignition of a reel of nitrate film which continued to burn even when fully submerged in water. Due to public safety precautions, the London Underground forbade transport of nitrate films on its system until well past the introduction of safety film. A cinema fire caused by ignition of nitrocellulose film stock (foreshadowed by an earlier small fire) was a central plot element in the Italian film *Cinema Paradiso*. Today nitrate film projection is usually highly regulated and requires extensive precautionary measures including extra projectionist health and safety training. Additionally, projectors certified to run nitrate films have many containment strategies in effect. Among them, this includes the chambering of both the feed and takeup reels in thick metal covers with small slits to allow the film to run through. Furthermore, the projector is modified to accommodate several fire extinguishers with nozzles all aimed directly at the film gate; the extinguishers automatically trigger if a piece of flammable fabric placed near the gate starts to burn. While this triggering would likely damage or destroy a significant portion of the projection components, it would prevent a devastating fire which almost certainly would cause far greater damage.

It was discovered decades later that nitrocellulose gradually decomposes, releasing nitric acid which further catalyzes the decomposition (usually into a still-flammable powder or goo). Low temperatures can delay these reactions indefinitely. It is estimated that the great majority of films produced during the early twentieth century were lost forever either through this accelerating, self-catalyzed disintegration or studio warehouse fires. Salvaging old films is a major problem for film archivists (see film preservation).

Nitrocellulose film base manufactured by Kodak can be identified by the presence of the word *Nitrate* in dark letters between the perforations. Acetate film manufactured during the era when nitrate films were still in use was marked *Safety* or *Safety Film* between the perforations dark letters. Letters in white or light colours are print-through from the negative.

Color negative film was never manufactured with a nitrate base, nor were 8 mm or 16 mm motion picture film stocks.

Other Uses

Depending on the manufacturing process, nitrocellulose is esterified to varying degrees. Table tennis balls, guitar picks and some photographic films have a fairly low esterification level and burn comparatively slowly with some charred residue. See celluloid.

In professional wrestling, nitrocellulose, in the form of a small piece of flash paper, is used to execute an illegal (and dangerous) move called a **Fireball**. The wrestler sets the piece of quick-burning paper alight (using a concealed lighter) and throws it at the opponent, giving the impression of a supernatural ball of fire emerging from their hand.

Nitrated cotton: Used as liftcharge for indoor fireworks, for hand flashers and magicians

Nitrated yarn: Used as fuse for indoor fireworks and to make things fall down on command

Nitrated paper: Mainly used by magicians to make paper disappear in a flash, but also indoor fireworks as comets

Nitrated cellulose: Ice fountains, indoor fireworks, making smokeless gunpowder, celluloid, paints

Collodion is a solution of nitrocellulose in ether or acetone, sometimes with the addition of alcohols.

Its generic name is pyroxylin solution. It is toxic and highly flammable. As the solvent evaporates, it dries to a celluloid-like film. It was discovered about 1846 by the French chemist and writer Louis Ménard. n 1851, the Englishman Frederick Scott Archer discovered that collodion could be used as an alternative to albumen on glass plates. This also reduced the exposure time when making the image. This became known as the wet plate Collodion or wet collodion method. Collodion was also grainless and colourless, and allowed for one of the first high quality duplication processes, also known as negatives. This process also produced positives, the Ambrotype and the Ferrotype (aka Tintypes).

Celloidin is a pure type of pyroxylin used to embed specimens which will be examined under a microscope.

Uses

- Collodion is widely used to glue electrodes to the head for electroencephalography.
- Pyroxylin with added pigments is used as a nitrocellulose lacquer.
- It was also added to nitroglycerine to stabilise it as blasting gelatine.
- Collodion is also used in theatrical makeup for various effects, such as simulating old-age wrinkles or scars.
- Similarly, it was sometimes used in boxing and other applications to cover up cuts. However, it has been generally replaced by the less toxic Liquid Bandage, which includes pyroxylin as an ingredient.
- Collodion also finds use in the cleaning of optics such as telescope mirrors. The collodion is applied to the surface of the optic, usually in two or more layers. Sometimes a piece of thin cloth is applied between the layers, to hold the collodion together for easy removal. After the collodion dries and forms a solid sheet covering the optic, it is carefully peeled away taking contamination with it.
- Collodion was also used by pathologist Thomas Stoltz Harvey to preserve Einstein's brain in 1955.